International Political Economy Series

Series Editor

Timothy M. Shaw, University of Massachusetts Boston, Boston, USA;
Emeritus Professor, University of London, London, UK

The global political economy is in flux as a series of cumulative crises impacts its organization and governance. The IPE series has tracked its development in both analysis and structure over the last three decades. It has always had a concentration on the global South. Now the South increasingly challenges the North as the centre of development, also reflected in a growing number of submissions and publications on indebted Eurozone economies in Southern Europe. An indispensable resource for scholars and researchers, the series examines a variety of capitalisms and connections by focusing on emerging economies, companies and sectors, debates and policies. It informs diverse policy communities as the established trans-Atlantic North declines and 'the rest', especially the BRICS, rise.

NOW INDEXED ON SCOPUS!

Ryan David Kiggins
Editor

The Political Economy of Robots

Prospects for Prosperity and Peace in the Automated 21st Century, 2nd edition

Second Edition

palgrave
macmillan

Editor
Ryan David Kiggins
Department of Political Science
University of Central Oklahoma
Edmond, OK, USA

ISSN 2662-2483 ISSN 2662-2491 (electronic)
International Political Economy Series
ISBN 978-3-031-95170-1 ISBN 978-3-031-95171-8 (eBook)
https://doi.org/10.1007/978-3-031-95171-8

Cover illustration: Rob Friedman/Stockphoto.com

This Palgrave Macmillan imprint is published by the registered company Springer Nature Switzerland AG
The registered company address is: Gewerbestrasse 11, 6330 Cham, Switzerland

Preface to the Second Edition

In the years since publication of the First Edition, artificial intelligence and associated robotic technologies have undergone notable innovations. An updated and expanded Second Edition seems appropriate. This Second Edition includes lightly revised original contributions and new updated contributions[1] to round out the collection considering recent technological innovations.

We sought to preserve the original text while offering reflection and critical analysis of the ways in which advances in artificial intelligence (AI) and robotics inform current perspectives. Some attention is given to Generative Artificial Intelligence without limiting ourselves solely to that innovation and additional chapters address AI as a threat analysis tool. By and large, many issues raised in the First Edition, are pertinent in this updated and expanded Second Edition, if not gaining urgency.

[1] Rogerson, K., & Sherman, J. (2021). AI in Public Education: Humble Beginnings and Revolutionary Potential. In *Towards an International Political Economy of Artificial Intelligence* (pp. 63–83). Cham: Springer International Publishing; Swed, O., & Chávez, K. (2021). Between Scylla and Charybdis: The Threat of Democratized Artificial Intelligence. In *Towards an International Political Economy of Artificial Intelligence* (pp. 177–194). Cham: Springer International Publishing; Yankoski, M., Theisen, W., Verdeja, E., & Scheirer, W. J. (2021). Artificial intelligence for peace: An early warning system for mass violence. In *Towards an international political economy of artificial intelligence* (pp. 147–175). Cham: Springer International Publishing.

Contributions by AI and robotics to peace and prosperity remain viable, as does, unfortunately, several perils. The First Edition sought to illuminate these potentialities as AI and robotics transform global political economy. This updated Second Edition is intended to do likewise in response to AI and robotics innovations including Generative AI.

Generative AI refers to a suite of AI techniques that create (generate) new content, code, large datasets, and data patterns. Generative AI learns from large datasets, then creates unique outputs resembling the training data. Techniques in this class of AI include Generative Adversarial Networks, Variational Autoencoder, and Transformer Models.

Generative Adversarial Networks use two neural networks (a generator and a discriminator) that compete against each other. The generator creates content, while the discriminator evaluates it against real data, pushing the generator to improve. This technique is widely used in creating realistic images and videos.

Variational Autoencoders are used in applications like image generation and data synthesis. They learn a compressed representation of input data and then reconstruct similar data from it.

Transformer Models like GPT (for text) and DALL-E (for images) use transformer architectures to predict the next piece of information, enabling them to craft fluent, coherent text, and highly detailed images.

Applications for Generative AI include:

- Text generation: Content creation, chatbots, translation, summarization, etc.
- Image and art generation: Creating art, generating realistic human faces, and transforming image styles.
- Audio and music generation: Composing music, generating human-like voices, and creating sound effects.
- Code generation: Automating parts of software development.
- Drug discovery and scientific research: Designing molecules for pharmaceuticals by predicting potential interactions.

These and other recent developments are energizing its practical applications in both civilian and military sectors, with increasing autonomy in logistics, vehicles, and military weapon systems. This will likely quicken shifts in cognitive and labor loads away from humans to AI and robots. With great power competition intensifying between the United States and

China, as both vie for superiority without either resolving important regulatory and ethical issues. The market power of companies based in Japan, South Korea, and Germany will continue to be important influences in the race for AI & robotics superiority in the global political economy. The mix of this competition and cooperation among sovereign and market actors will continue spilling over into other market economies, impacting human resources, industrial bases, and governance frameworks. The reach and implications of these trends portend to be substantial, requiring attention from scholars to industry to policymakers.

Development of institutions to harness AI and robotics potential, while limiting challenges will greatly impact prospects for prosperity and peace. Our hope is this updated and expanded Second Edition, will further deliberation necessary to reach consensus about how to govern AI, robots, and ourselves. The potential for prosperity and peace in the automated twenty-first century rests on such consensus.

Edmond, USA Ryan David Kiggins

Contents

Notes on Contributors

Ronald C. Arkin is Regents' Professor of Computer Science and Director of the Mobile Robot Laboratory at Georgia Tech University (Atlanta, GA). He also serves as the Associate Dean for Research and Space Planning in the College of Computing at Georgia Tech since October 2008. During 1997–98, Professor Arkin served as STINT visiting Professor at the Centre for Autonomous Systems at the Royal Institute of Technology (KTH) in Stockholm, Sweden. From June-September 2005, Prof. Arkin held a Sabbatical Chair at the Sony Intelligence Dynamics Laboratory in Tokyo, Japan and then served as a member of the Robotics and Artificial Intelligence Group at LAAS/CNRS in Toulouse, France from October 2005-August 2006. His research interests include behavior-based reactive control and action-oriented perception for mobile robots and unmanned aerial vehicles, hybrid deliberative/reactive software architectures, robot survivability, multiagent robotic systems, biorobotics, humanrobot interaction, robot ethics, and learning in autonomous systems. He has over 170 technical publications in these areas. Prof. Arkin has written a textbook entitled Behavior Based Robotics, published by MIT Press in May 1998, co-edited (with G. Bekey) a book entitled Robot Colonies published in 1997, and a book published in Spring 2009 entitled Governing Lethal Behavior in Autonomous Robots published by Chapman-Hall (Taylor & Francis). The following article is derived principally from that latter work.

Greg Austin is a professorial fellow at the East West Institute, where he served as a vice president from 2006 to 2011. Greg is concurrently a visiting professor in the Australian Centre for Cyber Security and a co-chair of the EastWest working group on Measures of Restraint in Cyber Armaments. He has held senior posts in the International Crisis Group and the Foreign Policy Centre (London). Other assignments include service in government, defense intelligence, academia, and journalism. He is the author of several books on China's strategic policy. He has a Ph.D. in International Relations and a Master's degree in international law. He is an Australian citizen. His most recent book, *Cyber Policy in China* (Wiley 2014), offers the first comprehensive analysis (military, economic, and political) of China's leadership responses to the information society. It explores the dilemmas facing Chinese politicians as they try to marry the development of an information economy with old ways of governing their people and conducting international relations. The book concludes that unless China's ruling party adapts more aggressively to the defining realities of power and social organization in the information age, the 'China dream' is unlikely to become a reality.

Elaine Carey is a professor of History and chair of the Department of History at St. John's University in Queens, New York, and she holds the Lloyd Sealy research fellow at CUNY's John Jay College of Criminal Justice. Currently, she is the vice president for the Teaching Division of the American Historical Association (AHA), and she is active in the AHA's Tuning Project. She is the author of *Plaza of Sacrifices: Gender, Power, and Terror in 1968 Mexico* (2005); co-editor with Andrae Marak, *Smugglers, Brothels, and Twine: Transnational Flows of Contraband and Vice in North America* (2011), the award winning *Women Drug Traffickers: Mules, Bosses, and Organized Crime* (2014), and the forthcoming textbook *Protests in the Streets: 1968 Across the Globe*(2016).

Kerry Chávez is Assistant Professor at the United States Air Force Academy, a nonresident research fellow with the Institute for Global Affairs, and an advisor for Project Air and Space Power at the Irregular Warfare Initiative. She was previously a two-time nonresident fellow with the Modern War Institute at West Point. Her research on the politics, strategies, and emerging technologies of modern warfare and international security has been published in several venues.

Ronald W. Cox is a professor in the Department of Political Science and International Relations at Florida International University. Dr. Cox has written extensively in the areas of US foreign economic policy, international political economy, American foreign policy, and the political economy of baseball. His books include *Power and Profits: US Policy in Central America*, University of Kentucky Press, 1994; *Business and the State in International Relations* (ed.), Westview Press, 1997; *US Politics and the Global Economy*, with Dan Skidmore-Hess, Lynne Rienner Press, 1999; and *Free Agency and Competitive Balance in Baseball*, McFarland Press, 2006, and *Corporate Power in American Foreign Policy* (ed.), Routledge Press, 2012. He is currently working on a new book for Lexington Books titled *Transnational Corporations and the New Globalization*, forthcoming 2014.

Berkeley Franz is a Post-Doctoral Health Policy researcher in the Department of Social Medicine at the Ohio University, Heritage College of Osteopathic Medicine.

Loren Gatch is a professor of political science at the University of Central Oklahoma where he researches and teaches on governing the American and Global economies, the political history of American and global monetary and financial systems, and political theory.

Ryan David Kiggins is currently an assistant professor of political science at the University of Central Oklahoma, Edmond, OK, USA. He has published on US Internet governance policy, US cyber security policy, and global security and rare earths. His broad research interest is the intersection and interaction of information technologies, international security, and global political economy.

Dr. Peter Lee is a University of Portsmouth reader in Politics and Ethics who specializes in the ethics and ethos of remotely piloted aircraft (drone) operations, the politics and ethics of war and military intervention, and the politics and ethics of identity. In November 2012, Peter transferred from King's College London after four years in the Air Power Studies Division and continues to lecture across a range of diverse subjects, from international relations to terrorism and insurgency. He is currently conducting research with the personnel of the two Royal Air Force Reaper squadrons, writes in various formats on the ethics of UAV operations and is regularly invited to lecture on this and other subjects to military, academic, political, religious, media, and wider audiences. In 2012 he published *Blair's Just War: Iraq and the Illusion of Morality*and in 2015 published *Truth Wars: The Politics of Climate Change, Military Intervention and Financial Crisis*, both with Palgrave Macmillan.

Andrae Marak is a professor of History and Political Science, a chair of the Division of Humanities and Social Sciences at Roosevelt University in Chicago, Illinois, and is an associate of the East Asian Studies Center at Indiana University. Currently, he is on the executive committee of the Mexico Section of the Latin American Studies Association and of the Midwest World History Association and is an contributing editor of *The Middle Ground Journal*. He is the author of *From Many, One: Peasants, Borders, and Education in Callista Mexico (1924–1935)*(2009); co-editor with Elaine Carey of *Smugglers, Brothels, and Twine: Transnational Flows of Contraband and Vice in North America*(2011); with Laura Tuennerman of *At the Border of Empires: The Tohono O'odham, Gender and Assimilation*(2013); and co-edited with Clarissa Confer and Laura Tuennerman of *Transnational Indians in the North American West*(2015).

Mary Manjikian is an associate dean of the Robertson School of Government at Regent University. She is the author of three books on national security, and has also authored articles on intelligence, gender aspects of emerging technologies, ethics of warfare and cybersecurity. Most recently, her manuscript *A Typology of Arguments about Drone Ethics*was published by the Strategic Studies Institute of the US Army War College.

Justin Quinn Olmstead, Ph.D. is a staff historian at Sandia National Laboratory in New Mexico, USA. He presents at national and international conferences on First World War and has acted as content consultant for several books on the subject. Currently, he is on the executive committee of the Midwest World History Association. His book *Acquiring America: The Diplomatic Battle for the United States, 1914–1917* is under review with the University of Kentucky Press. Dr. Olmstead's currently researching the impact of Lloyds of London on British policy during the First World War. Dr. Olmstead gained his Ph.D. from the University of Sheffield, England and is a member of the Royal Historical Society.

Kenneth S. Rogerson is Professor of the Practice at Duke University's Sanford School of Public Policy, and former Research Director of the DeWitt Wallace Center for Media and Democracy. He is currently the

University's faculty advisor for the BN Duke Scholars. He has served as chair of the American Political Science Association's Information Technology and Politics Section and the International Studies Association's International Communication Section.

Walter J. Scheirer, PhD is Associate Professor in the Department of Computer Science and Engineering at the University of Notre Dame. Previously, he was a postdoctoral fellow at Harvard University, with affiliations in the School of Engineering and Applied Sciences, Department of Molecular and Cellular Biology and Center for Brain Science, and the director of research & development at Securics, Inc., an early stage company producing innovative computer vision-based solutions. He received his PhD from the University of Colorado and his M.S. and B.A. degrees from Lehigh University. Dr. Scheirer has extensive experience in the areas of computer vision, machine learning and digital image forensics. His overarching research interest is the fundamental problem of recognition, including the representations and algorithms supporting solutions to it.

Bryant William Sculos is a lecturer at the University of Texas-Rio Grande in Corpus Christi, Texas, USA. His research deploys the social-psychological frameworks of Theodor Adorno and Erich Fromm to criticize contemporary theories of cosmopolitanism and post-Marxism.

Brett S. Sharp is a professor of Political Science and Public Administration at the University of Central Oklahoma where he also served as the founding director of Leadership Studies. He was the former employment manager for the City of Oklahoma City and former agency services coordinator for the State of Oklahoma Office of Personnel Management. The Oklahoma Political Science Association recently named him as Scholar of the Year. His publications include articles in *American Review of Public Administration*and *Review of Public Personnel Administration* as well as several books including his latest, *Managing in the Public Sector: A Casebook in Ethics and Leadership*.For the past few years, he has been serving as a visiting research professor at Swansea University in Wales, UK focusing on application of social media in political movements. His latest research is in data visualization of major administrative reform efforts.

Justin Sherman is Distinguished Fellow at the Center on Privacy & Technology. He is the founder and CEO of Global Cyber Strategies, a Washington, DC-based research and advisory firm. He is also an adjunct professor at Duke University's Sanford School of Public Policy; a nonresident senior fellow at the Atlantic Council; a fellow at the Starling Lab for Data Integrity at Stanford University; and a contributing editor at Lawfare. He co-hosts the "Power On, Power Off" podcast with Jenna Ruddock about technology, people, and power. Justin's focus on privacy and technology includes research and work on cybersecurity policy, data brokerage, advertising technologies, privacy protections for large-scale datasets, geolocation data, and emerging data analytic techniques.

Daniel Skinner is an assistant professor of Health Policy in the Department of Social Medicine at the Ohio University, Heritage College of Osteopathic Medicine (skinnerd@ohio.edu).

Dr. Ori Swed is Associate Professor in the Sociology Department at Texas Tech University. He is also the Director of the Peace, War, & Social Conflict Laboratory, where he leads multiple research projects on new technologies and emerging security threats. His research focuses on the role of violent non-state actors in contemporary conflict settings, among them private military and security contractors and mercenaries.

William Theisen is a PhD student working in the Department of Computer Science under the supervision of Dr. Walter Scheirer. His current research is studying the interplay between misinformation, propaganda, and internet memes. His recent work focuses on the 2019 Indonesian Presidential Election and associated online media.

Leslie Paul Thiele is a distinguished professor of Political Science, teaching political theory and sustainability studies at the University of Florida, where he serves as a director of the Center for Adaptive Innovation, Resilience, Ethics, and Science (UF CAIRES). His interdisciplinary research focuses on continental political thought, sustainability, technology, and the intersection of political philosophy and the natural sciences. His central concerns are the responsibilities of citizenship and the opportunities for leadership in a world of rapid technological, social, and ecological change. Dr. Thiele's work has been published through Cambridge University Press, Oxford University Press, and Princeton University press. His most recent book, *Sustainability*, 2013, published through Polity Press was selected for the list Outstanding Academic Titles for 2014 by *Choice*.

Ernesto Verdeja is Associate Professor of Political Science and Peace Studies at the Kroc Institute for International Peace Studies, University of Notre Dame. He has published on political reconciliation, transitional justice, the causes of genocide and mass atrocities, and mass atrocity early warning. He is currently the Executive Director of the Institute for the Study of Genocide, a non-profit founded in 1982 to promote research and policy analysis on the causes and prevention of genocide and political violence. He regularly consults with the government and human rights organizations on mass atrocity prevention and political reconciliation.

Sean Noah Walsh is a political theorist at Capital University in Columbus, Ohio. His areas of interest include ancient political philosophy, postmodernism, and Marxism and critical theory. He received his doctorate from the University of Florida in 2010.

Michael Wartenbe is a Ph.D. candidate in the Department of Political Science and International Relations at Florida International University.

Samuel Woolley is an associate professor in the communication department at the University of Texas-Austin. He currently serves as the project manager of two research projects focused on the study of politicized social bots: the European Research Council supported 'Computational Propaganda (CompProp)' project at the Oxford Internet Institute and the National Science Foundation supported 'Political Bots Project' at the University of Washington. He works on the Digital Activism Research Project and as the research and undergraduate community coordinator at the Center for Communication and Civic Engagement. His scholarship focuses on questions related to automation, politics, and digital culture.

Michael Yankoski earned his PhD from the University of Notre Dame's Kroc Institute for International Peace Studies and has since enjoyed two Postdoctoral Research Associate positions at Notre Dame's Department of Computer Science and Engineering as well as at the Davis Institute for Artificial Intelligence at Colby College. He joined the FAIS Lab at The College of William & Mary, as a Research Faculty helping steer research initiatives that utilize Large Language Models for HAT (Human and AI Teaming) experimentation as well as large scale population simulation research.

Acronyms

ADA	Americans With Disabilities Act
ADHD	Attention Deficit-Hyperactivity Disorder
AGI	Artificial General Intelligence
AGV	Automated Guided Vehicle
AI	Artificial Intelligence
AIG	American International Group, Inc.
A.L.I.C.E.	Artificial Linguistic Internet Computer Entity
AMA	Agricultural Marketing Act of 1929
ANT	Actor Network Theory
API	Application Programming Interface
ARPA	Advanced Research Projects Agency
AUV	Autonomous Underwater Vehicle
BBC	British Broadcasting Company
CEO	Chief Executive Officer
CIA	Central Intelligence Agency
DARPA	Defense Advanced Research Projects Agency
DDoS	Distributed Denial of Service Attack
DoD	US Department of Defense
DoS	Denial of Service Attack
DSB	Defense Science Board
ECM/s	Electronic Contract Manufacturers
EU	European Union
FHA	Federal Housing Administration
FHLB	Federal Home Loan Bank
FM	Field Manual
FOMO	Fear of Missing Out

GATT	General Agreement on Trade and Tariffs
GB	Great Britain
GDP/s	Gross Domestic Product
GPS	Global Positioning System
GSEs	Government-Sponsored Enterprises
GVC	Global Value Chain
HOLC	Home Owners Loan Corporation
HRI	Human Robot Interaction
HUD	Department of Housing and Urban Development
IBM	International Business Machines
ICT	Information Communications Technologies
ID	Identification
IFR	International Federation of Robotics
IMF	International Monetary Fund
INF Treaty	Intermediate-Range Nuclear Forces Treaty
IoT	Internet of Things
IP	Intellectual Property
IPR	Intellectual Property Rights
IR	International Relations
IRC	International Committee of the Red Cross
IRC	Internet Relay Chat
ISA	Ideological State Apparatus
IT	Information Technology/Technologies
JTAC	Joint Terminal Attack Controller
KGB	Komitet gosudarstvennoy bezopasnosti
LARs	Lethal Autonomous Robots
LAVs	Lethal Autonomous Vehicles
LAWS	Lethal Autonomous Weapons Systems
LIEO	Liberal International Economic Order
LMIC	Low-and-Middle-Income
MBSs	Mortgage Backed Securities
MITI	Ministry of International Trade and Industry
NAFTA	North American Free Trade Agreement
NATO	North Atlantic Treaty Organization
NBS	National Bank System
NLP	Natural Language Processing
NSA	National Security Agency
PFIAB	President's Foreign Intelligence Advisory Board
PLA	People's Liberation Army
RAF	Royal Air Force
RFC	Reconstruction Finance Corporation
RMA	Revolution in Military Affairs
RPA	Remote Piloted Vehicle

RSA	Repressive State Apparatus
S&Ls	Savings and Loan Banks
SCI	Strategic Computing Initiative
SEC	Securities and Exchange Commission
SMS	Short Message Service
SNS	Social Networking Services
SWAT	Special Weapons and Tactics
TIB	Transnational Interest Bloc/s
TRIPS	Trade-Related Intellectual Property Rights
UAV/s	Unmanned Areial Vehicles
UK	United Kingdom
UN	United Nations
US	United States
USAF	United States Air Force
USSR	Union of Soviet Socialist Republic
VA	Department of Veterans Affairs
WTO	World Trade Organization

List of Figures

CHAPTER 1

Robots and Political Economy

Ryan David Kiggins

Utopia or Reality?

The "Terminator conundrum" is gripping the attention of and provoking anxiety among policymakers, military officers, and academics. First articulated by vice chairman of the joint chiefs of staff, US air force General Paul Selva, during an address delivered at the Brookings Institution, in Washington, D.C., on January 22, 2016[1]; the concept expresses that human technological development has reached an inflection point where policymakers decide either to press forward with employing autonomous, artificial intelligence supported, smart robots, computers, and machines in war or, we choose not to cross the robot Rubicon. Roff and Singer (2016) echo this claim arguing that the next president of the United States will be faced with a decision on whether or not to press forward with autonomous and artificial intelligence enhanced weapons. Such technological developments have led some luminaries such as Stephen Hawking

R. D. Kiggins (✉)
Department of Political Science, University of Central Oklahoma, Edmond, OK, USA
e-mail: rkiggins@uco.edu

[1] Selva, Paul Gen. 2016. https://www.flightglobal.com/news/articles/dod-official-world-faces-terminator-conundrum-on-421094/. Last accessed 26 August 2016.

R. Kiggins (ed.), *The Political Economy of Robots*, International Political Economy Series, https://doi.org/10.1007/978-3-031-95171-8_1

to assert that, "The development of full artificial intelligence could spell the end of the human race" (Cellan-Jones 2014, no page no.). Others, including a founder of Google, Larry Page, are more sanguine, hopeful that recent and future technological developments in robot autonomy and artificial intelligence may benefit humankind (ibid.). Fears of artificial intelligence induced apocalypse and Terminator conundrums reveal an enduring puzzle concerning the prospects for human prosperity and security in an automated global political economy. This puzzle reflects human fear of the unknown and the unpredictability of technological change. This collection of essays constitutes brief interventions intended to provoke reflection, deliberation, and debate about practical implications to human affairs presented by the advent of autonomous robots, artificial intelligence, and other automated and autonomous information technologies.

We have in mind machines and computing systems that may be understood as functioning, completing a task, or otherwise performing an activity independent of human intervention (see Nof 2009). Going forward in this introduction, the term "robots" should be understood as referring to autonomous, automated, computer networked, and artificial intelligence supported machines and information technologies. Such robots include, for example, commercial and military drones, autonomous cars, ships, and planes, software "bots," viruses, malware, the Internet, and Internet connected devices, among other information technologies with consumer and military applications. This collection examines topics that include automated war, automated production associated with changing global trade patterns, artificial intelligence and big data supported policy decision-making, automated surveillance, terrorism, financial speculation, and automated transnational crime with an eye to providing insights into the effects of automation on political, social, and economic institutions, practices, and norms created to govern the global quest for prosperity and security. Intentionally, this collection, save for the reprinting of a primer on ethics and robotics by, Ron Arkin, is focusing on practical implications of the automated robot revolution to human affairs.

Some may quibble with the intentional divergence from solely investigating ethical dilemmas associated with automated robotics technologies, especially in war. Much has been undertaken scrutinizing ethical dilemmas of military grade robots engaging in target selection, engagement, and destruction that have necessarily brought such wicked problems into public consciousness and discourse (Arkin 2009; Beard 2013; Krishnan

2009; Roff 2014; Schulzke 2013; Sharkey 2007; Sparrow 2007, 2009, 2011; Walsh 2015). Pondering, expounding, and opining about Terminator conundrums certainly casts needed illumination on normative politics associated with the advent of robots. Such politics deserve attention and critique but, we risk overlooking other areas of human affairs where robots do replace, replicate, and respond in what have been or, currently are, solely human activities outside the narrow confines of war making. Baldy put, in the global political economy, the robot Rubicon has already been crossed and is rapidly fading in the distance. Supply chains (global value chains) are increasingly automated, we are on the verge of autonomous transportation technologies, and shopping via the Internet has long entered the consumer mainstream. Robotic surgery, online college courses, and egovernment-based policy decision-making likewise support the claim that humankind is fast leaving the age of humans and entering the age of robots. How will this new age interact with and effect human prosperity and security beyond the narrow confines of warmaking? Focus on this question is by no means a dismissal of ethical dilemmas, or utopia or, as Carr and Cox (2001) explains, what we think *ought* to be. Indeed, several contributions to this volume scrutinize militarized robots and attendant ethical dilemmas. Rather, it is to recognize that, "Politics are made up of two elements – utopia and reality. There is no greater barrier to clear political thinking than failure to distinguish between…utopia, and…reality" (Carr and Cox 2001, 87). Utopia, in the form of ethical ideals, has been identified, investigated, and re-investigated with little attention paid to reality—the direct and indirect implications of robots to human activity beyond war. This collection of essays addresses that oversight through the framework of political economy which enables diversification of interventions that investigate practical questions related to the robotics age in addition to new perspectives on the ethical dimensions of lethal and potentially autonomous robots.

The framework of political economy has been invoked to underscore that robots affect human affairs in addition to the discreet realm of military affairs and ethical dilemmas that obfuscate decisions to use violence against humans. Political economy has many meanings arising out of haphazard use of the term and owing to its contextual flexibility (Schofield et al. 2013). As an empirical matter, political economy may refer to the sum total and interaction of economic transactions, regulations, and political institutions within a specified geographic region such as a city, sub-national unit, national unit, or global unit. More abstractly, political economy may refer to the academic study of interactions between market, state, society, and individual; focusing on how those interactions distribute gains (see Gilpin and Gilpin 2001). Alternatively, one could view political economy as being fundamentally concerned with agency and how agency is affected by a clash between the "logic of the state" and the "logic of the market" (Gilpin and Gilpin 2001, 81). Of course, as one investigates human agency in the context of political economy, the shear complexity of understanding, describing, and explaining the emotional, psychological, and physical dimensions of human decision-making within the context of political economy, has led to the development of several theoretical approaches all pregnant with disparate ontologies, epistemologies, and methodologies (Wade 2012). This collection is necessarily ontologically, epistemologically, and methodologically heterodox in an effort to provoke some triangulated understanding of practical implications robots may present to human affairs.

The following interventions constitute insights and arguments that interrogate, investigate, and scrutinize potential affects that robot technologies may have on agents operating within political economy. We are less concerned with the academic study of political economy, being more so with practical considerations associated with the interaction of robots and humans in, for example, production, trade, finance, education, and, more broadly, public policy decision-making. Nevertheless, this collection of essays does not eschew ethical dilemmas that may arise as robots fill rolls in political economy previously filled by humans. Indeed, we attempt to strike a balance between reality and utopia, recognizing that investigating both is equally valuable, even essential, to clear understanding of political economy phenomena. There is a tradeoff to be made with such a broad focus. Adopting a political economy framework allows for more breadth in analysis and discussion of the different ways robots affect the prospects for human prosperity and peace. Making this tradeoff allows for a more broad-based view of the implications of robots to human affairs that may assist the identification of topics and issues that require more in-depth analysis in future research.

Human Prosperity and Peace

Political economy is uniquely suited, as a framework, for parsing through issues attendant with human prosperity and security. In this regard, human prosperity may be understood as to suggest a minimum pecuniary basis for sustaining human life. Through direct participation in political economy humans may be able to provide for their prosperity and, thereby, achieve a type of security. Ironically, under the current capitalist-free market structure of political economy, humans must experience a degree of insecurity as an incentive to participate in political economy (Buzan et al. 1998). Prosperity and security are dialectically entangled in the ongoing liberal international economic order (LIEO).

The LIEO is institutionally designed to pursue efficiency in use of resources and gains rather than justice or an equitable distribution of gains. The wage-labor contract is a classic example of this prosperity/security dialectic, incentivizing the selling of one's labor in the market in return for a wage on which one depends to sustain a basic living standard. The alternative is pecuniary insecurity risking starvation even death. Automation may contribute to that insecurity as it unravels the wage-labor contract that has remained at the core of the LIEO for nearly two centuries. Such a development directly supports the liberal ideal to do away with any barrier to efficient economic exchange and production. But at the cost of incentivizing political upheaval as economic insecurity for humans becomes acute. Controversies over fair wages for workers or, high wages that threaten employer profitability have contributed to rather nasty political upheavals within ostensibly liberal societies (Brown 2014; Marx and Engels 2002; Olzak 1989; Silver 2015; Tilly and Tarrow 2015). Automation benefits political leaders and employers by removing the risk of labor unrest which, serves as a barrier to efficient production, through replacing humans with robots. This replacement of humans by robots in production serves to enhance prospects for economic prosperity while potentially worsening economic inequality and, perhaps, political stability or, more directly, peace.

The notion human peace is intended to refer to a condition in which stability among potential belligerents prevails; that is, humans are free from fear of politically sanctioned violence. In the absence of violence, commerce, communication, and community may flourish. The advent of robots may worsen the prospects for human peace by threatening human security. As an analytical framework, human security is contested with critics noting its haphazard application and lack of conceptual rigor (Khong 2001; Paris 2001; Suhrke 1999). Others have attempted to refine, revise, and precisely delineate human security (Acharya 2001; Kfir 2015; Owens 2004). Among those scholars working with human security as an analytical, interpretive, and explanative framework, two schools shave emerged. One which argues for restricting human security to a focus on systemic violence against individuals, i.e., use of violence against individuals by governing institutions (Mack 2004). We might label this school as the freedom from fear approach to human security. The other argues for a more expansive conceptual application suggesting that for individuals to be secure, they must, in addition to being free from fear, also be free from want (Thakur 2004). The disagreement between the two human security approaches reflects a broader debate among international relations scholars regarding the conceptual boundaries of security as an analytical, interpretive, and explanative framework (Browning and McDonald 2013; Kiggins 2015; Smith 2005). All of this is to say that what counts as human security, for our purposes, is a recognition that in order for humans to be secure, freedom from want and freedom from fear must be experienced simultaneously. It is for this reason that contributions in this book speak to both the notion of human prosperity and human peace. This view situates the contributions in this book squarely in the expansive school of human security working within the broad framework of political economy. As new information technologies have developed, so have emerged benefits and detriments to human prosperity and peace.

Information and the Robotic Revolution

The information technology revolution can be traced to the invention of the transistor and networked computing (Abbate 1999). Combined these inventions spurred revolutions in military affairs altering the logic of national and global security and, regarding human prosperity, altering the very structure of global production, trade, and finance (Benkler 2006; Brynjolfsson and McAfee 2014; Castells 1996; Howard 2015; Mueller 2010; Morozov 2011; Singer 2009; Singer and Friedman 2014; Wu 2011). In other arenas of human activity, such as espionage, techniques, and tactics are being replaced by skilled hackers capable of gaining unauthorized entry into computer systems, at great distance, to acquire sensitive information stored on those very same computer systems. Network connected and armed military machines now fly skies over areas of the globe frequented by people labeled terrorists. These unmanned aerial vehicles or, drones, lower the risk of battlespace casualties for the owners of such robots while increasing the risk for death, destruction, and political instability in regions in which such robots fly. Through networked computing global finance has emerged facilitating the rapid transfer of capital to the furthest reaches of the global political economy in search of the highest return. The rapidity of capital transfer also brings with it the risk of near instantaneous capital flight with attendant economic catastrophe as local and regional capital markets come to a halt freezing economic activity. Networked computing has facilitated management of globalized value chains, the delicate logistical ballet, performed by global businesses, by which raw materials, components, and finished products are efficiently extracted, processed, manufactured, assembled, and shipped to retail locations around the globe for sale. And, with that globalization of production, have come employment dislocations as manufacturing in one locale has been replaced with assembling in another locale. We are experiencing significant structural change as robotics technologies reduce and, ultimately, replace human participation in political economy. How will human prosperity and peace change as a consequence? The short answer is no one knows for the simple reason that we are at the cusp of the robot political economy. What we can do, based on what we know is attempt to anticipate, illuminate, and scrutinize potential dislocations in an effort to prepare for the inevitable conclusion to the shift from human-based political economy to robot-based political economy.

Contributions grappling with implications of robots to human prosperity, begin with an examination by, Cox and Wartenbe, into how automation enhanced global value chains serve to strengthen a state power—corporate power symbiosis for the purpose of profit extraction. The advent of robots has spurred a shift in corporate organizational and production strategies that deepens, broadens, and enhances the profitability of globalized production. An important insight is that robot technologies, outsourcing, and the liberal ideal are all entangled in a dangerous game in which the winners reap windfall profits at the expense of the losers left with meager wages barely sufficient to sustain their basic survival needs.

Building off the concern that robots may place humans in a position of economic precariousness, Carey and Marak, investigate how automated drug interdiction strategies may ultimately worsen the drug epidemic.

They begin with critiquing the kingpin strategy employed by law enforcement during the last thirty-years. The logic underpinning this strategy is that by capturing, convicting, and incarcerating drug cartel heads, the organizational structure of drug cartels will crumble due to the vacuum of leadership that arises Carey and Marak note that the opposite has, in fact, occurred. Far from eradicating drug cartels, the kingpin strategy has proliferated cartels into smaller, more autonomous, and deadly drug trafficking organizations. In response, law enforcement has attempted to employ automated technologies at borders between sovereign entities to detect drug trafficking, direct law enforcement response, and predict future drug trafficking behavior. This focus on the supply-side of illicit drug trafficking neglects that as singularity approaches, more humans may find themselves economically vulnerable to illicit drug use, increasing demand for drugs. Robots may drive impoverished humans to self-medicate.

Brett Sharp continues investigating dimensions of potential human impoverishment through potential widespread replacement of human labor with automated labor. While acknowledging parallels between Luddite claims that mechanization would impoverish humanity, Sharp is less concerned about human labor being replaced by robots than the use of people analytics, supported by big data and artificial intelligence analysis, to parse through human job applicants. Most concerning for Sharp, is potential harm to the privacy of applicants from data culled from their personal, though public, social media, and web surfing behavior. Sharp advocates for government to develop regulatory regimes that prescribe what type of internet culled data, hiring organizations may use to assess applicant pools. These first three interventions bring to light the precarity human prosperity may experience during the shift from a human labor-based political economy to a robot labor-based economy. The next intervention focuses on the provision of human health reliant upon automated medicine.

An important feature of human prosperity is human health and well-being. Medical technology has advanced to include robotic surgery and diagnosis in an effort to assist healthcare providers offer services that enhance patient care, health, and well-being. At what cost? To what extent have new medical technologies contributed to a crisis in the American healthcare system through increased patient and health insurance costs? Might a partiality toward technology-based medicine shift financial resources of patient and insurer to expensive technology-based services and away from a principled non-technology-based healthcare provider-based medicine? Skinner and Franz investigate these questions and others through focusing on the politics of emergent technologies in the American healthcare system. Healthcare represents a multi-billion dollar industry within political economy with positive and negative spillover effects throughout other sectors. Focusing on the provision of primary health services, Skinner and Franz question the extent to which emergent technologies will assist primary health providers in efforts to improve overall health and well-being among their patient populations.

Striking a more confident tone, Walsh and Sculos offer a provocative analysis that draws attention to the unprecedented use of a law enforcement bomb disposal robot, to deliver and detonate a C4 explosive device within a kill radius of Micah X. Johnson. Johnson had been cornered by police after callously, brazenly, and without warning, shooting, wounding, and killing several Dallas, TX police officers. Who, in the course of their duty, were providing security for, ironically, a peaceful protest against police killings of unarmed black citizens in the United States on, July 8, 2016, in downtown Dallas, TX. While acknowledging the culpability of Johnson and the awful tragedy experienced by others in consequence of his violent crime spree, Walsh and Sculos critically assess the first use of an armed robot by US law enforcement to kill a criminal suspect. Relying on Althusser's repressive state apparatus concept, Walsh and Sculos connect the use of an armed robot with state agents, acting at the behest of the repressive state apparatus, enforcing and defending the class structure of late capitalism. Walsh and Sculos are most concerned with the growth in capability observed in the repressive state apparatus. Armed robots are now a tool of repression calling into question the narratives of democracy, constitutionality, and freedom employed by the repressive state apparatus to ensure working class fealty to capitalism. In closing, Walsh and Sculos rely on science fiction portrayals of robot resistance to highlight how resistance to the automation of state violence may be accomplished. Far from sounding a note of caution regarding singularity, Walsh and Sculos argue that such a development may finally precipitate the collapse of capitalism and class conflict, replaced by a robot labor-based political economy that frees human beings to pursue creative labor. Thus, the radical restructuring of political economy anticipated by Marx and Engels may yet be birthed through the advent of singularity. The first five contributions represent interventions that investigate some practical implications of robots to human affairs with a focus on human prosperity. The next four contributions occupy a middle ground between the practical and ethical implications of robots to human affairs.

Working for the US government, Department of Commerce, Office of Telecommunications, in 1977, Marc Porat published a prescient research report in which he coined the term "information economy." The focus of the report was to estimate the transmission of economically valuable information on the US telecommunications system of the period. Forty years later, to claim that the current epoch of liberal-capitalist political economy is structured by information is bordering on cliché. Nevertheless, widespread adoption, access, and use of the internet and internet connected devices has created a tsunami of information, economically valuable information on which the fortunes of founders of companies such as Microsoft, Google, Apple, and Facebook have been built. A recent technological development in the political economy of information is the political bot, an autonomous software program that, according to Woolley's contribution, may account for as much as 55% of internet traffic. Woolley investigates the political economy of political bots for the purpose of assessing the extent to which public opinion may be influenced for political or economic gain. His contribution also serves as a bridge to others that investigate the implications to human affairs of autonomous or artificial intelligence supported public policy decision-making and democracy.

Continuing the pursuit of autonomous decision-making, Gatch, investigates how autonomous financial innovation contributes to riskier financial speculation. Specifically, he describes a divide between the political economy of commodity money and that of fiat money, or what can be called "chartalism." A chartalist political economy is characterized by autonomous financial innovation, driven by private interests, that creates new forms of liquidity and embeds new forms of risk in the financial system. To the extent that private financial innovation is profitable, the supply of private financial assets can be increased and configured as information-insensitive, or "safe," serving in turn as a basis for new rounds of risk-taking. Public financial and monetary authorities aim above all to maintain the stability of the system, and toward that end seek to regulate and otherwise control this innovation. However, as part of the logic of chartalism, these regulatory interventions, undertaken to reinforce safety, themselves form the terrain upon which private interests create new strategies of risk-taking that assume, as their points of departure, sites of risklessness in the financial system. Autonomous financial innovation, ironically, arises out of regulatory efforts to control risks associated with autonomy potentially undercutting governance of autonomy initiatives and, possibly, market stability.

An important feature of democratic governance is ensuring stable, consistent, and informed political judgment. Lacking this feature, democratic governance is threatened with instability and whimsical ad-hoc policy decision-making. While policy experts, technocrats, and political leadership all contribute to stable informed political judgment, equally important is political participation of the citizenry. An engaged public that utilizes its reason, its capacity to make informed judgments, consistent with political ideals of their society, during the course of their participation in democratic processes, is of most concern to Thiele. Noting that there is much concern that artificial intelligence may prove the ultimate demise of humanity, Thiele shifts our focus to the implications of artificial intelligence to political judgment. Labeling the replacement of human-based political decision-making with artificial intelligence-based political decision-making engenders a wicked problem concerning the loss by humans of political judgment. Lacking political judgment, humans may do to democracy what artificial intelligence may do to their civilization. Shifting political decision-making to robots may undercut the capacity for humans to reason about the limits and possibilities of liberty, equality, and justice ironically rendering humans automatons. The remedy Thiele proposes is to cultivate practical wisdom, knowledge acquired through experience, in the hope that human cognitive capacities will remain undiminished in a future shared by humans, artificial intelligence, and robots.

Following the same conclusion, that artificial intelligence and robot-supported political decision-making is inevitable, Kiggins argues that theories of international relations may soon confront a crisis of explanative power. This crisis emerges from a shift in how policymakers determine policy decisions to effectuate outcomes consistent with global policy objectives. International relations theory is a composite of heterodox traditions many of which are in disagreement but nonetheless share a common objective: the explanation of human decision-making under certain conditions. The emergence of information technologies has inundated policymakers with data derived from tracking, recording, and analyzing information technology user behavior. This data deluge has spawned new data analysis techniques and technologies leveraged when making and automating policy decisions. International relations scholars must shift thinking away from human centricity to a framework that takes seriously the proposition that automated systems are consequential actors in global politics. Automated systems have agentive capacity in global politics, the power to influence outcomes, shape interests, and change relations among all actors in global politics. Incorporating reflexivity into theory building activities may enable international relations scholars to account for all consequential actors in global politics.

Echoing the notion that technological change has outpaced the explanative power of our analytical frameworks, Olmstead, looks to draw lessons from US diplomatic history that may prove useful to scholars and policymakers today. Comparing the US diplomatic response to the submarine with the US diplomatic narrative for use of drones, Olmstead focuses our attention on how technological revolutions in military affairs presents significant challenges for policymakers. For the last 100 years, US presidents have been confronted with moral and legal dilemmas arising out of new military technologies being deployed in and used by US armed forces to prosecute war. Much controversy concerning the use of drones for targeted killing harkens back to the controversy over German use of submarines for targeted sinking of US ships before US entry into World War I. The framework developed between the United States and Germany Olmstead, suggests, may be suitable for incorporating drone-based targeted killing into contemporary legal and moral norms and practices affecting the use of military technologies during armed conflict. The following two interventions shift our attention back to how emergent technologies with military applications present novel problems to political decision-makers. The first, explores the extent to which the incorporation of robots into military formations may alter military organization structure, practices, and, the second, investigates how adopting lethal autonomous weapons systems by China and the US armed forces may alter affairs between these states.

Manjikian draws attention to how adopting robots to fulfil military missions will radically alter the structure of military organizations. Military decision-making relies on information, leadership, and unit cohesion which includes, for Manjikian's purposes, discipline. Artificial intelligence supported robots may enhance military decision-making in terms of accessible, accurate, and timely information. However, two problems arise from this radical restructuring of militaries where leadership and unit cohesion are concerned. On the issue of leadership, Manjikian explores the dilemma of human soldiers being led by an algorithm; that is, receiving orders and directions in the battlespace by an automated system. It becomes questionable whether or not trust and inspiration that leaders must prompt in their soldiers can ever be achieved by an automated command and control system. Such a development undercuts the efficacy of militaries. This concern spills over to the issue of unit cohesion so necessary for successfully prosecuting military campaigns. As soldiers themselves become technologically enhanced, the loss of individual identity associated with being a member of a unit may undercut unit cohesion as such soldiers view themselves and are viewed by others as mercenaries effectively undermining the moral enterprise of war. The incorporation of robots into military units presents significant challenges going forward, some have pointed to potential adversaries of the United States and its allies as gaining a technological edge as they adopt militarized robots.

Assessing fears that Chinese military robots may grant a technological edge against potential adversaries, Austin suggests this view may be immoderate. Austin notes that China's industrial capability for robotics technology remains in development, that China's armed forces have yet to fully achieve an information-centric approach to warfare, and that within the leadership of China's armed forces there is lacking advocacy for developing robotic armies and navies. It is unclear that China is capable of achieving a technological edge where lethal autonomous weapons systems—armed robots—are concerned. Furthermore, lethal autonomous robots in China's armed forces may prove to be small in number and restricted for use by specialized military intelligence and security services rather than by frontline military forces. Any threat to the United States or its allies, posed by China's military robots, will be low. However, China will not face the same ethical dilemmas currently confronting Western Democracies, possibly heightening security dilemma[2] effects for US policymakers. US policymakers may be forced to press forward with developing, deploying, and using lethal autonomous robots as a check against other states, like China. The price paid will be less emphasis on the ethical dilemmas associated with employing lethal autonomous robots.

[2] The security dilemma, within international relations theory, is the choice to arm, risking provocation; or, not to arm, risking vulnerability.

Opening our shift in focus to ethical dilemmas presented by lethal autonomous robots, Lee scrutinizes autonomy, ethical decision-making, and ethical responsibility. Autonomy is a contested concept making it difficult to develop consistent and widely accepted norms governing the use of lethal autonomous robots by state actors. The contested nature of autonomy serves to heighten the need for the incorporation of ethical decision-making into the computer code on which autonomous decision-making is based. Crucially, for Lee, humans are solely capable for developing ethical-responses. Computer coders, engineers, military leaders, and policymakers, consequently, all share ethical responsibility for ensuring ethical decision-making by lethal autonomous robots.

Arkin closes our collection with his case for continuing the development of lethal autonomous robots. Noting that humans appear unrelenting in their penchant for resorting to violence and war to resolve disputes, Arkin suggests that developing lethal autonomous and intelligent robots may prove to limit intended and unintended casualties in war. Pushing still further, Arkin argues that widespread adoption of lethal autonomous and intelligent robots on the battlefield may also eliminate war atrocities given that humans have a poor record of ethical decision-making during combat. The possibility that robot-based war may prove more ethical than human-based war is enough for Arkin to offer a full-throated endorsement for continuing the development of autonomous and intelligent robots.

Works Cited

Abbate, J. (1999). *Inventing the Internet*. Cambridge, Mass, MIT Press.

Acharya, A. (2001). "Human security: East versus west." *International Journal* 56(3): 442–460.

Arkin, R. (2009). *Governing lethal behavior in autonomous robots*, CRC Press.

Beard, J. M. (2013). "Autonomous Weapons and Human Responsibilities." *Geo. J. Int'l L.* 45: 617.

Benkler, Y. (2006). *The wealth of networks: How social production transforms markets and freedom*, Yale University Press.

Brown, C. (2014). *Racial Conflicts and Violence in the Labor Market: Roots in the 1919 Steel Strike*, Routledge.

Browning, C. S. and M. McDonald (2013). "The future of critical security studies: Ethics and the politics of security." *European Journal of International Relations* 19(2): 235–255.

Brynjolfsson, E. and A. McAfee (2014). *The second machine age: work, progress, and prosperity in a time of brilliant technologies*, WW Norton & Company.
Buzan, B. Wæver, O. and J. De Wilde (1998). *Security: A new framework for analysis*. Lynne Rienner Publishers.
Carr, E. H. and M. Cox (2001). *The twenty years' crisis, 1919–1939 : an introduction to the study of international relations*. Basingstoke, Palgrave.
Castells, M. (1996). *The rise of the network society*. Cambridge, Mass., Blackwell Publishers.
Cellan-Jones, R. (2014). "Stephen Hawking warns artificial intelligence could end mankind." Retrieved July 10, 2015, from http://www.bbc.com/news/technology-30290540.
Gilpin, R. and J. M. Gilpin (2001). *Global political economy : understanding the international economic order*. Princeton, N.J., Princeton University Press.
Howard, P. N. (2015). *Pax Technica: How the Internet of things may set us free or lock us up*, Yale University Press.
Kfir, I. (2015). "NATO's Paradigm Shift: Searching for a Traditional Security–Human Security Nexus." *Contemporary Security Policy* 36(2): 219–243.
Khong, Y. F. (2001). "Human security: A shotgun approach to alleviating human misery?" *Global Governance* 7(3): 231–236.
Kiggins, R. D. (2015). The strategic and security implications of rare earths. *The Political Economy of Rare Earth Elements*, Springer: 1-19.
Krishnan, A. (2009). *Killer robots: legality and ethicality of autonomous weapons*, Ashgate Publishing, Ltd.
Mack, A. (2004). "A signifier of shared values." *Security Dialogue* 35(3): 366–367.
Marx, K. and F. Engels (2002). *The communist manifesto*, Penguin.
Morozov, E. (2011). *The net delusion: The dark side of Internet freedom*. PublicAffairs.
Mueller, M. L. (2010). *Networks and states: The global politics of Internet governance*. MIT press.
Nof, S. Y. (2009). Automation: What It Means to Us Around the World. i. S. Y. Nof, Springer Berlin Heidelberg: 13–52.
Olzak, S. (1989). "Labor unrest, immigration, and ethnic conflict in urban America, 1880–1914." *American Journal of Sociology*: 1303–1333.
Paris, R. (2001). "Human security: Paradigm shift or hot air?" *International Security* 26(2): 87–102.
Porat, M. U. (1977). The information economy: definition and measurement. O. o. T. (DOC). Washington, DC.
Roff, H. M. (2014). "The strategic robot problem: Lethal autonomous weapons in war." *Journal of Military Ethics* 13(3): 211–227.
Roff, H. M. a. P. W. Singer. (2016). The Next President Will Decide the Fate of Killer Robots—and the Future of War. *Wired*. 2016.

Schofield, N., et al. (2013). *Advances in Political Economy*, Springer.
Schulzke, M. (2013). "Autonomous weapons and distributed responsibility." *Philosophy & Technology* 26(2): 203–219.
Sharkey, N. (2007). "Automated killers and the computing profession." *Computer* 40(11): 124–123.
Silver, B. (2015). "The Global Restructuring of Labor Movements." *Journal of World-Systems Research* 4(1): 8–9.
Singer, P. W. (2009). *Wired for war: The robotics revolution and conflict in the 21st century*, Penguin.
Singer, P. W. and A. Friedman (2014). *Cybersecurity: What Everyone Needs to Know*, Oxford University Press.
Smith, S. (2005). "The contested concept of security." *Critical security studies and world politics*: 27–62.
Sparrow, R. (2007). "Killer robots." *Journal of applied philosophy* 24(1): 62–77.
Sparrow, R. (2009). "Predators or plowshares? Arms control of robotic weapons." *IEEE Technology and Society Magazine* 28(1): 25–29.
Sparrow, R. (2011). "Robotic weapons and the future of war." *New wars and new soldiers: Military ethics in the contemporary world:* 117–133.
Suhrke, A. (1999). "Human security and the interests of states." *Security Dialogue* 30(3): 265–276.
Thakur, R. (2004). "A political worldview." *Security Dialogue* 35(3): 347–348.
Tilly, C. and S. G. Tarrow (2015). *Contentious politics*, Oxford University Press.
Wade, L. L. (2012). *Political Economy: Recent Views*, Springer Science & Business Media.
Walsh, J. I. (2015). "Political accountability and autonomous weapons." *Research & Politics* 2(4): 2053168015606749.

CHAPTER 2

The Politics of Global Value Chains

Ronald W. Cox and Michael Wartenbe

Introduction

Over the past thirty years of the global economy, there has been a steady expansion, consolidation, and integration of global value chains across the borders of states.[1] In these value chains, networks of actors are linked in the production and distribution of products and services. This trend has established a hierarchy of relationships that is dominated by a small number of firms at the top of the value chain, characterized by their ownership of patent rights, their control of branding and marketing, and their linkage with powerful retail outlets that sell (and privilege) their

[1] There is a vast literature on global supply chains and "value" chains. For one of the best recent summaries, complete with a good bibliography and several timely articles in the same journal issue, see Jeffrey Nielson, Bill Pritchard, and Henry Wai-chung Yeung, "Global Value Chains and Global Production Networks in the Changing International Political Economy: An Introduction," *Review of International Political Economy*, 21:1, 1–8.

R. W. Cox (✉) · M. Wartenbe
Florida International University, Miami, FL, USA
e-mail: coxr@fiu.edu

M. Wartenbe
e-mail: michael.wartenbe@fiu.edu

R. Kiggins (ed.), *The Political Economy of Robots*, International Political Economy Series, https://doi.org/10.1007/978-3-031-95171-8_2

branded products on their store (or virtual) shelves. At the bottom of the value chain, subcontractors engage in fierce competition with each other to be able to offer the top-tier firms the lowest production and delivery costs for a component part or a finished product (Nolan et al. 2007). In between these extremes are firms that engage in more sophisticated technological design and production to meet the specifications of high-end computer and electronic production processes that have become central to the new system of transnational production and accumulation (Applebaum 2008; Starosta 2010). Financial networks are also becoming more prominent in connecting the different actors in the value chain with each other (Serfati 2008). Corporations at the top of the value chain look toward institutional financial actors to help establish the political and economic linkages essential to the establishment of the value chain. They also look to governments and multilateral trade and financial organizations to facilitate the terms and conditions necessary for the establishment and functioning of global value chains (Cox 2008). All of these actors form what we refer to as a Transnational Interest Bloc that is led by the most powerful transnational firms that secure the highest profit from the value chain, but also engage in relationships with a wide range of other actors positioned at different levels of power and influence within the value chain (Cox 2012).

The Transnational Interest Bloc includes numerous corporate, business, and political actors that are linked by this process of global production. This interest bloc is led by the structural and instrumental power of large-scale transnational corporations that occupy a hegemonic position within the bloc. The modern value chain provides the linkage between production processes that involve multiple corporate actors, whose position in the value chain is shaped by the degree of power, profit, and ownership rights distributed throughout the value chain. A production profile of the current global economy starts with the ascendancy of both the high-tech computer and information sectors and the financial sector. The integration of advanced computer networks in modern manufacturing and the incorporation of these networks in retail operations allow for a "just-in-time" delivery system for the sale of finished goods. As such, the high-tech sector occupies the most important strategic location in modern global capitalism (Harris 2008: 19–39). High-tech firms engaged in the production of computers, operating systems, microprocessors, and information systems are central to the new economics and politics of global production. This is due to the steady expansion of the high-tech sector over the past 30 years, measured in rising corporate profits (relative to other sectors) and dispersed location of production. The electronics sector was a pioneer of producing component parts in a wide variety of locations as early as the 1960s. The high-tech sector has also been crucial for the success and expansion of other sectors of the global economy, including manufacturing and retail, giving it enhanced strategic importance for global capitalism.

The creation of a global value chain involves considerable expressions of political and economic power wielded by dominant firms and their partners located at different geostrategic and spatial positions within the value chain. The central purpose of this chapter will be to examine the power relationships embedded within Transnational Interest Blocs, especially in the high-tech sectors of the global economy, and the political economy of the extraction of profit within the global value chain. We will also address the increasing consolidation of corporate power within the value chain by examining the intersection of market power and political power with the incorporation of high-technology production methods and its impact on the global workforce.

There has been extensive scholarship over the past two decades on the economics of global value chains. But much of this literature does not fully capture the political aspects of how these supply chains came into existence. Dominant firms, or "system integrators," have been able to secure their position at the top of global value chains by using their economic and political power within the market and within the state (Levy 2008). Firms such as Microsoft, Apple, and Intel did not simply emerge as powerful market players through their own innovation and productivity. They have been able to take advantage of considerable public financing of research and development to secure low-cost access to technological innovations that preceded their very existence as corporations (Perelman 2003). The manner in which public goods have been socialized into private profits is central to understanding the politics of corporate power, and has been true in every era of capitalism.

Neoliberalism as ideology wants to strip this history from our knowledge base in order to proclaim the unfettered virtues of private accumulation, but such accumulation has long rested on a close relationship between corporate power and the state. What has changed in recent decades is that governments are stripping away more of their public functions for private profit, especially compared to the Bretton Woods era of capitalism from 1945 to 1973. This has contributed to a corporate-driven process of accumulation that has simultaneously increased the gap between rich and poor on a global scale while delegitimizing the public functions of governments (Kotz 2015).

Transnational Interest Blocs have emerged as a powerful political force in a wide range of locations throughout the global political economy. However, their emergence is still conditioned and mediated by the institutional, socioeconomic, and class histories of nation-states. In other words, corporate political actors have extended their influence on a global scale, but not under the circumstances of their choosing. They have had to advance their political agenda in a wide range of political and economic contexts, and often compete with other corporate blocs for access to state and market power. This means that any account of Transnational Interest Blocs cannot stop with an analysis of the dominant firms at the top of global value chains. We must also examine the politics of how value chains are constructed at various locations within the system. This means examining which actors within the value chain control the most profitable activities, how costs are pushed downward within the value chain through advanced technology and automation, and how workers are segmented within the value chain through a geographical/spatial configuration that reinforces exploitation and extraction of profit from the bottom of the chain to the top.

Origins and Scope of the Global Value Chain

Transnational firms that locate much of their research and development activities in the U.S., Western Europe, and Japan have increasingly relied on global value chains from the 1980s to the present. The most extensive manifestation of this trend has occurred within firms based in the U.S. and Japan, with Western European firms lagging behind in the timing, pace, and quantity of reliance on global value chains. Transnational corporations with research and development activities concentrated in the U.S., with different intensities depending on sectoral position, began to restructure their operations as early as the 1980s to concentrate on ownership of high-value activities that included research and design, branding and marketing, and ownership of intellectual property rights or patents (Prechel 2000: 20). These firms then began to sell corporate divisions that were involved with the manufacturing of finished products in an effort to lower the costs of production. Instead of producing goods "in-house," transnational firms subcontracted or outsourced production to companies in low-cost locations where production methods had to meet the specifications of the contractor. The process has grown more complex over time, and often involves a range of actors that facilitate the transaction between the transnational firm that establishes the terms of production, the intermediaries that are paid to facilitate the logistics of the production operation, the often large-scale "turnkey" companies that manage the production of the product from start-to-finish, and the numerous small-scale producers that produce inputs needed as part of a "just-in-time" delivery system that links to the finished product. Workers at every phase of the production process are subject to rules, regulations, and restrictions on their job classification that makes it easy for contractors to maintain low wages and poor working conditions, reinforced by the reliance on temporary work contracts and part-time classification schemes (Sealy 2010).

The politics of the global value chain rests on a hierarchy of power that is made possible by the segmentation of design and manufacture, which is endemic to the value chain. The most highly profitable actors within this chain are those transnational firms, narrowly concentrated at the top of the value chain hierarchy or pyramid, that own the intellectual property rights to the design, branding, and marketing of the product. Reinforced by the politics of national and international patent law, corporations at the top of the value chain pyramid are able to leverage their ownership of high-value activities so that costs are pushed further down the value chain, impacting the profit margins of a range of other actors who compete with each other to move up the value chain ladder, or to prevent themselves from falling off the ladder entirely (Nolan and Zhang 2010). It is technological innovation that has made this system possible in the first place, with advanced telecommunications systems and computer hardware and software being integrated within the coordination of production activities that occurs across countries and regions of the global economy. The high-tech sector, broadly defined as including telecommunications, computer software and hardware, semiconductors, electronic component parts, and information systems, occupies a central strategic location in the creation and maintenance of global value chains. Therefore, we will focus on this high-tech sector to provide examples of the politics of global value chains as we work through the history and implications of power relationships within these chains.

At the top of the global value chains are the transnational corporations that own the rights to intellectual property that provide these firms with both economic advantage and political power relative to subordinates within the supply chain. Baruch Lev's analysis of the value of corporations on the Standard and Poor's 500 from the 1980s through the 1990s concluded that intangible assets such as intellectual property rights have become much more important than tangible assets as a percentage of a firm's value. By 1998, intangible assets accounted for 85% of a corporation's value, while tangible assets accounted for only 15%. This was a significant reversal from 1982, when tangible assets accounted for 62% of a firm's value while only 38% of that value was derived from intangible assets. This shift in corporate value tracks the changes in corporate organizational structure toward a concentration of ownership of high-value activities focused around product design and a selling off of corporate divisions that produce goods. This shift in corporate organizational structure was made possible by the high-tech revolution, which enabled corporations to segment design from production across state borders in a more spatially compartmentalized fashion.

Concretely, this shift in corporate organizational structure has meant that corporations at the top of the global value chain own the intellectual property rights that generate a disproportionate share of the revenue and the profits from the value chain. Corporations acquire ownership of intellectual property rights through a variety of mechanisms, which include research and development; purchasing patents from other inventors; mergers and acquisitions; and reliance on publicly funded research and development programs which privatize innovations at subsidized costs. By concentrating their investments on research and design and the strategic acquisition of patents, corporations based in the U.S., Japan, and Western Europe have been able to steadily increase their reliance on the foreign production of intermediate goods whose labor and final assembly is divided across multiple platforms across countries that the World Bank classifies as low-and-middle income (LMIC). So in assessing this shift in global production strategies, a common statistic used by researchers is the extent to which intermediate goods are becoming a more important part of the overall percentage of imports from LMICs for high-income countries. As expected, imports from LMICs have increased from the 1991 to 2010 in the U.S. (from 20 to 50%), Japan (from 21 to 52%), France (from 10 to 21%), Germany (8% to 23%), and the U.K. (9% to 22%) (Milberg and Winkler 2013: 38).

These percentage increases reflect a greater reliance on global value chains by the dominant transnational corporations based in these five countries. The transnationalization of production has occurred across all sectors of the global economy, but is especially pronounced in the category of high-tech products, whose dependence on global offshoring of both sectoral materials and servicing is at the high end compared to other sectors of global production (Milberg and Winkler 2013: 43–48). The reason is very straightforward: the new high-tech revolution has enabled a concomitant shift in corporate organizational and production strategies by making possible segmentation of production on a global scale. High-tech products are central to such a shift, impacting not just the high-tech industry but all other industries that rely on these new high-tech platforms, from traditional manufacturing to retail services. Those firms that have intellectual property rights ownership of these high-tech sectors therefore have emerged as very strategically powerful, both within the marketplace and within governments, where favorable tax laws have facilitated corporate re-organization on a global scale.

Intellectual Property Rights and Corporate Hegemony

Transnational corporations at the top of global value chains have leveraged ownership of intellectual property rights to wield power relative to other actors in the value chain. The reliance of transnational firms on intangible assets for a majority of revenues and profit margins is part of a global corporate restructuring endemic to the current phase of global capitalism. Facing a declining or stagnating profit rate from the mid-1960s to the early 1980s, transnational corporations that comprise the Global 500 looked toward corporate reorganization to reverse this trend (Prechel 1997: 414; Cox 2012: 17). The decade of the 1980s saw the first steps toward a process of corporate reorganization that occurred unevenly depending on the country, the sector, the particular circumstances of individual firms, and the political and economic opportunities that existed to facilitate corporate reorganization. Nonetheless, we can identify the political economy of transnational corporate restructuring by examining several factors that help us understand the power dynamics of contemporary global value chains.

First, transnational corporate lobbies worked with governments in the U.S., the European Union, and Japan to make protection of Intellectual Property Rights (IPR) a priority. Firms in the computer, electronics, chemical, pharmaceutical, and information technology sectors were disproportionately active in lobbying for stronger IPR protection in both domestic law and international trade agreements. Faced with mounting global competition, high costs, and declining rates of profit, firms looked toward restructuring their operations to regain their competitiveness or in some cases to fend off bankruptcy. Such restructuring involved selling off corporate divisions and concentrating on the most profitable activities centered around ownership of IPR. But for such a strategy to be successful, corporations enlisted their governments to commit to changing domestic law to enable greater protection of IPRs. In the U.S., corporate lobbying contributed to Congressional legislation strengthening patent law, alongside additional administrative and judicial bodies created for the purpose of protecting patent rights. In the EU, the European Business Roundtable took the lead in lobbying for patent protection as part of the push for a strengthened "competition policy" in the 1990s. Transnational firms based in the U.S., the EU, and Japan cooperated in pursuit of a Trade-Related Intellectual Property Rights agenda that would form an important part of the Uruguay Round of GATT, and would be institutionalized in the creation of the World Trade Organization (Sell 2003).

Second, the corporate lobbying for increased protection of IPR was a product of corporate restructuring on a global scale. Corporations sold off less profitable divisions that involved manufacturing or producing goods in favor of high-value activities such as ownership, consolidation, and protection of IPR. This restructuring of the corporate firm was made possible by the lucrative returns available to owners of intellectual property rights, which allowed corporations to reverse declining rates of profit by restructuring around high-value activities. The process of corporate restructuring was the product of a political relationship between corporations and states in the U.S., Western Europe, and Japan, where corporations lobbied aggressively for the passage of favorable laws pertaining to taxation policy and foreign direct investment that facilitated the growth of profitable supply chains. In the U.S., corporate lobbying contributed to the Congressional passage, and the Reagan Administration support, of the U.S. Economic Recovery Act of 1986, which enabled U.S. corporations to sell off corporate divisions and re-direct the profits back into the global restructuring of the firm at low rates of taxation (Prechel 1997: 420). In Japan, the corporate lobby Keidanren succeeded in securing legislation that facilitated corporate foreign direct investment, allowing Japanese firms to begin to restructure by legalizing, encouraging, and facilitating foreign direct investment that would be crucial to establishing more flexible global supply chains that would begin to supplant the domestic supply chain relationships that long embedded Japanese firms (Yoshimatsu 1999). In Western Europe, corporate lobbies led by the European Business Roundtable lobbied successfully for the creation of a European Union that would facilitate the growth of European foreign direct investment that would lead to corporate restructuring through lower costs and greater reliance on supply networks.

Third, transnational corporations have used the acquisition of patents to create quasi-monopoly power within the market. This is made possible by corporate purchase of patents just to keep competitors from accessing particular technologies or to discourage market entry by using patents to drive up the costs for start-up businesses. Corporations acquire patents not simply due to their own innovations, but more often due to the ability to purchase patents from publicly funded institutions at low costs, or the ability to use mergers and acquisitions to acquire control over patent technology held by a competitor or potential competitor. Therefore, the acquisition of patents is not primarily a process to reward corporate innovation. It is instead a political process through which corporate actors acquire patents from other entities, often public institutions, so that the costs of creating innovative products are socialized while the profits become privatized (Perelman 2003; May and Sell 2005). This process of socializing the costs and privatizing the profits can be seen most clearly in the case of the U.S., where military spending, funneled heavily through research universities during the Cold War and through private contractors in locations such as Silicon Valley as early as the 1970s, provided the technological innovations and infrastructure that was later developed and patented by dominant high-tech corporate firms. The computer and telecommunications sector owes much of its economic success to the ability to "tap into" publicly funded institutions such as the research university, whose ability to enter licensing arrangements with corporations was enhanced in the U.S. by the passage of the Bayh-Dole Act of 1980, which allowed public universities to license and patent technologies that were developed as a result of federal funding.

The result of the Bayh-Dole Act was an exponential increase in the appropriation by private corporations of federally funded research through exclusive licenses with public universities. Previously, such research was often (though not always) shared freely and widely available to the public through organizations such as the National Institutes of Health. The legislation expanded private–public partnerships that patented federal research dollars for the benefit of entrepreneurial interests who claimed that such partnerships would spur greater research and would be mutually beneficial to businesses and universities. Proponents of Bayh-Dole justified the patent system by emphasizing the monetary opportunities for universities able to effectively utilize these market relationships. However most universities have not been able to afford the high capital outlays necessary to even enter the patent game, let alone make money from it. So the idea that patents can somehow replace the shortfall from reductions in public finances has been patently false (pardon the pun). Instead, Bayh-Dole helped usher in a process that "began to alter the internal standards and procedures of academia". The Texas A&M University System, for example, began to allow "patents and the commercialization of research" to count toward the criteria for tenure. As a former Dean of the Graduate School of the Arts and Sciences at Emory University, Donald Stein, described the situation created by Bayh-Dole, "publish or perish" has been supplanted by "publish, patent or perish". Since Bayh-Dole, there has been exponential increases in universities securing patents in the U.S., from a relatively modest 436 university-issued patents in 1981 to more than 3,500 in 2001, although patent ownership has been concentrated among a relatively small group of research universities.

In contrast to the U.S., where universities license patent agreements with private firms, in many Western European countries universities are prohibited from holding patents. Instead, professors are allowed to retain patent rights and therefore to bargain directly with private corporations or companies to commercialize their product. At the same time, unlike in the U.S. where federal funding filters to universities which can then license and patent the innovations that are derived from federal funding, European national agencies carry out their own research in publicly funded labs, instead of sending more of those research dollars to European universities. Furthermore, European professors in Germany, Austria, Denmark, and Sweden, enjoy a "professors' privilege" which allows them to retain property rights over their research findings. As a result, there has not been a huge increase in European universities' patent activities from the early 1980s to the present, in contrast to the U.S. However, commercial firms own most of the patents which are derived from university research in Europe, while in the U.S. universities mostly retain patent ownership and license their products to private commercial businesses.

Despite the differences between the U.S. and European commercialization of publicly funded knowledge through patents, there has been a commonality that researchers have noted: especially in the high-tech, biotech, and pharmaceutical sectors, publicly funded university research has been more important in creating the basis for innovation in the private sector (Montobbio 2012: 201). What this has meant is that the ability of corporations to either license with universities or to directly own the products of publicly funded innovation, has grown increasingly significant for corporate research. These relationships are most robust in the developed countries, where public–private partnerships have increased the ability of a privileged strata of high-tech firms to consolidate ownership of innovations that have been initially developed with public funding. So instead of making those innovations available to the public at lower costs, particularly in sectors that would seem to have broad public utilization needs—broadband access, healthcare and biotech innovations, etc., the public–private relationships tends to lower corporation's cost of production (at public expense) while privatizing the benefits of public research.

This leads to our fourth point: there are clear power dynamics involved in this socialization of costs and privatization of profit that involve linkages between privileged transnational corporations, their host governments, and their universities in the U.S., Western Europe, and Japan. The embedded structure of these relationships provides leading transnational firms with strategic market power in the global market and in their dealings with developing countries. This is illustrated by the extent to which transnational firms, working closely with their home states, can leverage this market power through international trade and investment agreements and/or through global institutions such as the World Trade Organization. As the leading research on this topic has demonstrated, transnational firms in the high-tech, chemical, and pharmaceutical sectors have worked closely with their governments in the U.S., the European Union, and Japan to protect and codify their ownership of intellectual property rights on a global scale in the negotiations of the Uruguay Round of GATT that led to the establishment of the World Trade Organization (Sell 2003; May and Sell 2005).

Prior to the 1980s and 1990s, there were few successful or lasting attempts to expand and enforce patent rights on a global scale. But by the 1980s, sectoral pressure from corporate interests based in the U.S., Western Europe, and Japan, led to a strengthening of domestic patent laws in the developed countries and a new set of policies that would be incorporated into the Uruguay Round of GATT and made enforceable under the World Trade Organization. The corporate sectors that were represented in both domestic and global lobbying efforts included biotech, chemical, pharmaceutical, entertainment, and software industries. Thus the high-tech sectors referenced in this chapter were crucial players in the corporate coalitions that formed to lobby for IPR protections during the 1980s and 1990s. There were three outcomes of the establishment of a Trade-Related Intellectual Property Rights (TRIPS) that strengthened corporate power within global value chains. The first is that patents secured in developed countries were immediately given wider protection within the markets of all WTO member states. Developing countries were required to develop administrative and legal procedures necessary to meet the conditions of establishing a patent system that would guarantee some form of protection for patents that were overwhelmingly granted by governments in developed countries. Whereas countries could previously deny patent rights to corporate investors in order to spur technological innovation or to help create start-up industries, now they were required to grant patent protection across all sectors of the world economy. Second, whereas countries could previously offer patent protection for short-time spans to help spur spin-off research and development and local innovation, they were now required to extend patent protection for 20-year periods in most cases. Third, governments were now limited in how they regulate patent holders, so that there can be no requirement that patent holders share technology with domestic firms, or disseminate patented technologies for wider domestic policy or societal utilization (Shadlen 2009: 8–9).

There are exceptions that are written into these global patent rules in the areas of health and safety, but governments that seek to override patent protection requirements have to go through considerable costs and legal hurdles to justify such a move. As such, the global institutionalization of patent rights has epitomized the increasing instrumental and structural power of transnational corporations and governments in the developed world to write rules within the WTO that disproportionately privilege the profits associated with ownership of intellectual property rights. Still, the WTO does give some leeway to developing countries in the type of patent regime that they may choose to develop. Within this framework, countries are allowed what is often referred to as a "dynamic" patent regime which makes the transfer of patents acceptable under particular circumstances to help promote technology transfer and certain development objectives. Recognizing this loophole in the global patent regime, transnational corporations in high-technology industries have pushed for stronger patent and "investment" protections in bilateral or regional trade agreements. This has meant that the trade agreements being negotiated outside the boundaries of the WTO are primarily "investment" agreements, rather than trade agreements, with extensive and substantial political, legal, and enforcement protection for intellectual property rights as the centerpiece of these deals whose provisions go beyond the WTO (Shadlen 2009: 10).

In short, international trade agreements lock in privileges for dominate transnational corporations whose political influence and market power has enabled them to secure patents by tapping into publicly funded sources of innovation or by purchasing patents from other firms or through mergers and acquisitions. The transfer of patent rights is a process that delivers a quasi-monopoly privilege to a single actor, despite the fact that high-technology innovations are often the product of what some have called "universal labor," meaning the cumulative work of a multitude of actors, both public and private, who have been responsible for developing an "innovation," often over a period of two decades or more. The transfer of "universal labor" to private ownership has been crucial for reversing a declining rate of profit among transnational capitalist firms. Therefore it's noteworthy to point out that "universal labor" represents a mechanism that was identified by Marx as both a source of potential surplus value extraction by capitalists but also as a potential source of liberation for workers. In other words, to what extent "universal labor" is recognized in the form of compensation to all who are involved in the creation of a product versus the expropriation of "universal labor" by a narrow strata of the capitalist class illustrates an ongoing structural contradiction of the system of capitalism. Capitalism is supposed to reward innovation with material benefits, but it appears that innovation that is occurring over a lengthy period of time, and therefore is contingent on the contributions of multiple actors, many of whom do not see adequate compensation for their labor value (Perelman 2003).

The contradictions become apparent when one understands that this privatization of "universal labor" and publicly generated knowledge comes at a significant price. First, private ownership comes with extensive privileges, including a patent system that makes wider dissemination of knowledge subject to restrictions based on patent rights that have increased in duration over the decades and are now locked in for a 20-year duration in global markets. Second, scientific advances that are dependent on open exchange of publicly funded knowledge becomes subject to greater secrecy as innovators work with corporate patrons to keep the details of innovative work from being taken by rival firms or rival investors. Thus considerable efforts are placed on legal issues that detract considerably from scientific work. Third, corporations, especially in the high-tech sector, wage patent wars that involve purchasing patents merely to prevent competitors from gaining access. These patent wars are also often accompanied by litigation, further legislation protecting patent ownership rights, and efforts to privilege secrecy and market position to satisfy the short-term profit-making interests of owners and shareholders (Perelman 2003).

Transnational Corporations and Contract Manufacturers

As seen, the neoliberal global market rests upon many mechanisms that are necessary for the development and empowerment of Transnational Interest Blocs. These include the strengthened Intellectual Property Rights (IPR) regime, transnational production, and closer links with governments, to name a few. Essential to this structure has been the growth of the Global Value Chain (GVC) and the role of Contract Manufacturers. This has allowed for lead firms to offshore factories, utilize labor of the global South, and segment and retain a higher value-added sector of production. On the positive side, this has allowed for a greater efficiency of manufacturing, as lead firms focus on their core competency, thus reducing their overhead of expensive factories, and generating greater economies of scale.

However, this important role of Contract Manufactures within transnational production must be examined through the lens of Transnational Interest Blocs (TIB) and North/South divisions. With a concentration of intellectual property creation in the global North and dispersion of low-end manufacturing in the global South, the Contract Manufacturer plays the role of mediator within the TIB hierarchy. Again, this means profits flow up this structure while risk and costly adjustment flows down. Lead firms, such as Intel or Apple, are thus able to expand their profits due to their relationships with their Contract Manufacturers, while simultaneously strengthening their relationships with Northern governments, who provide IP protection, subsidies, and subsidized research. At the bottom of the hierarchy low-skilled labor in the global South are placed in a precarious position as wages, hours, and job security are necessarily vulnerable to changes and adjustment.

While Contract Manufactures operate in numerous industries, from textiles to automotive, we will be focusing on Electronic Contract Manufacturers (ECMs). An analysis of ECMs will be illustrative to not only the structure of TIBs, but also the larger topics of transnational production, GVCs, and North/South relations. As the name suggests, ECMs specialize in electronic components, from circuit boards and semiconductors, to finished products of computers, tablets, and smart phones. As such, ECMs manufacture for the leading firms in the high-tech sector. These include large corporations such as Apple, Microsoft, and Cisco, who play a leading role in the Western TIB structure. These high-tech firms have been essential in developing the global IP regime, but have also been essential in developing the GVC and transnational production we see today.

Despite the immense size of leading ECMs, the hierarchical relationship between lead firms and Contract Manufactures persist within the high-tech industry. Illustrative of this is the relationship between Apple, which has seen record profits in recent years, and its primary manufacturer, Taiwanese ECM Foxconn. By focusing on research, design, and advertisement, American lead-firm Apple has achieved monumental success, retaining as much as 30–60% of the profits from products like the iPad and iPhone (Golson 2012; Barboza 2010). By contrast, Foxconn has amassed a fraction of profits per product produced. This amounts to as little as $8 per product (Mishkin and Palmer 2012), or even manufacturing some products for free, just to retain Apple as a client (Chan and Ngai 2010). Despite the razor-thin profit margins, Foxconn and other ECM heavyweights, such as American firms Flextronics and Jabil Circuit, have been able to generate profits in the billions (HHPI 2015; Flex 2016). This is thanks to their participation in developing a transnational production chain that has generated transnational exploitation of labor.

As such, these manufacturers have been able to develop a truly transnational network of production. This has allowed them a leading part in developing the international division of labor with geographically dispersed factories specializing in appropriate modules along the production chain. North American ECMs have utilized this division, having the lowest value-added portion of production done by cheap labor, today heavily within China, while the bulk of high-value-added portions, such as engineering or design, reside largely within the global North (Lüthje et al. 2013). While providing foreign investment, factories, and jobs into developing regions, it has also accelerated the global race to the bottom. This has large implications for possible industrial upgrading for developing states, and generates a system of vulnerability and precariousness among labor.

ECM, Labor, and the Race to the Bottom

As lead firms, by definition, have retained the highest value-added portion of the supply chain, along with the largest means for growth through research and development of IP, Contract Manufactures have had to find means of retaining a degree of profit. They have done this in numerous ways, from offering related services in logistics and product testing to producing for numerous lead firms. This broadening of supply chain operations and multiple manufacturing contracts allows for greater economies of scale, which in turn allows for a cheaper product and thus greater competitiveness. Despite an expansion of services, the primary cause of ECM success can be found in their exploitation of cheap labor. As mediators in transnational production, these corporations have found and developed such labor in abundance, strengthening not only their dominant position, but also the structure of the GVC itself. Illustrative to this is the Chinese labor market, heavily organized by the state, and its partnership with the leading ECM, Foxconn.

Foxconn's great success is in the utilization and consolidation of sweatshops and Chinese labor. There are largely two aspects to the use of Chinese labor that will be essential for understanding how adjustment is managed within the industry. One is the exploitation itself; the relationship that is developed with low-paid workers who hold very little collective power. The second is the relationship to the Chinese government, which has resulted in assistance to the Taiwanese ECM in the form of subsidies, supply and pacifying labor, and participating in the race to the bottom both within China and the global labor force.

Perhaps most telling of the labor control practices Foxconn has developed is the existence of so-called "Foxconn City." This manufacturing compound in Shenzhen lives up to its description as a "city," with hundreds of thousands of Chinese workers. Being primarily migrants from other areas of China, they are bused into the compound to work and live within the corporate dormitories (Chan and Ngai 2010). This movement and isolation is no accident, as it is a common strategy of developing worker control and obedience. This represents the race to the bottom for labor, whereby worker vulnerability is both found and developed to depress wages and labor power, as well as generate corporate dependence. For a company with razor-thin profits on manufacturing this power relationship is essential.

The Chinese government itself has been very receptive to participating in this race to the bottom and for utilizing sweatshops as a means to development. On top of facilitating an internal race to the bottom among provinces, China also assists with infrastructure, recruiting workers, such as through vocational schools, and the development of the infamous *Hukou* system, whereby a regional registration generates a vulnerable class of internal migrants (van Liemt 2016). Without support of government programs these laborers provide much flexibility to the ECM by accepting decreased working hours or expanding overtime, as required by fluctuations in the market (Lüthje et al. 2013).

This utilization of artificially cheap labor and state subsidized development is not unique to Foxconn, nor is it unique to China. Indeed, there are numerous Contract Manufacturers that make up the transnational system, with the next largest ECMs being American firms Flextronics and Jabil Circuit. These, in turn, work with and utilize numerous locations to reach and generate the cheapest cost in production. Foxconn itself, heavily concentrated in China, has made plans to enter India and Indonesia as Chinese wages become too expensive (van Liemt 2016). This competition has allowed for the value chain hierarchy that privileges those at the top, and pushes risk down.

In conjunction with the aforementioned squeeze on labor and the exploitation of the race to the bottom among states, there has been a larger trend of ownership consolidation within this sector (Applebaum 2008). This competitive monopolization is a natural affect of capitalism as a strategy to deal with asymmetric industry power. Thus, while competition continues to exist within the industry/sector it does so in a more oligopolistic form. Naturally this allows for greater negotiation power, but also expands the ability to push risk farther down the Value Chain, to the most vulnerable and dependent actors. By expanding direct and indirect control to local factories and junior partners, ECMs are able to develop a large network of actors to participate in cost reduction (van Liemt 2016).

Automation, Change, and Adjustment

In conjunction with being by far the largest ECM, Foxconn is a useful example as it is also one of the more ambitious ECMs to adopt to the so-called "Industry 4.0." This is seen as the fourth industrial revolution, which will call for even greater flexibility of production than we saw in the digital revolution (bmbf.de 2016). Rather than simply an expansion of robotics and production line automation, this fully incorporates a labor automation, logistics, and planning by developing a global network of factories, cloud computing, and autonomous decision making. This broad project of reshaping how production is done will reshape both the industry and the broader market and society.

The expansion of automation and networks will have large implications for all actors involved in the GVC. Those benefiting most will again be lead firms with a specialization in intellectual property. ECMs will continue to have a large role in mediating the adjustment, which means labor will continue to find itself in its precarious position. It has been seen numerous times that during downturns, labor has necessarily been forced to take the largest adjustment, meaning loss of wage, hours, and employment (Lüthje et al. 2013). This makes up for the large overhead and the ability to push risk down for ECMs. This is because of the race to the bottom and can be seen in the examples of weak labor laws, weak enforcement, and lack of labor collection and power in host states. This weak enforcement is sometimes reserved just for export sectors, industrial parks, and transnational corporations, even if strong laws exist for domestic producers (ibid.).

While adoption of global automation and networked robotics will provide ECMs with greater leverage over labor, it is unlikely that this high-tech development will drastically change the structure of the GVC. The immediate future is unlikely to see much increase in automation, as the abundance of cheap labor negates the high cost of such equipment.

But even aspects of networked factories and autonomous decision making will only reinforce the system as intellectual property remains the dominant sector in production. ECMs will continue to compete for contracts as they take on the cost of technological upgrading, with much of the cost–benefit being filtered to the lead firm. As a result, we will likely see greater labor vulnerability, greater costs for ECMs to compete, and cheaper manufacturing for lead firms.

Earlier in Taiwan's development, Foxconn was one of many benefactors from the state's participation and assistance in the economy's industrial upgrading (Lüthje et al. 2013). Following the "Asian Model" of development, these policies of assistance, such as directing production, organizing industry, and offering finance and subsidization, follows a more corporatist organization of the economy by the state. As such, coordination and cooperation with the government is essential for leading firms. The later push by mainland China to utilize the GVC facilitated the regional focus of Foxconn (Ibid). Not only did this allow Taiwan and China to find relative success in development, and massive growth for Foxconn, but it also developed a regional interest bloc, which not only continues to assist local corporate power, but also reinforces the global structure of TIBs.

To reach or maintain a dominant position in the global market, states throughout the GVC have assisted transnational corporations and facilitated the exploitation of labor. For successful semi-peripheral states like Taiwan, this has been in relation to the more recent developmentalism of China and utilization of cheap labor there. For a state like Singapore this is in its relations to Malaysia, and Malaysia, in turn, in its utilization of cheaper labor from Indonesia (Bormann et al. 2010). What this means, however, is that industrial upgrading and successful development requires not only the exploitation of the most vulnerable population of labor, but also strong corporate-state cooperation.

Japan is one of the better known states in providing extensive subsidies to its corporations, both to promote research and development and to foster greater competitiveness in the global market. Through a lengthy history starting most prominently with the Ministry of International Trade and Industry (MITI), Japan has organized and supported its leading corporations to become dominant actors within the global market (Johnson 1982). As the name suggests, MITI has used this as a trade (and investment) strategy, which has helped Japan become a world leader within the high-tech industry, but also enriched specific actors. The automation company FANUC received great benefit from this intervention, as standardization of machine tools were developed around their products (Kende 1991). But the old national developmentalism in Japan has long since given way to a Japanese state that is linked to transnational firms that have greatly expanded their relationships with foreign transnational interest blocs. In the process, the growth of transnationalized production, in Japan and elsewhere, has lessened the identification of corporations with their "home" territory in favor of a transnational identity focused on maximizing global profits. As documented by a wide range of scholarly work, the role of transnational corporations in Japan has pushed the Japanese state toward the adoption of neoliberal policies that have steadily eroded the basis for the old "developmentalism." This has meant considerable Japanese government support for foreign direct investment and trade agreements that have promoted the linkages of transnational firms in Japan to GVCs (Lechevalier 2014), and away from previously entrenched domestic supply networks.

Like Japan, the U.S. state has actively supported leading transnational corporations to become dominant globally (Block 2008). Outside of the better known assistance, such as direct subsidies and infrastructure, the US has provided both major political support and high-tech research to a wide range of transnational firms. This includes strong corporate leadership in the development of the TRIPS regime within the WTO (Sell 2003), and the inclusion of "investment" provisions in U.S.-led bilateral and regional trade agreements, which have locked in privileges for transnational corporate investment that go beyond those mandated in the WTO. Investment agreements, often mistakenly referred to as "trade" agreements, lock in privileges for transnational capital that make traditional "developmentalist" approaches more difficult for states. Instead, states have more incentives to subsidize trade and investment networks that contribute to the growth of GVCs, while organizing their population into cheap labor for transnational corporations. In research, the U.S. has acted as a mediator to unite universities and corporations in partnership. Known as the "triple-helix model" (Etzkowitz 2003), this has made the U.S. economy the world leader in high-tech products, but at the cost of hidden research subsidization and a corporatization of the University system (Giroux 2009). In summary, a transnational corporate lobby has had considerable success expanding its influence in U.S. foreign economic policymaking during the neoliberal era, further providing incentives for an expansion of GVCs within U.S. tax and investment legislation (Dreiling and Darves 2011).

China's ascendancy has been closely linked to GVC networks that transcend nation-states, with the U.S., the EU, and China promoting policies that have helped to establish the conditions and the growth of transnational production networks. These networks involve considerable cooperation between transnational firms based in the West and the Chinese state, despite ongoing geopolitical and economic tensions between the West and China. China has seen spectacular growth due to its inclusion in transnational production, while actively assisting and subsidizing the growth and consolidation of transnational ECMs. This is especially true for the stellar success of Foxconn, as China's globalization has allowed Taiwanese corporations access to abundant cheap labor and cemented Taiwan's position as a middle-income state. While this has brought much into the coffers of the Chinese government, its own subsidization of high-tech IP has been on seemingly contradictory paths. Some factions within the Chinese government want to provide further assistance to Chinese-owned high-tech businesses that can compete effectively with transnational interest blocs dominated by Western transnational corporations. At the same time, highly profitable Chinese firms are linked to a transnational interest bloc that includes Western corporations like Apple. Thus the Chinese state is divided into competing corporate factions, with each using its political influence within the Chinese Communist Party to steer government policy in its preferred direction (Hung 2009).

Conclusion

As the previous cases illustrate, in this age of the GVC, transnational interest blocs often exert power across a range of states, including states with seemingly distinct political cultures and institutions. Transnational interest blocs exert both structural and instrumental power relative to states, ECMs, and labor within an increasingly global production system. These interest blocs are often able to use their extensive political and economic ties to leverage their power in a wide range of circumstances. Many of the most important future battles over the distribution of profits, wages, and resources will be fought within the context of a transnational production system that has disproportionately privileged dominant transnational firms.

Within this transnational interest bloc, the role of labor could become increasingly important as a potential vehicle for transformative battles over allocation of profits, wages, and resources. The rapidly expanding architecture of the modern GVC has required the increasing centralization and consolidation of workers within what some researchers have called "logistical" nodes of coordination and distribution (Sealy 2010). This has meant an increasing concentration of workers whose activities in facilitating transportation and delivery of products connected to GVCs are crucial to a functioning system. There are now huge concentrations of workers in warehouses that have become indispensable "clusters" facilitating the coordination of rail, truck, air, and water transportation. In just three of these clusters in the U.S., there are concentrations of over 100,000 workers that provide a crucial logistics function for global value chain networks. To the extent that these workers can be mobilized politically, they represent a potential challenge to the current skewed distribution of profits at the top of GVCs, a political and economic system that has thus far been dominated by the preferences of the lead transnational firms that control transnational interest blocs (Brooks and Moody 2016).

Addendum[2]

The race among dominant transnational corporations to achieve a monopoly or quasi-monopoly position by acquiring the latest technological advances is a defining characteristic of the latest phase in the "new globalization." This phase accelerated after the global capitalist crisis of 2008, which has resulted in several trends that have been driven by a systemic crisis in neoliberal capitalism. The first is the stagnation of global value chains reflected by global trade statistics, as the volume of global trade dramatically fell in the aftermath of the 2008 capitalist recession and has failed to rebound to pre-crisis levels (Timmer et. al. 2016). The biggest factor behind the global trade slowdown has been the stagnation of trade within global value chains, which has intensified after 2014 after having rebounded from 2011 to 2014 (Constantinescu et al. 2015). In response, transnational corporations are reducing the complexity of global value chains (incorporating fewer suppliers) in favor of greater

[2] This section borrows from the recent work of Cox (2019, 2023).

capital-intensity and greater consolidation of those value chains around the most important global markets (Degain et al. 2017). As part of this process, transnational corporations have reorganized their global capitalist employees more rigidly around two categories: production supervisors and managers that oversee the logistical and supply operations of the global value chains and are relatively well-paid, and workers who produce within these chains at wages that continue to stagnate and decline relative to capitalist revenues (Timmer et. al. 2013).

Second, the gap between the profits of the dominant transnational corporations and the wages of the global capitalist workforce has grown wider, as more wealth is concentrated at the top. This has resulted in a further "hollowing out" of the industrial workforce in the core capitalist states, often expressed as a shrinking "middle class," alongside the increased exploitation of the global working class (Milanovic 2016). These class tensions have resulted in further deligitimacy of capitalist governments in the West, contributing to the rise of Donald Trump in the U.S. and other quasi-fascist political figures and parties in Western Europe. The increased class tension is a long-term byproduct of the contradictions of a neoliberal capitalist political project that has steadily increased the power of transnational interest blocs while reducing the political influence of workers and ordinary citizens, whose political marginalization and alienation have steadily increased.

Third, the global value chains that were previously dominated by North–South linkages, with the dominant transnational corporations concentrated in the Northern states, have increasingly been challenged by the dramatic rise of China and the policies of the Chinese state, particularly in the area of high technology, which has helped to shift global value chains a bit more in a South-South direction, with countries of the global South, led by China, reorienting global value chains (to some extent) toward production, marketing, and consumption within the global South (Horner and Nadvi 2018). The Chinese state has enacted policies that have contributed to a greater emphasis on domestic production and accumulation and to more stringent requirements for foreign transnationals investing in high-technology manufacturing in China in an attempt to assert greater control over these value chains.

The business and political coalitions in U.S. foreign policy advocating political and economic cooperation with China have been eclipsed from the mid-2010s to the present. Global capitalist crises, starting with the

2008 global recession, have sharpened the tensions between U.S. corporations who have long benefitted from locating their global value chains within the China market and those corporations who have long felt threatened by the competition posed by Chinese exports (Cox 2019). Meanwhile the Chinese government has sought to exert more concessions from foreign investors and to increase the scope and scale of state intervention toward an upscaling of Chinese capacity in 5G technology and artificial intelligence. An acceleration of Chinese nationalist political strategies in recent years, combined with the global political and economic crises of the COVID pandemic, have increased the costs of foreign direct investments for Western corporations. Corporations that have long been in competition with China, especially in the steel and aluminum sectors, have aggressively led the campaign for sanctions against China (and were major contributors to President Trump's first campaign). At the same time, interest groups within the military-industrial complex, which have long sought to identify China as a threat justification for higher military budgets, have identified an expansion of Chinese military capacity and utilization, especially in the South China Sea, but also throughout East Asia, as a threat to U.S. extra-regional hegemony in Asia (Grazier 2021). The rise of China's economic and military capacity in Asia and its long-term claim to the island of Taiwan, with newly rising military maneuvers and tensions in recent years, has accelerated an increasing confrontational posture between the U.S. and China (Grazier 2022).

The heightened competition between the U.S. and China, filtered through transnational interest blocs in the China market, reflects the current contradictions within global capitalism. The Chinese state has long been an important player in mediating the political and economic conditions necessary to help transform and expand global capitalism into a vast global value chain network. Indeed, as many scholars have documented, China has become the epicenter of global production activities, which includes high-value activities centered around production of computers, electronics, telecommunications, or, taken as a single category: information technology products. The battle over the extraction of surplus value from the China market is at the center of the latest crisis of neoliberal capitalism. Competition among transnational interest blocs within the China market has intensified under the current dynamics of global capitalist accumulation. The key battles revolve around how surplus value will be extracted from contemporary global value chains, which transnational interest bloc coalitions will benefit the most from the political economy of value extraction and, most importantly, how will ownership and control over high-technology processes be determined in this phase of global capitalism?

Bibliography

Applebaum, Richard. 2008. "Giant Transnational Corporations in East Asia: Emergent Trends in Global Supply Chains," Competition and Change, Vol. 12, No. 1, 69-87.

Barboza, David. 2010. "Supply Chain for iPhone Highlights Costs in China". http://www.nytimes.com/2010/07/06/technology/06iphone.html?_r=0 (July 20, 2016)

Block, Fred. 2008. "Swimming against the current: The rise of a hidden developmental state in the United States." *Politics & society* 36, no. 2: 169-206.

bmbf.de. "Zukunftsprojekt Industrie 4.0". www.bmbf.de/de/zukunftsprojekt-industrie-4-0-848.html **(August 2, 2016)**

Bormann, Sarah, Pathma Krishnan, and Monika E. Neuner. 2010. *Migration in a Digital Age: Migrant Workers in the Malaysian Electronics Industry; Case Studies on Jabil Circuit and Flextronics*. WEED.

Brooks, Chris and Moody, Kim. 2016. "Busting the Myth of a Workerless Future." *Labor Notes*, July 26 Newsletter.

Chan, Jenny, and Pun Ngai. 2010. "Suicide as Protest for the New Generation of Chinese Migrant Workers: Foxconn, Global Capital, and the State." *The Asia-Pacific Journal* 8, no. 37: 2.

Constantinescu, Cristina, Aaditya Matoo, and Michele Ruta. 2015. "The Global Trade Slowdown: Cyclical or Structural?" Washington D.C.: IMF Working Paper.

Cox, Ronald W. 2008. "Transnational Capital, the U.S. State and Latin American Trade Agreements," *Third World Quarterly*, Vol. 29 (8): 1527-1544.

Cox, Ronald W., ed., 2012. *Corporate Power and Globalization in U.S. Foreign Policy*. London: Routledge.

Cox, Ronald W. 2019. *Corporate Power, Class Conflict and the Crisis of the New Globalization*. Lanham: Lexington Books.

Cox, Ronald W., "Capitalism and Class Power in the Military-Industrial Complex in the U.S." In Ronald W. Cox, ed., 2023. *Capitalism and Class Power*. Boston: Brill, 2023, 11–42.

Degain, Christophe, Bo Meng and Zhi Wang. 2017. "Recent Trends in Global Trade and Global Value Chains." In "Measuring and Analyzing the Impact of GVCs on Economic Development," World Bank Group. Washington, D.C.: World Bank Publications.

Dreiling, Michael and Darves, Derek. 2011. "Corporate Unity in American Trade Policy: A Network Analysis of Corporate-Dyad Political Action." *American Journal of Sociology*, Vol. 116, No. 5, 1514-63.

Etzkowitz, Henry. 2003. "Innovation in innovation: The triple helix of university-industry-government relations." *Social science information* 42, no. 3: 293-337.

Flextronics International (Flex). 2016. Annual Report

Gao, Haichao, Ruipeng Guo, and Pengcheng Li. 2015. "Development Situation and Prospect of Chinese Industrial Robots." In *2nd International Conference on Civil, Materials and Environmental Sciences*. Atlantis Press.

Giroux, Henry A. 2009. "Democracy's nemesis: The rise of the corporate university." *Cultural Studies↔ Critical Methodologies*.

Golson, Jordan. 2012. "Foxconn Profit Margins Remain Tight as Apple Flourishes". January 5, 2012. http://www.macrumors.com/2012/01/05/foxconn-profit-margin-remains-tight-as-apple-flourishes/ (August 12, 2016).

Grazier, Dan. 2021. "The China Threat Is Being Inflated to Justify More Spending." Project for Government Oversight, Feb. 17, Washington, D.C.

Grazier, Dan. 2022. "Chinese Threat Inflation and America's Nonsensical Plans." Project for Government Oversight, Dec. 7, Washington, D.C.

Harris, Jerry. 2008. *The Dialectics of Globalization: Economic and Political Conflict in a Transnational World*. London: Cambridge Scholars Publishing.

Hon Hai Precision Industry (HHPI). 2015. Annual report.

Horner, Rory and Khalid Nadvi. 2018. "Global Value Chains and the Rise of the Global South: Unpacking Twenty-first Century Polycentric Trade." *Global Networks: A Journal of Transnational Affairs* 18, Issue 2: 207-237.

Hung, Ho-Fung. 2009. "America's Head Servant." *New Left Review*, Vol. 60, Nov.-Dec., 5–24.

Jabil Circuit (Jabil). 2015. Annual Report

Johnson, Chalmers. 1982. *MITI and the Japanese miracle: the growth of industrial policy: 1925-1975*. Stanford University Press.

Kende, Michael. 1991. *Gains from standardization: the case of numerical controls*. INSEAD.

Kotz, David. 2015. *The Rise and Fall of Neoliberal Capitalism*. Harvard University Press.

Lechevalier, Sebastien, ed. 2014. *The Great Transformation of Japanese Capitalism*. London: Routledge.

Levy, David. 2008. "Political Contestation in Global Production Networks." *Academy of Management Review*. 33 (4): 943-962.

Lüthje, Boy, Stefanie Hürtgen, Peter Pawlicki, and Martina Sproll. 2013. *From Silicon Valley to Shenzhen: global production and work in the IT industry*. Rowman & Littlefield.

May, Christopher, and Susan Sell. 2005. *Intellectual Property Rights: A Critical History*. Boulder: Lynne Rienner.

Milanovic, Branko. 2016. *Global Inequality: A New Approach for the Age of Globalization*. Boston: Harvard University Press.

Milberg, William and Deborah Winkler. 2013. *Outsourcing Economics*. New York: Cambridge University Press.

Mishkin, Sarah and Palmer, Maija. 2012. "Foxconn survives on thin slices of Apple". http://www.ft.com/cms/s/0/170a225c-0356-11e2-a284-00144feabdc0.html#axzz4HJrUuX9R (August 21, 2016)

Nolan, Peter, Zhang, Jin, and Liu, Chunhang. 2007. *The Global Business Revolution and the Cascade Effect*. New York: Palgrave Macmillan.

Nolan, Peter and Zhang, Jin. 2010. "Global Competition After the Financial Crisis." *New Left Review*, Vol. 64: 97-108.

Perelman, Michael. 2003. *Steal This Idea: Intellectual Property and the Corporate Confiscation of Creativity*. New York: Palgrave Macmillan.

Prechel, Harland. 1997. "Corporate Transformation in the Multilayered Subsidiary Form: Changing Economic Conditions and State Business Policy." *Sociological Forum*, Vol. 12 (3): 405–439.

Prechel, Harland. 2000. Big Business and the State: Historical Transformation and Corporate Transformation, 1880s-1990s. Albany: State University of New York Press.

Sealy, Roger. 2010. "Logistic Workers and Global Logistics: The Heavy Lifters of Globalisation." *Work Organization, Labor and Globalisation*, Vol. 4, No. 2, 25–38.

Sell, Susan K. 2003. *Private power, public law: the globalization of intellectual property rights*. Cambridge University Press.

Serfati, Claude. 2008. "Financial Dimensions of Transnational Corporations, Global Value Chain and Technological Innovation," *Journal of Innovation Economics*, Vol. 2, 35–61.

Shadlen, Kenneth C. 2009. "The Politics of Patents and Drugs in Brazil and Mexico." *Comparative Politics*, Vol. 42 (1): 41–58.

Starosta, Guido. 2010. "The Outsourcing of Manufacturing and the Rise of Giant Global Contractors: A Marxian Approach to Some Recent Transformations of Global Value Chains," *New Political Science*, 15:4, 543–563.

Timmer, Marcel P., Abdul Azeez Erumban, Bart Los, Robert Stehrer, and Gaaitzen J. de Vries. 2013. "Slicing Up Global Value Chains." Paper for European Commission, Research Directorate General. Portions of this were later published in the *Journal of Economic Perspectives* 28, Number 2 (Spring 2014): 99–118.

Timmer, Marcel P., Bart Los, Robert Stehrer, and Gaaitzen J. de Vries. 2016. "An Anatomy of the Global Trade Slowdown Based on the WIOD 2016 Release." University of Groningen: GGDC Research Memoranda, No. 162.

van Liemt, Gijsbert. 2016. "Hon Hai/Foxconn: which way forward?." In *Flexible workforces and low profit margins: electronics assembly between Europe and China,* edited by Drahokoupil, Jan, Rutvica Andrijasevic, and Devi Sacchetto, 45–66. ETUI.

Yoshimatsu, Hidetaka. 1999. "The State, MNCs, and the Car Industry in ASEAN." *Journal of Contemporary Asia*, Vol. 29 (4): 495–515.

CHAPTER 3

The Safe Hand: Liquidity, Money, and Financial Innovation

Loren Gatch

Safe Assets, Financial Innovation, and Financial Technology

The financial crisis of 2007–2008 has invigorated a debate over the role placed by "safe assets" in the financial system. "Safe assets" refer to a class of financial instruments whose properties and contexts allow them to be considered riskless, in the sense that their owners have no need or obligation to acquire information about them. Another way of describing this property is that, as long as no owner suspects that another owner possesses private information about the asset, each owner has no incentive to avoid the asset out of fear of being exploited. To that extent, the possibility and presence of safe assets renders financial investment automatic in the specific sense that, by eliminating the need to collect information and otherwise engage in cognitive efforts to assess risk, investors can behave

L. Gatch (✉)
Dept. of Political Science, University of Central Oklahoma, Edmond, OK, USA
e-mail: lgatch@uco.edu

R. Kiggins (ed.), *The Political Economy of Robots*, International Political Economy Series, https://doi.org/10.1007/978-3-031-95171-8_3

formulaically. Money is the fundamental example of an information-insensitive asset, but public authorities and private actors coordinate to produce a variety of other assets that approximate safety.

Safe assets arise out of innovations in financial technologies that construct safety so as to allow certain assets to play important roles in the financial system. Following Goetzmann (2016), financial technologies in general take two forms: "hardware" and "software". Hardware refers to various physical abilities to transmit and receive information faster and over longer differences, as well as the computational capacity to process large amounts of information at ever-quicker speeds. Hardware in this sense need not be unique to financial markets. For example, the introduction of the telegraph (1844) and the laying of the first transatlantic cable (1866) led rapidly to the narrowing the price of securities across different markets (Garbade and Silber 1978). Similarly, information technology has driven financial disintermediation in the competition among payments systems (Winn 2000); increases in computational power have also facilitated the securitization of financial assets and thus the rise of the "shadow banking system" (Lengwiler 2016). These innovations do not occur in a regulatory vacuum. Improvements in both information and computational technology enable the high-frequency trading of securities. However, for these trading strategies to be realized, the Securities and Exchange Commission had to countenance certain changes in market rules and practices—the decimalization of stock prices, abolition of the "uptick" rule, rate deregulation—as well as changes in institutions—the demise of the specialist system of market-making and the decentralization of the stock exchange—to make this trading practice possible (Traflet and Gruver 2015).

In contrast to financial hardware, software refers to instruments that are created and traded in financial markets. Unlike hardware, innovations in financial software are specific to the interests and purposes of market actors. Although their deployments may be abetted by the prevailing state of hardware, financial software can arise independently of hard technologies. For example, while problematic mortgage securitizations were at the heart of the 2007–2008 crisis, previous (and similarly problematic) episodes of mortgage securitization in American history date back to the 1880s (Simkovic 2013). More generally, financial crises have unfolded with characteristic regularity across centuries, quite apart from the particular level of hard technology (Reinhart and Rogoff 2009).

However they emerge, all innovations in financial software seek to make illiquid assets more liquid. This chapter seeks to provide perspective on the safe assets debate by examining the relationship between safety and liquidity. This chapter begins by introducing the contemporary debate on safe assets, as it has arisen out of the recent financial crisis. Various kinds of safe assets share the common property of being liquid assets. How does an asset acquire the property of liquidity? The chapter then turns to the nature of liquidity, and how social processes and institutions generate it. Liquidity is a kind of financial privilege that accrues to those who possess it, or control how it is made.

Money is the ultimate form of liquidity, but the meaning of money and its relationship to other liquid assets is also elusive. In particular, the historical shift from commodity monetary systems to those based upon abstract fiat has created new possibilities for a politics of liquidity. While the financial crisis of 2007–2008 illustrates such a politics in abundance, this chapter seeks to put this politics in historical context by sketching out the political economy of liquidity as it has emerged at certain junctures of American financial history.

In particular, this chapter describes a divide between the political economy of commodity money and that of fiat money, or what can be called "chartalism". A chartalist political economy is characterized by autonomous financial innovation, driven by private interests, that creates new forms of liquidity and embeds new forms of risk in the financial system. To the extent that private financial innovation is profitable, the supply of private financial assets can be increased and configured as information insensitive, or "safe", serving in turn as a basis for new rounds of risk-taking. Public financial and monetary authorities aim above all to maintain the stability of the system, and toward that end seek to regulate and otherwise control this innovation. However, as part of the logic of chartalism, these regulatory interventions, undertaken to reinforce safety, themselves form the terrain upon which private interests create new strategies of risk-taking that assume, as their points of departure, sites of risklessness in the financial system. As a result, the quest for safety in financial assets may produce uncontrollable eruptions of its opposite.

1. The Financial Crisis and the Safe Assets Debate

The financial crisis of 2007–2008 threatened the first globally systemic financial collapse since the 1930s. Long- and short-term factors created the crisis, some emanating from the American epicenter, others from outside. Long-term factors included a bubble in American home prices, driven by public policies that pushed home ownership rates to unsustainable levels (Wallison 2015), and which treated higher levels of home equity as a palliative for stagnant living standards and rising income inequality (Rajan 2010). Responding to their stagnating incomes, Americans ran down their savings and increased their debt loads (including mortgages and home equity loans) to maintain their standard of living (Drennan 2015, Chapter 4).

Rising housing prices after the late 1990s were fed by global interest rates kept low by excessive saving by Asian countries (Bernanke et al. 2011). These savings stimulated the production of mortgage-backed securities (MBSs) to meet the enhanced demand for private investment-grade securitized products perceived to be as safe as U.S. government debt. Conversely, chronic U.S. current account deficits were sustainable only so long as foreign investors funded it (Reinhart and Rogoff 2009, Chapter 13). For its part, the Federal Reserve enabled the bubble by refusing to cool the housing market via higher interest rates (Taylor 2009).

On a more technical level, the bubble in housing was exacerbated an originate-to-distribute model of mortgage securitization that reduced the incentives of loan originators to undertake due diligence. Competition among lenders drove underwriting standards downward (Simkovic 2013). Securitization itself proved to be a flawed technology. Specific features of subprime mortgage securitizations depended upon a continued rise in home prices, and the refinancings that appreciation made possible. These securitizations were repackaged and tranched in ways that concealed risks from investors. The lack of information arising from the opacity of subprime securitization later created the flight by institutional investors out of this asset class. The resulting stampede toward liquidity unfolded, in essence, like the banking panics of the pre-Federal Reserve era (Gorton 2010, Chapter 3). Indeed, despite the widespread conceit among top policymakers that the sophistication and resilience of the American financial system rendered crises unlikely, the events of 2007–2008 unfolded essentially like any emerging-market financial collapse (Reinhart and Rogoff 2009, pp. 171–173).

Across these interpretations, a common theme is actors' incentives to engage in reckless behavior. Government-sponsored enterprises (GSEs) like Fannie Mae and Freddie Mac pushed securitization to enhance their profits, and to accommodate political pressures to increase home ownership. Though private companies, the GSEs' political salience made their rescue inevitable (Calomiris and Haber 2015, Chapters 7–8). Ratings agencies were corrupted by the prospect of the continued fees from mortgage securitizers. Overleveraged banks and investment companies badly miscalculated the riskiness of mortgage-backed securities, as did companies like the American International Group (AIG) who sold insurance hedges for these portfolios. Funding these portfolios with short-term funding put investors in the perilous position of having to sell into a falling market when funding dried up. Finally, government oversight of securitization, and of shadow banking more generally, was hobbled by fragmented regulatory jurisdictions, forum-shopping by financial institutions, inadequate expertise and outright incompetence on the part of regulators, and bureaucratic jealousies that prevented a coordinated response to the unfolding disaster (Financial Crisis Inquiry Commission 2011).

While these explanations emphasize different factors, none disputes the salience of those specific events that marked the acute phase of the crisis in September 2008. These included the takeover of Fannie Mae, Freddie Mac, and AIG; the failure of Lehman Brothers; and the government's temporary backstopping of money market mutual funds and the commercial paper market. After some hesitation, Congress voted regulators the authority to purchase distressed assets from banks and investment companies on an immense scale. In the course of barely two weeks, federal interventions in the nation's financial markets reached a scope not seen since the 1930s.

Policy responses to these events embodied one common element: they substituted discredited private-sector assets with public assets deemed "safe" by dint of the full faith and credit of the government. At its acute phase, the crisis was marked by a widespread and uncontrollable shift in expectations concerning the safety of a large class of privately generated assets, resulting in the financial system's "sudden financial arrest" (Caballero 2009). In addition to sparking political controversy, the sheer magnitude of this substitution has also engendered a debate about the nature of safe assets, their role in the financial system, and how private actors and public authority generate these assets. The International Monetary Fund (2012, Chapter 3) has been prominent in arguing that, with the collapse of private triple-A rated securitized assets in the United States and the downgrading of some European sovereign debt as a result of the Eurozone crisis, there exists a shortage of safe assets that is destabilizing for the global financial system.

Although examples of safe assets vary, an asset is deemed "safe" if it is designated risk-free to the extent that financial actors have no reason to inquire into the riskiness of the asset, making it "information insensitive" (Gorton 2010, p. 5), thus permitting financial actors to treat different examples of the same asset class as interchangeable for a variety of purposes. Alternately, "a safe asset can be defined as any debt asset that promises a fixed amount of money in the future with no default risk" (Gourinchas and Jeanne 2012, pp. 5–6). Safe assets serve as the collateral that facilitates routine transactions in overnight lending markets. Their high quality makes them a sufficiently stable store of value or assured income stream that they underpin a variety of short- and long-term investment strategies. They serve as a capital cushion for financial institutions in a variety of regulatory frameworks (IMF 2012, p. 88).

Though treated as information insensitive themselves, safe assets are nonetheless vital for the assessing the riskiness of all other financial assets. Positing that there is a risk-free asset with a particular yield creates a benchmark against which the costs of borrowing can be ascertained. The risk-free yield also represents a discount rate for calculating the time value of money. Fundamental tools of portfolio construction like the capital-asset pricing model assume that there is a risk-free rate against which the riskiness of other investment alternatives can be measured, or by which hedging strategies can be constructed. Finally, public safe assets generated by central banks serve as the base asset for the banking system, enabling such basic phenomena as the money multiplier and the very possibility of monetary policy (Fisher 2013).

Although their forms have changed with increasing financial disintermediation, safe assets have maintained a rather constant presence in the economy. The rise of "shadow-banking" institutions since the mid-twentieth century has increased the supply of financial assets fourfold to 400% of GDP by 2010. Over the same period, the proportion of the economy's total assets that are financial has risen from 25 to 40%. Yet the percentage of financial assets that are considered "safe" to total assets has remained relatively stable, suggesting that, whether provided by private or public sources, safe assets play an important and invariant role in economic life more generally (Gorton et al. 2012).

Properties of a safe asset include a large supply, tradeability, and a high credit rating, but ultimately the status of a safe asset persists as a matter of self-reinforcing expectation: "It is of the essence of a safe asset that it cannot become unsafe" (Gourinchas and Jeanne 2012, p. 4). The circularity of this definition underscores how safety is not an intrinsic property of assets. Indeed there is no such thing as absolute safety. Rather, assets are designated safe by processes of financial innovation, institutional development, and shared social understandings of what constitutes risk. They are, in effect, useful "fictions" that are created at the intersection of the private interest and public power (Gelpern and Gerding 2016).

While, apart from government debt, the creation of most safe financial assets is endogenous to the private sector, government also plays a role structuring how the private safe assets are produced. Following Gelpern and Gerding (2016), governments influence the creation of private safe assets in three ways: by (1) "making" them safe; (2) "labeling" them as safe; and (3) "guaranteeing" their safety. *Making* an asset safe entails balance sheet regulation that limits the portfolio risks that institutions can take on. *Labeling* an asset safe doesn't per se render an asset safe, but signals to financial actors which kinds of assets regulators consider appropriate for their investment horizons. *Guaranteeing* an asset safe more directly involves government interventions to either redeem the asset at par (bank deposit insurance), or to rediscount it in exchange for central bank liabilities. Private guarantees like collateralized credit default swaps or surety bonds mitigate risk in a similar way, though public guarantees have an intrinsic advantage in that they are backed by the taxing and money-creating powers of the state (see also Diamond and Dybvig 1983; Holmstroem and Tirole 1998; Gourinchas and Jeanne 2012).

From a macroprudential perspective, the basic problem of safe assets is that there may not be enough of them, especially in light of the post-crisis drop in private securitizations, and the increased demand for safe assets as a consequence of reforms that require financial institutions to hold greater collateral for different purposes. Additionally, central bank policies of quantitative easing aggravate the problem by soaking up available supplies of sovereign debt, increasing the price of public safe assets, and pushing all investors into riskier assets in a search for yield (IMF 2012). As benchmark interest rates converge toward zero, yield compression among riskier assets leads safe assets to lose their ability to generate information about risks elsewhere in the financial system.

In contrast, what this chapter seeks to highlight are the *political* aspects of the safe asset problem, particularly insofar as what constitutes a safe asset has shifted over time, thanks to developments in financial hardware and software, as well as in their regulatory and institutional environments. From this dynamic point of view, the significance of safe assets inheres not merely in their specific functions, but in the fact that they represent *structures of privilege* assembled by economic interests that succeed in shaping the financial environment at historically crucial junctures. Safety means different things to financial actors with different investment horizons. Yet a common feature of safe assets is that, to be safe, they must possess the property of *liquidity*.

II. The Nature of Liquidity and Money as a Means-in-Itself

In modern finance, "liquidity" has a number of nuances (including a simple synonym for money), but is typically understood in the sense Keynes defined it in 1930: a financial asset is liquid if is "more certainly realizable [i.e. convertible into cash] at short notice without loss" (quoted in Hicks 1962, p. 789). While "liquidity preference" with respect to investor motives (transactional, precautionary, speculative) underpinned Keynes' theory of interest and his explanation of capitalist market failure, Keynes was disdainful of liquidity per se: "Of the maxims of orthodox finance none, surely, is more anti-social than the fetish of liquidity, the doctrine that it is a positive virtue on the part of investment institutions to concentrate their resources upon the holding of 'liquid' securities. It forgets that there is no such thing as liquidity of investments for the community as a whole" (Keynes 1936, Chapter 12).

Liquidity cannot be a property of the financial system itself because investments only produce economic value across space and over time under probabilistic or even uncertain conditions: an economic determinism, by banishing space and time, contradicts the essential unpredictability of economic life. Likewise, liquidity cannot be an intrinsic property of any particular financial asset; rather, liquidity is conferred by the institutional framework within which the asset functions, above all within a *portfolio* of other financial assets. By construction, the portfolio acquires the property of liquidity that its constituent elements may not possess to the same degree (Hicks 1962).

At a minimum, conferring liquidity requires three elements: an organized market with a density of transactions that generate asset prices which are parametric with respect to any particular transaction; specialized actors ("market makers") whose role is to match buyers and sellers; and the creation of standard units of the asset, either through protocols of measurement, manufacture, or the creation of legal rights that enhance the fungibility and alienability of the asset (Carruthers and Stinchcombe 1999). Indeed, the first two elements (markets and market makers) depends upon creation of the third, which essentially solves a problem of social knowledge. Without a supply of standardized assets which buyers and sellers can regard as interchangeable, liquidity will not exist. "[L]iquidity presumes assets that are knowable by a large group of potential buyers and sellers. Idiosyncratic, private, or highly asymmetric knowledge undercuts liquidity. The creation of liquidity therefore becomes a problem in how to create generalized impersonal knowledge out of idiosyncratic personal knowledge" (Carruthers and Stinchcombe 1999, p. 356).

If liquidity entails the transformation of one financial asset in terms of another, then is money "the purest embodiment of liquidity" (Shackle 1992, p. 160) and if so how does money acquire liquidity? If money has merely because other people accept it, then a "generalized impersonal knowledge" of money's value that would imbue it with liquidity derives from the concept itself of money. This circularity inheres in "the proposition that money is 'liquid' because of a belief in its power to retain its exchange-value, and that this belief is due to its 'liquidity'" (Shackle 1992, p. 219). Having no value other than in facilitating exchanges and otherwise intermediating economic value, money "is the purest reification of means, a concrete instrument which is absolutely identical with its abstract concept; it is a pure instrument" (Simmel 1990, pp. 122, 211).

In addition to overcoming information asymmetries, money also solves trust problems by replacing personal trust with system trust, or "trust in trust". Such social reflexivity disrupts the "ontological opposition of being and appearance... appearance is reality if it is used as a premise for further experience and action" (Luhmann 1979, p. 67). The difference between personal trust and system trust resembles the relationship between a single financial asset and the larger portfolio in which it resides. It is inevitable that, in any social relationship, "to trust is to risk" (Earle and Cvetovich

1995, p. 107). Whereas personal trust emerges from inductive, interpersonal experiences that create the knowledge upon which personal trust is based, system trust is a self-validating heuristic that acknowledges the reality of social artifice because of its pragmatic results—it represents "fictions which work" (Luhmann 1979, p. 68) Similarly, evaluating any single financial asset imposes more information costs than evaluating the larger portfolio. To the extent that holding financial assets in a portfolio combines together uncorrelated risks, diversification lessens the need to evaluate the risks of any specific component of the portfolio.

By extension, money may be viewed in this analogy as that financial simulacrum with the most diversified portfolio of all. At any given time, the value of any amount of money is some fractionated reciprocal of the price level: As the universally accepted asset, money's utility in acquiring anything hedges its possessor against fluctuations in the prices of any particular things, or in the shifting urgencies of the possessor's own needs. Another way of rendering this relationship is to posit that money is that form of debt which a single debtor (the state) owes to an indefinite number of creditors (Crump 1981, p. 82). Or conversely, "money is a claim on society... [it] appears, so to speak, as a bill of exchange from which the name of the drawee is lacking..."(Simmel 1990, p. 177).

Through money, markets generate information (prices) as a condition for allocating resources. Yet the presence and use of money creates a distinctive tension insofar as money itself presupposes ignorance. For money to function in economic exchange requires that transactors be spared the need and burden of knowing anything about money, other than to trust in it. Just as the liquidity-premium of money (in the Keynesian sense) derives from the ability it gives holders of money to defer decisions about consumption and investment, liquidity as a property of any asset serves as a palliative for ignorance: "Liquidity is, in some sense and degree, a *substitute* for knowledge" (Shackle 1972, p. 216). Likewise, the liquidity of money minimizes the burden placed upon transactors to have knowledge of each other, since money substitutes for the need to know or trust other people. "Money is a substitute for the hypothetical system of a social organization where perfect information and unspoiled reciprocal confidence prevail" (Spahn 2001, p. 11). A completely efficient and frictionless barter economy populated by omniscient transactors, by avoiding the problem of the "double coincidence of wants" (Jevons 1875), would render money unnecessary. Money thus reduces social complexity, but at the cost (so to speak) of draining away contextual knowledge of interpersonal relationships. In these senses does Keynes's "antisocial" epithet characterize the depersonalized market relationships that money makes possible, and for which money serves as a template. For any financial instrument to be liquid, it must be easily convertible into money. As financial innovations convert less liquid assets into more liquid ones, those assets approach "moneyness" in sense of acquiring the characteristics of a generalized medium of exchange.

III. Money and Financial Innovation: Some Historical Episodes

The specific mechanisms by which liquidity has been created reflect historical contingencies, and the prevailing pattern of interests. That financial innovations invariably involve the transformation of less liquid assets into more liquid ones can be seen in episodes of monetary experimentation throughout American history. Prior to independence, Americans circumvented mercantilist restrictions imposed by Great Britain by creating their own circulating media (Sylla 1982). Dissatisfaction with those experiences led to a constitutional framework in which American states could no longer print paper money, and could only make gold or silver a legal tender. Conversely, federalism assured that the chartering and regulation of banks remained at the state level. Given the revenue benefits states derived from the chartering and taxing of their banks, they were loath to grant operating privileges to out-of-state banks; the resulting prohibitions on interstate branching ensconced a decentralized pattern of unit banking that endured into the late twentieth century (Calomiris and Haber 2015, Chapter 6).

This combination of monetary restriction and financial opportunity encouraged the proliferation of banknotes issued by state-chartered banks. Though not legal tender and often of uncertain value, state banknotes were the chief form of monetary liquidity during the antebellum era. Private bank clearinghouse associations developed procedures for the more efficient settling of debits and credits, as well as for the provision of emergency liquidity during periodic financial stringencies. In this way, clearinghouses substituted for a central bank (Timberlake 1984; Gorton 2010, pp. 33–38).

In American financial history, the establishment, during the Civil War, of the National Bank System (NBS) illustrates nicely how innovations in liquidity provision can at once link the interests of public and private actors, liquefy disparate financial assets, and yet create constraints on liquidity that incentivize further innovation to circumvent those very constraints. Established in order to fund war debts and provide for a uniform currency, the NBS tied the support of government debts to a new currency that, unlike state bank notes, circulated at par and would remain valid even if the issuing bank failed, rendering them safe assets (Friedman and Schwartz 1963, Chapter 2). By tying the financing of the Civil War to a new circulating medium, the NBS became the linchpin of a Republican-dominated political economy that discriminated against agrarian regions both in terms of the distribution of national banking facilities and the use of high tariffs on manufactured products to pay down the debt, tariffs which effectively transferred wealth from the agriculture to the manufacturing sector (Bensel 1990, Chapter 5). State banks were encouraged to switch to national charters by a prohibitive tax placed on state bank notes. Initially most state banks did become national banks, but high minimum capital requirements to join the NBS, a prohibition on mortgage lending, and statutory limits on the overall supply of national bank notes created liquidity constraints (especially in the South and West) that encouraged a revival of state banks, which by the end of the century far outnumbered their national counterparts.

While state banks could no longer profitably issue paper currency, they did expand another form of liquidity, namely checkable deposits, and postbellum banking evolved as a dual system in which state banks competed with their national counterparts. State-level unit banking systems were protected by the legal inability of national banks to engage in interstate branching (Sylla 1969). The prevalence of unit banking interacted with the reserve rules of the NBS to stimulate further innovations. The NBS tiered banks into country, reserve city, and central reserve city banks, with increasingly strict reserve requirements; it also allowed banks to hold a fraction of their reserves as interest-bearing bankers' balances with institutions at the next level. This "pyramiding" of reserves created a framework for a pattern of correspondent banking by which both state and national banks channeled into the securities markets (James 1978, Chapter 4). Not only did the United States lack a central bank, but its financial system looked different than those of other industrializing countries. The country sustained thousands of state and national banks, many with miniscule capitalizations. Unlike other countries, its correspondent banking practices inhibited the development of discounting as a means of obtaining liquidity (White 1983, Chapter 1).

This financial system proved both unstable and, for certain interests, insufficient in terms of how much liquidity it provided. Though the NBS pooled reserves in big-city banks, this arrangement fell short of the centralization of reserves that would deal with financial stringencies before they broke out into outright panics. Moreover, deployment of reserves in stock market speculations meant that any widespread call for the return of bankers' balances could precipitate a crash in securities prices, as was spectacularly the case during the Panic of 1907.

This cursory sketch of nineteenth-century American finance highlights how limited was the capacity, not to mention the willingness, of the federal government to ameliorate sectoral or sectional conflicts by creating public liquid assets or conferring liquidity upon private ones. Most forms of liquidity were generated and maintained by private institutions. Above all, it was the political commitment to maintaining a commodity money standard that placed fundamental limits upon the potential for government to provide for, or otherwise manipulate flows of, financial liquidity to propitiate various interests.

IV. The Chartalist Turn: How American Farmers Became a Constituency of Credit

Up through 1907, debate percolated among bankers as to how best to secure financial stability and elasticity of the currency. Yet divisions among them, particularly over the need for a central bank, prevented the Republicans from committing to comprehensive reform. (Friedman and Schwartz 1963, Chapter 4; McCulley 1992, Chapter 3 esp. p. 97). The Panic of 1907 reinforced the need to do something, and by 1913 legislation established the Federal Reserve System. Two new sources of government-sponsored liquidity created by the legislation—the Federal Reserve note and the rediscount facilities of the regional reserve banks—reflected traditional notions about lending only against short-term, productive assets. As before, agrarian interests were again slighted in that much of their short-term paper was not eligible for rediscounting. In effect, initially at least, the Fed was committed to the passive, "real bills" principle that it was only to accommodate the liquidity demands generated by real economic activity, and that the new central bank would not pursue discretionary policies other than to defend gold convertibility (Friedman and Schwartz 1963, Chapter 5).

Very quickly, the creation of the central bank occasioned the first examples credit constituencies that would become ubiquitous over the next century. As early as 1914, the disorganization of agriculture arising out of World War I led to passage of the Cotton Futures Act, which created a fund to help Southern cotton farmers, a precedent that prefigured other measures. The Grain Standards Act and Warehouse Act of 1916 established federal standards for defining different grades of farm output, and imposed rules for the operation of warehouse facilities. The latter act also provided for a standardized warehouse receipt that farmers could then use as collateral in bank borrowing. Following the model of the Federal Reserve System, the Federal Farm Loan Act of 1916 established twelve land banks and joint-stock banks. Capitalized with federal money, these banks would originate farm mortgage loans, financing them by issuing debentures that enjoyed tax-free privileges and which were collective liabilities of the land banks (Holt 1924).

Farmers, who had enjoyed higher prices for the last two decades and had especially prospered during the war, were crushed by the short but sharp depression of 1920–1921. A group of Congressmen known as the "Farm Bloc" argued that farmers had been victimized by government policies that favored industry over agriculture (Capper 1922). In particular, farm leaders blamed the Fed's higher interest rates for precipitating the price collapse, and for starving agriculture of credit. Although Congressional hearings ultimately exonerated the central bank, the inquiries did identify a need for some kind of credit facilities for agriculture between the short-term paper eligible for Fed rediscount, and the long-term mortgage loans available through the Federal Farm Loan banks. Accordingly, in 1923, the Agricultural Credits Act established a parallel system of twelve intermediate credit banks to provide agricultural financing over periods beyond the maturity of the three-month commercial paper which the Fed was authorized to rediscount and use for its note backing. Agricultural paper eligible for hypothecation included warehouse receipts, bills of lading, and titles to livestock. Like the system of land banks, capital stock of the intermediate credit banks was owned by the government and its financing provided by tax-advantaged debentures sold to investors (Benner 1925a; Wooddy 1934).

The effects of this emerging credit infrastructure for agriculture were mixed, and hardly made up for the farmers suffering in the depression of 1920–1921. Its limited success reflected the Farm Bloc's own premise that their economic problems were fundamentally a matter of inadequate financing and ineffective marketing, and not the actual overproduction of farm commodities. The conceit even prevailed among supporters that federal credit programs did not actually provide farmers with "government loans", but merely helped them establish their own credit (Wiprud 1921). Other evidence suggested that the programs crowded out private lenders that had traditionally financed farm mortgages, and led to lower lending standards on short-term farm paper (Wright 1923; Benner 1925b, 1926).

The politics of agricultural credit as it unfolded up to 1929 illustrated the potential for public policies that conferred benefits upon private interests by creating a financial liquidity to support the value of their real assets. From the standardization of commodity grading and creation of negotiable warehouse receipts to the establishment of institutions that would lend against agricultural land and short-term paper, farm policies transformed the agricultural sector into the first modern constituency for credit. This new arrangement was made possible above all by the existence of the Federal Reserve, and its powers to create liquidity via rediscounting and open-market purchases of the liabilities of land banks and intermediate credit banks.

Herbert Hoover, who was an activist President by Republican standards of the day, represents a liminal figure in the evolving framework of government-sponsored liquidity. Hoover increased the government's commitment to agriculture by signing the Agricultural Marketing Act in 1929. This measure established a Federal Farm Board that sought to promote the merchandising of agricultural products by various methods, including the creation of farmers' cooperative associations to which the government would loan funds, enabling the associations to keep crop surpluses off the market for the sake of higher prices. By late in the year, marketing corporations were set up for grains and wool; other crops soon followed. These organizations served as channels for the loans provided under the act. Despite the desire of farmers in getting higher prices, Hoover insisted on treating the AMA as a measure to improve agricultural marketing, rather than as an exercise in price stabilization, even though funds were increasing devoted to a futile attempt to support farm prices as the depression deepened (Fausold 1977; Hamilton 1991, Chapter 3).

Hoover's "associational" vision of public policy stressed the voluntary organization of private interests rather than direct government interventions in the economy. This restrained view of public power also informed the implementation of his administration's other major financial initiative, the Reconstruction Finance Corporation (RFC). As a weakening banking system retrenched on lending with little support from a timid Federal Reserve, Hoover made a futile attempt to organize private capital into a National Credit Corporation before signing into law the RFC. An independent government agency, the RFC was given the resources to lend to against collateral to almost any private financial or nonfinancial entity; in July 1932 its remit was extended to include agricultural lending as well as municipal governments. That same month Hoover also signed into law the Federal Home Loan Bank Act, creating a similar facility to channel funds to support home mortgages by borrowing from the Treasury and issuing its own tax-free bonds (Barber 1985, pp. 130–131).

Characteristically, Hoover supported this extension of federal credit policy even as he espoused the fiscal orthodoxy of balanced budgets; borrowings of the RFC were placed off-budget. The RFC's significance for the present-day debate about safe assets is that it represented the first general-purpose vehicle for transforming private financial assets into public liquidity. Doing what the Federal Reserve System at the time couldn't, or wouldn't, the RFC embarked on a recapitalization program similar to that undertaken by the Federal Reserve after 2008. To be sure, elements of the latter rescue effort were lacking in 1932. For one, the earlier Federal Reserve did not yet take the view that it was possible or desirable to undertake massive open-market operations; gold convertibility was still regarded as an operational restraint. Nor did the federal government regard as legitimate fiscal policies—let alone deficit spending—that would have provided direct relief to citizens, or have stimulated aggregate demand.

Roosevelt's New Deal swept away many of these inhibitions, and the country's modern architecture of economic governance dates from the Depression era. Once the gold standard no longer represented a policy constraint, deficit spending created open-ended fiscal possibilities. The pre-Depression edifice of agricultural credit was reinvigorated under Roosevelt as the Farm Credit Administration. Price support and production controls finally became official policy (Romasco 1983, Chapter 9). With similar continuity, Hoover's Federal Home Loan Banks served as the basis of a federal effort to revive the construction industry. In 1933 the Home Owners Loan Corporation (HOLC) was added to the FHLB system. The HOLC's particular task was to minimize foreclosures by refinancing existing mortgages into longer-term, self-amortizing loans, a form of mortgage finance that hitherto was not widely used. Traditionally, mortgage lending committed funds only up to five years, at the end of which borrowers were expected to return the principal in a balloon payment; loan-to-value ratios were typically below 50%. In contrast, the HOLC popularized a form of lending that amortized the principal over a much longer period, allowing for much higher loan-to-value ratios (smaller down payments). To manage the risks of this new form of financing, the Federal Housing Administration (FHA) was created in 1934 to provide mortgage insurance to approved lenders; Fannie Mae, established as a government agency in 1938, gave birth to a secondary market in mortgages by purchasing mortgages from private lenders for its portfolio and (initially, at least) originated its own (Babcock 1939).

Federal housing policy dating from the 1930s had the same goal as agricultural policy after 1916: create constituencies for credit in order to liquefy financial assets connected to the economic fortunes of different interests or industries. Achieving this liquidity required new financial software. Just as the Grain Standard and Warehouse Acts of 1916 created instruments that allowed farmers to hypothecate their crops, so too did housing policy under the FHA popularize a form of mortgage which, combined with federal mortgage insurance and a government-provided secondary market, opened up home ownership to a much wider swathe of the American public. Indeed, these housing policy initiatives transformed a risky and illiquid asset into a uniform financial product that could be disposed of in an active and national secondary market.

V. Financial Innovation in a Chartalist World

If the financial crisis of 2007–2008 was a deluge that nearly washed the financial system away, the housing policy innovations of the 1930s represented the original head waters where a new form of financial liquidity first emerged. It should be stressed that the appearance of the modern mortgage predates their securitization by more than three decades. Financial assets are not per se stable or unstable; rather, it is the evolving systemic context that bestows this quality. The financial architecture established by the New Deal consisted of a few basic conceptual and institutional elements. With the jettisoning of the domestic gold standard, money was reconfigured as legal tender fiat. After 1935, a reconstituted Federal Reserve System, its powers now centralized in

Washington, D.C., embraced the premise that monetary policy should be subordinate to, and accommodative of, fiscal policy (Friedman and Schwartz 1963, Chapter 8; Stein 1969, Chapter 3). Continued prohibitions against branching protected unit banking, while deposit insurance effectively transformed their demand liabilities into safe assets. For their part, commercial banks were forbidden from paying interest on checking accounts. Savings accounts were subject to the Fed's Regulation Q, which placed limits on the interest rates which with which banks could entice depositors. In response to the perceived abuses of financial speculation and fraud, the Glass-Steagall Act of 1933 separated commercial from investment banking, leaving the former a highly regulated yet fragmented industry in which local bankers, facing little competition, enjoyed steady profits in exchange for limits upon their investment activities. As long as these profits continued, the "charter value" of a traditional banking institution muted the incentives to increased risk-taking and "shadow banking" financial innovation (Gorton 2010, pp. 54–55; Calomiris and Haber, p. 194).

That this architecture remained stable for nearly forty years is reflected in the fact that bank deposits represented most of the "safe assets" generated by the financial system until the late 1970s (Gorton et al. 2012). For some three decades after its implementation, Regulation Q represented no constraint upon banks' activities, since the market rates banks paid on various deposits were consistently below the prevailing regulatory ceiling (Gilbert 1986). Commercial banking remained at the core of the financial system up through the 1960s. Although Bretton Woods did re-establish an international dollar-gold exchange standard, for the first fifteen years of its existence global demand for dollar liquidity ensured that gold convertibility did not represent a constraint upon American monetary or fiscal policy. A relatively closed international financial system characterized by significant exchange and capital controls blunted speculative pressures on fixed exchange rates (Reinhart and Rogoff 2009, p. 205). Only by the late 1950s did the world supply of dollars, managed in the growing Eurodollar market, begin to impinge upon credibility of the gold-dollar link.

The "Great Inflation" after 1964 created incentives for financial innovation that ultimately undermined the New Deal architecture. Monetary policymakers proved unwilling to incur the political costs of correcting for their previous accommodations of mounting fiscal deficits. Of key significance was the breakdown in the expected relationship and tradeoff between inflation and unemployment—the so-called Phillips Curve—that was fundamental to Keynesian economic policies (Meltzer 2005). Deficit-driven inflation also served to erode the postwar bipartisan "fiscal consensus" that had reconciled a traditional solicitude for balanced budgets with Keynesian countercyclical policy (Stein 1969, p. 240). Those deficits, and the inflation they ignited through the 1960s, both undermined international commitment to Bretton Woods and stimulated financial software innovations that sought to circumvent interest-rate limits. In addition, trends toward deregulation where reinforced by hardware innovations that increased the data processing capabilities of banks, narrowing the profit margins of their traditional businesses (Horvitz 1986; Kane 1986). As Bretton Woods promoted international financial liberalization, private actors dissatisfied with regulated domestic interest rates in an inflationary environment now had the motive and means to proliferate a parallel shadow-banking system. This liberalization, and the financial innovations it facilitated, set the stage for further instability and crises (Reinhart and Rogoff 2010, pp. 155–156). High inflation was critical to undermining the domestic policy equilibrium that had sustained unit banking since the early years of the Republic. Facing a squeeze between their portfolio incomes and costs of funds, bankers acquiesced to a lifting of banking restrictions that effectively introduced interstate branching by the early 1990s (Calomiris and Haber 2015, pp. 195–202).

The demise of Bretton Woods after 1971 left a global fiat monetary system and expanded the potential for a political economy of chartalism. A political economy of commodity money placed fundamental convertibility limits on government's capacity to generate and distribute financial liquidity. In contrast, the advent of chartalism has enabled government to sublimate political conflict by dramatically expanding its ability to shape the generation of financial assets. Liquidity policies are redistributive in character: they invariably involve a transfer of resources from one part of the population to the other. Yet unlike direct policies of taxation and appropriation, liquidity policies typically do not involve upfront fiscal commitments. When government insures bank deposits or implicitly guarantees the debts of GSEs, the costs of these actions do not appear as budget items; rather, they represent contingent liabilities that may, or may not, be invoked depending upon the circumstances.

The political economy of chartalism required the abandonment of commodity money standards. A world of monetary nominalism empowers public authority to embark on any number of policies that distribute resources to favored groups by enhancing the liquidity of the assets they control or the liabilities they have incurred. If money is no longer a "thing", but a property of the matrix of prevailing debt relations, then public policies apart from monetary policy in the conventional sense can, in effect, create money by targeted interventions that impart to financial assets greater "moneyness".

The absence of convertibility restraints supplanted the hardness of rules with the softness of a shifting politics of credibility whereby government triangulates its macroeconomic policies with a set of flexible constraints—exchange rate volatility, budgetary and balance of payment financings; income, growth, and inflation indicators—all of which will create fields of discretion that vary from country to country. The United States, as the world's largest economy and with the deepest finance markets, enjoys the privilege of borrowing in its own currency and thus the greatest policy discretion. Yet this discretion is counterbalanced by the possibility that financial liberalization and innovation interact so as to outstrip the regulatory capacity of even the largest economy and the most capable public authorities. Key here are intensified episodes of securitization which, by liquefying disparate assets, link their values in systematically unpredictable ways (Cerny 1994). While securitization has been justified as a more efficient way of managing aggregate risk, it encourages the banking system to take on *more* risk as well (Rajan 2005) which, in the absence of capital requirements analogous to those in traditional banking, only increases the system's fragility (Stein 2010).

The possibilities of chartalism were not immediately realized with the end of Bretton Woods, since the expansion of chartalist liquidity policies has occurred in a characteristically reactive fashion—problems must first emerge in order for the new policy muscles to be flexed. One early feature of the new environment was the change in what it meant for a core economy like the United States to conduct monetary policy. With the demise of the international gold standard and the accompanying surge in global capital flows, monetary policy confronted a new and disturbing conceptual reality. If relentless and open-ended innovations in deregulated financial markets blurred the distinction between money and other financial instruments, then what operational criteria would monetary authorities use to target their desired macroeconomic outcomes? Ironically, just as the breakdown of Phillips-Curve relationships between inflation and unemployment discredited the Keynesian paradigm, so too by the 1980s was Monetarism confounded by the apparent breakdown in the empirical relationships between the money stock, income, and prices

(Friedman 1988; Goodhart 1989). In an environment of financial innovations that continually seek to render all manner of assets more like money, monetary policy itself assumes esoteric characteristics. This is not to suggest that monetary policy is ineffective. Rather, any intervention short of automatic rule-following may have uncertain or even perverse consequences.

After Bretton Woods also began the sustained decline in the proportion of liabilities associated with traditional forms of financial intermediation (banks, insurance companies, etc.) relative to shadow disintermediation. Between 1971 and 2007, traditional intermediation fell from 80 to 40% of all financial liabilities (Adrian and Ashcraft 2012). The first money market mutual fund, established in 1971, sought to circumvent Regulation Q by paying interest on checkable, stable net-asset-value shares. High inflation also gave a fillip the securitization of mortgages since it allowed loan originators to offload interest-rate risks onto third-party investors. The end of Regulation Q's interest rate caps between 1981 and 1986, also driven by the impact of inflation upon an increasingly deregulated and competitive financial marketplace, triggered the first domestic disruption of the postwar financial era, the Savings & Loan debacle. While the collapse of the thrifts preceded the financial crisis by nearly twenty years, it was the first event whose elements prefigured those of 2007–2008. In both cases, financial deregulation increased incentives for risk taking which, combined with inadequate oversight and excessive regulatory forbearance, produced a festering moral hazard problem that eventually provoked a socialization of private losses—the conversion of private assets into public safe assets.

Unlike many of the actors in 2007–2008, S&Ls were traditional institutions of financial intermediation. However, their risky responses to a high interest-rate environment underscore how the traditional advantages of a financial charter within a highly regulated industry had eroded. Just as the business model of thrift-based mortgage lending had been rendered obsolete by the 1980s, so too had the diminished attractiveness of unit-based commercial banking fostered both consolidation and burgeoning shadow-banking disintermediation. The S&L experience also revealed that public authorities face an imperative to rescue the financial system, no matter what the cost might be or the consequences for future bouts of instability. In the context of increased mobility of global capital that drove both financial deregulation in the United States and various shadow-banking innovations, policymakers in the Treasury and especially the Federal Reserve have no choice but to intervene when financial disruptions of a certain magnitude threaten the integrity of the system.

Since each bout of financial turbulence has its own antecedents and etiology, the modalities from these public interventions will differ from case to case. Those crises that do not have systemic significance may be resolved through some combination of private investor losses and IMF-orchestrated restructuring. Yet the increasingly autonomous quality of the interactions between financial crises and public liquidity policies can be appreciated in the sequential overlap across crises and responses. Responses become preconditions for future crises. Freed of Bretton Woods, the inflationary accommodation of oil price shocks in the 1970s enabled the United States to run ever-larger balance of payments deficits, which were financed by petrodollar recycling. These funds also found their way into the borrowings by Latin American countries, which culminated in their debt defaults of 1982. In response, the United States attempted, with mixed results, to liquefy that debt via Brady bonds. A combination of tight American monetary policy and loose fiscal policy in the early 1980s pushed the value of the dollar to unacceptable levels, leading to the Plaza Accord interventions of 1985. A resulting stronger yen encouraged a property bubble in Japan, which burst in 1991, ushering in its "lost decade" and the first experiments with unconventional monetary policies. American responses to its own problems with thrift institutions (brought about by the same monetary policies of the early 1980s) included pushing Fannie Mae and Freddie Mac into giving greater support to the mortgage market; resolving that crisis also stimulated the securitization of mortgages and the growth of shadow banking, setting the stage for America's housing bubble. Efforts later in the decade to address Japan's economic sluggishness—now by *devaluing* the yen—kept American monetary policy easy enough to fuel asset bubbles elsewhere in Asia. In the aftermath of the Asian financial crisis of 1997, Asian countries built up their foreign exchange reserves with dollar-denominated assets, thus fueling America's housing bubble. Easy money responses to Russia's sovereign default (1998) and Long-Term Capital Management's collapse and bailout (1998) fed the internet stock bubble, which deflated by 2001.

This litany of financial episodes is by no means exhaustive. Each episode has its own complicated explanation, and all are hardly attributable to a single bugbear like "easy money". Rather, it is in the totality of these episodes' confluence that the contours of a political economy of chartalism become perceptible. The events of 2007–2008 have made it abundantly clear that the American authorities will go to any lengths to address system-threatening instabilities. This is comforting, albeit in an ominous way. Public financial authorities represent the ultimate backstop to the financial system insofar as they possess the ability to create public safe assets (central bank liabilities, government debt) and act as a lender of last resort. Unconstrained by any definition of money that would limit these abilities, public authorities have no choice but to rescue the system when private risk-taking creates unacceptable levels of instability. *Every private actor knows this*. But because they know that authorities possess the ability to rescue the system by socializing private risks, the level of aggregate risk acquires an upward momentum. In effect, the political economy of chartalism gives rise to a state of "meta' moral hazard" (Miller et al. 2001) that is profoundly corrosive of the information-producing capacity of the financial system. If all private risk can, in principle, be rendered publicly safe to an indefinite extent, then a determined financial system becomes essentially disengaged from the irreducible uncertainty of the real economy.

The ultimate governor of this engine of contradiction is the capacity of private financial actors to trust in the validity of what public authorities define as safe assets—above all, to trust in the fiat money generated by those authorities. Participants in the financial system, big and small, have no choice in the matter. Yet in an environment where monetary strategies of quantitative easing create indefinite amounts of safe assets, there will be a point when that very plenitude flips into its opposite. If and when interest rates fall into negative values, any financial asset, no matter how safe, will become hostage to the very distinction between investment and speculation. To own a financial asset with a negative yield will only produce positive returns if that asset can be sold to someone for a higher price. In these circumstances, the very pursuit of safety creates its opposite.

References

Adrian, Tobias and Adam B. Ashcraft. 2012. "Shadow Banking: A Review of the Literature" Federal Reserve Bank of New York *Staff Report No. 580.*

Babcock, Frederick M. 1939. "Influence of the Federal Housing Administration on Mortgage Lending Policy" *The Journal of Land and Public Utility Economics* 15: 1–5.

Barber, William J. 1985. *From New Era to New Deal. Herbert Hoover, the Economists, and American Economic Policy*. Cambridge University Press.
Benner, Claude. 1925a. "Credit Aspects of the Agricultural Depression" *Journal of Political Economy* 33: 94–106.
Benner, Claude. 1925b. "Credit Aspects of the Agricultural Depression" *Journal of Political Economy* 33: 217–233.
Benner, Claude. 1926. *The Federal Intermediate Credit System*. NY: The Macmillan Co.
Bensel, Richard Franklin. 1990. *Yankee Leviathan: The Origins of Central State Authority in America, 1859–1877*. Oxford University Press.
Bernanke, Ben, Carol Bertaut, Laurie Pounder Demarco, and Steven Kamin. 2011. "International Capital Flows and the Returns to Safe Assets in the United States, 2003–2007" Board of Governors of the Federal Reserve System. International Finance Discussion Papers Number 1014.
Caballero, Ricardo G. 2009. "Sudden Financial Arrest" Mundell-Fleming Lecture, International Monetary Fund.
Calomiris, Charles W., and Haber, Stephan H. 2015. *Fragile by Design: The Political Origins of Banking Crises and Scarce Credit*. 1st paperback edition. Princeton University Press.
Capper, Arthur. 1922. *The Agricultural Bloc*. NY: Harcourt Brace.
Carruthers, Bruce G. and Arthur L. Stinchcome. 1999. "The Social Structure of Liquidity: Flexibility, Markets, and States" *Theory and Society* 28: 353-382.
Cerny, Philip G. 1994. "The Dynamics of Globalization: Technology, Market Structure, and Policy Response" *Policy Sciences* 27: 319–342.
Crump, Thomas. 1981. *The Phenomenon of Money*. Routledge & Kegan Paul.
Davis, Lance E. 1965. "The Investment Market, 1870–1914: The Evolution of a National Market" *The Journal of Economic History* 25: 355–399.
Diamond, Douglas W. and Philip H. Dybvig. 1983. "Bank Runs, Deposit Insurance, and Liquidity" *Journal of Political Economy* 91: 401–419.
Drennan, Matthew P. 2015. *Income Inequality: Why it Matters and Why Most Economists Didn't Notice*. Yale University Press.
Earle, Timothy C., and Cvetkovich, George T. 1995. "Social Trust: Toward a Cosmopolitan Society". Westport, CT: Praeger.
Fausold, Martin. 1977. "President Hoover's Farm Policies, 1929–1933" *Agricultural History* 51: 362–377.
Financial Crisis Inquiry Commission. 2011. *The Financial Crisis Inquiry Report*. Washington, D.C.: GPO.
Fisher, Peter. 2013. "Reflections on the Meaning of 'Risk-Free'" in *Sovereign Risk: A World Without Risk-Free Assets*. Bank for International Settlements BIS Paper 72: 65–72.
Friedman, Benjamin M. 1986. "Lessons on Monetary Policy from the 1980s" *Journal of Economic Perspectives* 2: 51–72.

Friedman, Milton and Anna Jacobson Schwartz. 1963. *A Monetary History of the United States 1867–1960.* Princeton University Press.

Garbade, Kenneth D., and Silber, William L. 1978. "Technology, Communication and the Performance of Financial Markets: 1840–1975" *Journal of Finance* 33 (3): 819–832.

Gelpern, Anna and Erik Gerding. 2016. "Inside Safe Assets" *Yale Journal on Regulation* 33:

Gilbert, Alton R. 1986. "Requiem for Regulation Q: What it Did and Why it Passed Away" Federal Reserve Bank of St. Louis *Review*: 22–37.

Goetzmann, William N. 2016. *Money Changes Everything: How Finance Made Civilization Possible.* Princeton University Press.

Goodhart, Charles. 1989. "The Conduct of Monetary Policy" *The Economic Journal* 99: 293–346.

Gourinchas, Pierre-Oliver, and Olivier Jeanne. 2012. "Global Safe Assets" Bank for International Settlements. Working Paper Number 399.

Gorton, Gary B. 2010. *Slapped by the Invisible Hand: the Panic of 2007.* NY: Oxford University Press.

Gorton, Gary, Stefan Lewellen, and Andrew Metrick. 2012. "The Safe-Asset Share" *The American Economic Review* 102: 101–106.

Hamilton, David. 1991. *From New Era to New Deal. American Farm Policy from Hoover to Roosevelt.* University of North Carolina Press.

Hicks, J. R. 1962. "Liquidity" *The Economic Journal* 72: 787–802.

Holmstroem, Bengt, and Jean Tirole. 1998. "Private and Public Supply of Liquidity" *Journal of Political Economy* 106: 1–40.

Holt, W. Stull. 1924. *The Federal Farm Loan Board.* Johns Hopkins Press.

Horvitz, Paul. 1986. "Technological Innovation. Implications for Regulation of Financial Institutions" in Colin Lawrence and Robert Shay, eds. *Technological Innovation, Regulation, and the Monetary Economy.* Cambridge, MA: Ballinger Publishing: 111–123.

International Monetary Fund. 2012. *Global Financial Stability Report: The Quest for Lasting Stability.*

James, John A. 1978. *Money and Capital Markets in Postbellum America.* Princeton, NJ: Princeton University Press.

Jevons, William Stanley. 1875. *Money and the Mechanism of Exchange.* D. Appleton & Co.

Kane, E. J. 1986. "Technology and the Regulation of Financial Markets," *in Technology and the Regulation of Financial Markets: Securities, Futures, and Banking.* A. Saunders and L. T. White eds. Lexington, MA: Lexington Books: 187–193.

Keynes, John Maynard. 1936. *The General Theory of Employment, Interest, and Money.*

Krippner, Greta R. 2011. *Capitalizing on Crisis: The Political Origins of the Rise of Finance*. Harvard University Press.
Lengwiler, Martin. 2016. "Risky Calculations: Financial Mathematics and Securitization Since the 1970s". *Historical Social Research* 41 (2): 258–279.
Luhmann, Niklas. 1979. "Trust", in *Trust and Power*. John Wiley & Sons.
McCulley, Richard T. 1992. *Banks and Politics During the Progressive Era: The Origins of the Federal Reserve System, 1897–1913*. New York, NY: Routledge.
Merton, Robert C. and Zvi Brodie. 1995. "A Conceptual Framework for Analyzing the Financial Environment", in *The Global Financial System: A Functional Perspective*. Boston, MA: Harvard Business School Press: 3–31.
Meltzer, Allan. 2005. "Origins of the Great Inflation" Federal Reserve Bank of St. Louis *Review* 87: 145–175.
Miller, Marcus, Paul Weller and Lei Zhang. 2001. "Moral Hazard and the U.S. Stock Market: Analyzing the 'Greenspan Put'" CSGR Working Paper No. 83/01, University of Warwick.
Miller, Merton H. 1986. "Financial Innovation: The Last Twenty Years and the Next" *The Journal of Financial and Quantitative Analysis* 21: 459–471.
Podolski, T. M. 1986. *Financial Innovation and the Money Supply*. Basil Blackwell.
Rajan, Raghuram G. 2005. "Has Financial Development Made the World Riskier?" in *Economic Policy Symposium Proceedings*. Federal Reserve Bank of Kansas City: 313–369, August.
Rajan, Raghuram G. 2010. *Fault Lines: How Hidden Fractures Still Threaten the World Economy*. Princeton University Press.
Reinhart, Carmen M. and Kenneth S. Rogoff. 2009. *This Time Is Different. Eight Centuries of Financial Folly*. Princeton University Press.
Reinhart, Carmen M., and Kenneth S. Rogoff. 2010. "Growth in a time of debt" *American Economic Review* 100 (2): 573–578.
Romasco, Albert U. 1983. *The Politics of Recovery*. Oxford University Press.
Shackle, G. L. S. 1972 [reprinted 1992]. "Liquidity: Its Nature" in *Epistemics and Economics: A Critique of Economic Doctrines*. EBook edition. Routledge/Taylor & Francis.
Shackle, G.L.S. 1992. *Epistemics and Economics*. New Brunswick, NJ: Transaction Publishers.
Simkovic, Michael. 2013. "Competition and Crisis in Mortgage Securitization" *Indiana Law Journal* 88: 213–271.
Simmel, Georg. 1990. *The Philosophy of Money*. Translated by Tom Bottomore and David Frisby, 2nd edition. London and New York: Routledge.
Spahn, Heinz-Peter. 2001. *From Gold to Euro: On Monetary Theory and the History of Monetary Systems*. Berlin: Springer-Verlag.
Stein, Herbert. 1969. *The Fiscal Revolution in America*. University of Chicago Press.

Stein, Jeremy C. 2010. "Monetary policy as financial stability regulation" NBER Working Paper No. 16883. National Bureau of Economic Research.

Sylla, Richard. 1969. "Federal Policy, Banking Market Structure, and Capital Mobilization in the United States, 1863–1913" *The Journal of Economic History* 29: 657–686.

Sylla, Richard. 1982. "Monetary Innovation in America" *The Journal of Economic History* 42: 21–30.

Taylor, John B. 2009. *Getting off Track: How Government Actions and Interventions Caused, Prolonged, and Worsened the Financial Crisis.* Stanford, CA: Hoover Institution Press.

Timberlake, Richard H. Jr. 1984. "The Central Banking Role of Clearinghouse Associations" *Journal of Money, Credit, and Banking* 16: 1–15.

Traflet, Jan M., and Gruver, William R. 2015. "The Quixotic Quest for Fairness: The SEC's Role in the Rise of High Frequency Trading" *Essays in Economic and Business History* 33 (1): 144–189.

Wallison, Peter J. 2015. *Hidden in Plain Sight: What Really Caused the World's Worst Financial Crisis and Why it Could Happen Again*. Encounter Books.

White, Eugene Nelson. 1983. *The Regulation and Reform of the American Banking System, 1920–1929.* Princeton University Press.

Wiprud, A. C. 1921. *The Federal Farm Loan System in Operation*. NY: Harper & Brothers.

Wooddy, Carroll. 1934. *The Growth of the Federal Government 1915–1932.* NY: McGraw-Hill.

Wright, Ivan. 1923. *Farm Mortgage Financing*. NY: McGraw-Hill.

CHAPTER 4

The New Algorithms Reshaping Human Labor: Policy Options for the Automated Workplace

Brett S. Sharp

Introduction

Recent technological breakthroughs are revolutionizing the modern workplace. These disruptive innovations challenge traditional roles of human workers within organizations. The promises offered by such advancements are extraordinary but so are the inherent risks. This chapter focuses on three emerging forces that are transforming the nature of work: (1) algorithmic management, (2) artificial intelligence (AI), and (3) robotics. While these technologies overlap, they represent only a portion of the numerous innovations impacting organizations today. Although

Prepared for *Political Economy of Robots* (2nd Edition) edited by Ryan David Kiggins.

B. S. Sharp (✉)
Department of Political Science, University of Central Oklahoma, Edmond, OK, USA
e-mail: BSharp@uco.edu

R. Kiggins (ed.), *The Political Economy of Robots*, International Political Economy Series, https://doi.org/10.1007/978-3-031-95171-8_4

workplace automation has been evolving for decades, recent technological breakthroughs are occurring at a pace that far exceeds what organizational leaders and policymakers are prepared to handle. Swift and informed policy responses are now critical to address the unintended consequences that may arise from these rapid development. The implications for the future of work are profound.

For some, serious consideration of the potential downside of these technologies is difficult. An interesting dynamic is at play in which many contemporary observers who are now expressing concern about these technologies seem to be echoing some of the wildest flights of fancy within science fiction. In other words, our experiential base on these matters comes as much from Hollywood as it does Silicon Valley. On the other side, the Luddites and their progeny have been prematurely expecting machines to replace human workers for well over a hundred years now. Yet, evidence continues to mount that we already are experiencing major shifts in the labor market as a result of relentless unfolding of the automated workplace. The most well-known and established instance is the automation occurring on the factory floor over the past several decades. As businesses very visibly outsourced to less expensive labor markets in the developing world, such industrial globalization eclipsed any displacement identifiable as due to new technologies. Manufacturing is now returning to the developed world as closer proximity to consumer markets reduces shipping costs (Sauter and Stebbins 2016). This reshoring has intensified in the wake of the global pandemic (Akbari 2024). Yet, reopening domestic production facilities is not generating a significant number of new jobs in the near term (Chung and Lee 2023). Why? Increasingly, the most efficient alternative for cheap labor is deployment of more and more versatile industrial robots (Lee 2016; Prettner 2023). Likewise, other segments of the economy are replete with the mobilization of various automation strategies such as self-service kiosks and gig apps in the blue-collar service sector as well as sophisticated analysis software in the upper reaches of white-collar professions.

Major innovations in the past may have dislocated whole economic sectors (e.g. mechanization of agriculture), but the overall trend has been one of increasing both productivity and employment overall. As one sector weakens demand for employees, another sector takes up the slack with new jobs. This new phase of automation may be qualitatively different. The industrial age introduced incredible new ways to extend the productivity of the human *body*. The information age is now extending the productivity of the human *mind*. In doing so, this new era of automation affects the entire workforce spectrum. A new generation of sophisticated software programs guide robots and less visible computer functions to not only augment work tasks, but possibly replace entire job categories altogether. As software improves its ability to take on more cognitive tasks, the go-to policy prescriptions of retraining and educating displaced workers loses force. Likewise, other economic sectors are likely to not be as readily available as employment safety nets. To further complicate matters, parts of the world and certain segments of society may be experiencing the impact of these accelerating technological applications at differential speeds (Acemoğlu 2021; Brollo et al. 2024). Existing disparities and inequities are likely to deepen (Glover 2024).

At least so far, our policy leaders have been able to contain the most extreme hazards of emerging technologies, but of course not without 100% success. Airlines and aviation regulations for example, set admirable standards for passenger safety over many decades and yet have not prevented all crashes. Likewise nuclear energy has been implemented worldwide supplying relatively clean power (as compared to say, coal-fired plants) with relatively few complications. Even so, there have been high-profile failures such as Three Mile Island, Chernobyl, and Fukushima. A remaining problem is still disposal of radioactive waste. Governments have had to remain vigilant to prevent terrorists from gaining access to radioactive materials useful for building an atomic or dirty bomb. Storage of these substances is also a struggle as the half-life of some radioactive waste stretches beyond several generations. Other streams of scientific research are also inherently fraught with risks and therefore governments have responded with policy interventions and ethical guidelines. For example, genetic modification is governed through international cooperative agreements such as the Cartagena Protocol on Biosafety.

The key to producing the right kind of policies in relation to the latest advances in algorithmic management, AI, and robotics is simple in theory: (1) minimize regulation so that new innovations and industries may flourish, (2) provide effective oversight to prevent and mitigate negative externalities, and (3) implement policies to correct for abuses, imbalances, and labor dislocation. To try and hold back emerging technologies until they are guaranteed to be risk-free would be foolhardy and likely futile. Identifying current technological trends and projecting their probable impact is necessary to keep policy leaders from being caught totally off guard by major innovative breakthroughs. Organizations face dramatic new changes in already turbulent environments. The pace of change limits the capacity of government and industry leaders to make decisions that both optimize organizational strength and minimize human costs. We have arrived once again at a major crossroads in human existence that pits the high aspirations of applied science with serious risks and unwelcome consequences. Both short-term decisions and long-term strategies will determine how and under what conditions we work and, in fact, whether or not we have the opportunity to work at all.

Algorithmic Management

Analysis of work and the workforce has been a major concern for organizations since at least the early twentieth century when Frederick Taylor and other efficiency experts began championing scientific management (1911/1997). The attempt to rationalize the workplace continued through the 1930s and beyond with the principles of administration movement (Gulick and Urwick 1937). The 1950s era of the "Organization Man" (Whyte 1956) ushered in three decades of efforts among industrial-organizational psychologists to refine personnel assessment instruments to help identify the right mix of work talent among candidates for selection and promotion. In the wake of increased litigation based on Title VII of the 1964 Civil Rights Act and other civil rights legislation and regulation, this type of standardized testing all but went away. By the 1990s, the batteries of tests so commonly used in public and private organizations were "replaced by a handful of ad hoc interviews, with the questions dreamed up on the fly" (Peck 2013). The payoff for investing in expensive assessment processes also seemed much more remote as the tradition of lifetime employment waned in the face of job shifting and layoffs.

Algorithmic management is the latest iteration of longstanding efforts to rationalize the workplace. It refers to the emerging use of data-driven technologies to select employees and then monitor, control, and optimize their performance (Lee et al. 2015). These systems rely on vast amounts of real-time data, such as location tracking, performance metrics, and customer feedback, to assign tasks, set productivity goals, and make decisions traditionally handled by human managers. This form of management can lead to heightened efficiency and cost reduction but has also raised concerns about worker autonomy, privacy, and fairness in decision-making processes (Kellogg et al. 2020). The rise of remote work during and after the pandemic, as well as the increased application of gig and platform-based industries is now being coupled with new and legacy datasets to process a wide variety of evaluative information. Algorithmic management comes in the forms of people analytics, algorithmic wage setting, and performance metrics.

Flagship corporations such as Google have been rolling out "people analytics" as an initial algorithmic strategy for managing employees at work (Bryant 2011). In this newer incarnation of scientific management, sophisticated statistical methodologies are used to analyze so-called "big data" to make decisions in areas such as talent management and performance appraisal. At first, digital traces of employees drawn from work email and available social media sites are used to begin compiling multi-dimensional portraits of worker potential. These sources of information may be augmented by tracking GPS locations of employees in fleet vehicles or through business cell phones or through employee identification badges used for passage through an organization's various workplaces. In other words, instead of using big data in its original manner to analyze customers and clients to devise new marketing schemes, it is now used to refocus an organization's attention ***inward*** upon its own core processes.

Specific applications might be incorporating retention data to refashion the types of employees recruited and selected and even adjust existing managerial practices to help keep good workers. Networking among employees—both professionally and socially—also can be traced to facilitate teamwork. For example, the Bank of America incorporated people analytics to approach its longstanding problem of employee turnover in its call centers. Up until its analysis, Bank of America managers would intentionally space out the 15-minute breaks among employees to maximize the number active at any time with customers on the phone and online. Through people analytics, they were surprised to discover that the leading indicator for success of its call center workers was inter-office collaboration. Almost all interactions took place in the small window of partially overlapping lunch breaks. The results indicated that the bank should allow teams to take breaks at the same time. When this change was introduced, subsequent measurements revealed that they "were handling calls 23% faster, and cohesion was up by 18%" saving the company an extra $15 million (Griswold 2014).

The power of people analytics lies not only in harnessing vast amounts of heretofore untapped data, but also bypassing inherent cognitive biases of human decision makers. The concept was first popularized by the book (and subsequent movie starring Brad Pitt), *Moneyball* by Michael Lewis (2004). It recounted how the down-and-out Oakland A's, having lost several star players to rival, more profligate baseball teams realigned its strategy for player recruitment in accordance with principles of economic analysis. General Manager Billy Beane and his staff employed statistical methods to bypass traditional gut-feeling approaches based on surface appearances for assessing player worth. As such, he and his administrative team identified several undervalued players and worked to get them to join the *Athletics*. Subsequently, they even adjusted game strategies and player field assignments based on their statistical analyses.

McAfee and Brynjolfsoon (2012) distinguish the big data approach from previous analytical formulations by its *volume* (billions of gigabytes of new digitized data are created every day), *velocity* (organizations are able to access data points in real time), and *variety* (new forms of digitized information are streamed from a variety of sources that weren't even available a few years ago). At the same time, computer hardware, processing speeds, and data storage capabilities are becoming more and more economical. The capacity to both store and analyze incredibly vast quantities of information across platforms is just now being realized.

A precursor to contemporary people analytics was the biodata movement fashionable among industrial-organizational psychologists in the 1990s. "Biodata" is a truncation of "biographical data." It generated information points based on employee and candidate responses to broad sets of questions on biographical surveys. The wide variety of questions ranged from favorite colors, choice of hobbies, tv program selection, relationship experiences, and any other item that the instrument developers imagined might create a fresh new data point. Responses were then correlated with performance measures. Questions revealing the highest correlation to job performance were retained for future administration. Biodata was a purely statistical approach. As such, it did not rely on any underlying theory except the well-accepted premise that past behavior might help predict future job success. In this way, it was a "black box" that did not even attempt to identify linkages among its predictions and identifiable areas of job relatedness. As with other sophisticated assessment tools (e.g. standardized testing, assessment centers, etc.), biodata fell out of favor due to high costs. Its lack of face validity invited legal challenges, and in this particular case, the newly passed American with Disabilities Act (ADA) surfaced a fresh, new class of civil rights concerns

(Sharf 1994). Many of the questions embedded within these biographical questionnaires could indirectly identify applicants or employees with disabilities in violation of ADA provisions. For example, a question about whether or not a job candidate played team sports in high school or college might positively correlate with performance on the job, but it might also have a disparate impact on other individuals who may have coped with a disability during that same time period. That disability might be totally unrelated to whether the candidate could successfully perform the essential functions of a job with or without reasonable accommodation which is the standard required under the ADA.

People analytics started with digital screening for employment applications using software-coded rubrics that identify and weight specific buzzwords. Now it incorporates analysis of narrative gleaned from social media sites and other digital traces before employment. For current workers, it has evolved into the more fully developed algorithmic management which takes metrics such as customer satisfaction responses, email correspondence, and location tracking to extrapolate meaning that benefits the overall organization.

Harnessing the power of algorithmic management promises to unleash enormous benefits to organizations in such areas as talent acquisition, employee development, innovation, organizational culture alignment, performance appraisal, productivity improvement, retention, risk management, teamwork, wellness, and workforce planning (Isson & Harriott 2016). Algorithmic management also attempts to better match employees with jobs specifically suited to their unique skill sets. Advocates (e.g. Peck 2013; Waber 2013, 6) claim that the people analytics style of employee selection for new hiring, promotions, project assignments, and development opportunities comes without the normal baggage of human bias and prejudice.

Algorithmic management certainly seems to set the table for a win–win situation. Organizations can boost productivity through better empirical analysis of its own workforce data. Employees would enjoy careers better matched to their skill sets, aptitudes, and interests. Training would be customized for individual needs. Profits and cost savings would be maximized. And the peculiarities of human bias could be contained. Furthermore, the latest advancements in data visualization techniques will give managers unprecedented insight into the internal work processes of their organizations to optimize decision-making.

Like biodata in the past, the algorithms employed by algorithmic management to discern workforce patterns is its own "black box." Really, only a handful of sophisticated statisticians have any chance of understanding the process. This invites the fallacy of misplaced precision. Differences among candidate scores may well be in the margin of error but even the experts will have a built-in motivation to display outsized confidence in the results. Assume that people analytics evolves in a direction that gives near perfect predictions of who will be successful and who will fail in any given circumstance. Such knowledge within a competitive environment is its own dynamic. People analytics might result in candidates prematurely withdrawing from the long-term set of steps necessary to progress in a career. After all, if the analytics suggest you won't be successful, why even try in the first place?

Since all the digital traces are correlated with job performance, what happens if the job performance measures themselves are imperfect? Imagine that within an organization, the performance appraisal scores of racial or ethnic minorities—or women—were just slightly below their white male counterparts due to subtle, hidden prejudices of organizational managers. When the algorithmic management process begins tying the digitized behavioral traces of the workforce to these performance measures, it would instill discriminatory algorithms deep within the programming. It may not be evident to anyone. It certainly would not have to be intentional. Remember that back in 2015, Google algorithms inadvertently identified African Americans in its Photos app as gorillas and later, Microsoft's Twitterbot emulated human conversation by imitating racist tweets it found in social media (McWhorter 2016, 23). It's a well-worn cliché in programming circles but very true: Garbage in is garbage out. The statistics may be sound, but the source information may be flawed. Or the inferential assumptions within the formulaic model may be wrong. The complexity of these models makes it prohibitive for the average person to comprehend and diagnose latent problems. As Barocas and Selbst (2016) suggest, determining legal liability under Title VII and other civil rights legislation may be difficult given the complexities of the data mining process and algorithms used for analysis.

The post-COVID era has seen a significant acceleration in the use of performance metrics across various industries, driven largely by the shift toward remote work, digital transformation, and the need for organizational agility. During the pandemic, companies had to adapt quickly to disruptions in traditional work environments, which led to an increased reliance on metrics to track employee productivity, operational efficiency, and business outcomes. For example, the adoption of real-time performance tracking tools and agile methodologies enabled organizations to make faster, data-driven decisions, particularly in sectors like technology, healthcare, and finance (De Smet et al. 2020; Faraday 2021).

In addition to this, organizations have redefined the metrics that matter most, focusing not only on financial performance but also on employee well-being, customer satisfaction, and digital engagement. As more businesses moved to hybrid and remote working models, the focus shifted to evaluating team collaboration, innovation, and resilience. Performance management systems were revamped to support continuous feedback and agile responses to change, making organizations more adaptive to external shocks like the pandemic (De Smet et al. 2020; Faraday 2021). This shift has emphasized the importance of flexibility and the ability to measure success beyond traditional financial indicators.

Another major potential problem in algorithmic management is its promise of near-perfection rationalization of the workplace. Even back in Frederick Taylor's day, steel workers would frequently stage strikes in protest of scientific management methods. Algorithmic management promises to deliver a new stage in workplace rationalization. Metrics continuously govern behavior in the work environment. The potential is to rob workers of their autonomy in performing their job functions. If algorithmic management determines that there is a better way to perform a particular job task, the managerial imperative will of course be to deploy that task methodology across all similar positions in the organization. Even if it truly is a better way, it will by definition limit the freedom of humans to work according to their own styles and preferences. As noted by Unruh et al. (2022), "Algorithms can enable ubiquitous surveillance and far-reaching control of workers" that "makes the value of autonomy particularly salient in the context of algorithmic management" (p. 753). Organizational productivity is maintained but at the cost of limiting individual worker autonomy and perhaps stifling innovation.

Beyond the quest to continuously improve performance through algorithmic management is the overwhelming potential to invade personal privacy. The danger lies in the exploitative potential of this kind of unprecedented intrusion into employee behavior. Amassing and organizing vast amounts of information on individuals creates great risks to personal privacy. We all understand that work emails are not really private, but we are comforted by the fact that few managers are willing to cull through the hundreds if not thousands of electronic messages as a means to monitor our work performance. Algorithmic management, on the other hand, promises to do all of that and more. The natural language processing available with the latest software enables electronic eavesdropping on work conversations. Ben Waber (2013), perhaps the leading advocate of people analytics, points to the latest generation of the worker ID badge which has the capability of processing audio information in real time. In order to ensure employee privacy, he suggests that such recordings only include major sound characteristics such as emphasis, pace, pitch, tone, and volume rather than substantive, conversational content (p. 15).

Formal restraints for many of these situations may not yet even be existent. Policy prescriptions might include anonymization of data used for analysis at the macro-organizational level. Waber's recommendation to avoid transcribing actual dialogue could be extended to other areas of communication among workers such as emails, texting, and phone calls. Standard operating procedures within organizations and policies from external regulatory, professional association, or accreditation agencies should enforce such standards.

As in studies conducted on human subjects in academics and health sciences, imperatives for informed consent (including effective protections against those not willing to participate) should be codified through law and regulation. The U.S. Congress should strongly consider assignment of related regulatory responsibilities to established oversight agencies such as the Department of Labor. Similar approaches can be tried at state and local levels and among other nations. Much experimentation will probably be needed to determine optimum regulatory strategies to avoid strangling business productivity and still prevent the workplace from developing into a corporate version of the panoptic prison.

Algorithmic management provides managers with incredible information processing tools that can maximize worker productivity at the individual, team, and organizational levels. The resulting boost in productivity might come with some disadvantages to privacy and other areas, but algorithmic management still assumes that the employee is the primary unit of value. Artificial intelligence and robotics, on the other hand, have the potential to extend productivity well beyond the highest levels of human capacity—and ultimately even to replace workers.

Artificial Intelligence

Artificial intelligence (AI) is much in the popular press these days as high-profile tech and science giants like Sam Altman, Bill Gates, and Stephen Hawking warn of the possibility of an AI getting out of control. In the scenario often painted, a software program approaching human intelligence is finally achieved. This advancement relies on what is known as deep learning. That is, a computer system designed to run autonomously analyzes all available digital knowledge. It can run through a massive series of successive comparisons or even generate "genetic algorithms." The latter are numerous software codes written by the program itself and applied to a series of problems or scenarios. Only those sets of coding elegant and effective enough to resolve the situation are allowed to "survive" through the next iteration of development. The system "learns" and uses this newfound knowledge to revise its own code.

This deep learning involves an algorithm "first trained on known data and ... then unleashed to solve similar problems with new information" (Ford 2015, 89). Large language models, facial recognition software, email spam filters, online language translation programs, and web-based product recommendation programs are but just a few examples of this deep learning process already at work.

Unlike normal evolution for biological creatures that may take millions of years, there is no such time limitation for synthetic systems. Once a software program cycles through a sufficient number of iterations in this fashion, it could theoretically become a "super intelligence" and function in ways far beyond human understanding. The futurist theorist, Ray Kurzweil (1999), refers to this point in time as the "singularity." He is using the metaphor of the event horizon of a black hole through which the laws of the known universe break down. Inevitably, human existence will be overwhelmingly changed in ways completely unknowable after a super intelligence emerges. According to those who take this possibility seriously, the world beyond that point would be utterly unrecognizable to those of us on this side of that event. Over 90% of AI experts predict human-level AI "before the end of this century" (Barrat 2013, 25; Roser 2023). Every major government and certainly some of the most capital-rich corporations on the planet are working on various forms of AI applications at this very moment. What sounded far-fetched just a few years ago, now seems very probable after the introduction of popular programs such as ChatGPT and the other competing generative AI applications have gone mainstream.

The potential doomsday scenario—even if the possibility is extremely distant and remote—deserves some serious policy analysis of its own. But for purposes of this chapter, one does not need to look decades into the future to see how the leveraging of sophisticated software programs is already altering the shape of employment. And it's not just at the low end of the pay scale. For example, Wall Street brokers, perhaps the most class-conscious segment of society, have seen their employment numbers drop precipitously even as profits soar stratospherically. In the year 2000, Wall Street companies employed about 150,000 financial professionals, but "by 2013, the number was barely more than 100,000" (Ford 2015, 115). In 2014 the industry saw another 1.2% drop in the number of people employed (Alden 2014). Even in 2024, Wall Street faces anemic growth coupled with continuing layoffs (Mayhew 2024). Meanwhile, the stock market is hitting record highs crossing over 5,850 in October 2024. Much of the work previously performed by top financial analysts has steadily been delegated to automated systems that are now responsible for the majority of trades in the stock market. The management consulting firm Accenture predicts an even more dramatic shift in the banking industry, suggesting that AI could potentially replace up to 75% of all working hours in the financial sector (Sor 2024). It's now really a business war among rival software codes and progressively more powerful servers that shave off microseconds to gain the competitive edge.

So the effect of automation has moved beyond Rust Belt-style manufacturing and is hitting directly into the heart of white-collar professional jobs. A story in the *The Atlantic* headlined, "Technology will soon erase millions of jobs" and described this potentiality as "the end of work." The question addressed in Derek Thompson's (2015) cover story is, "Could this be a good thing?" The story cites a notable forecast by Oxford researchers suggesting that within the next twenty years, automation may claim up to half of all jobs in the United States. According to the model developed by the Oxford researchers, the workers most at risk are those employed in the transportation, logistics, and administrative support occupations (Frey and Osborne 2013, 44). These bleak predictions have yet to be in evidence over a decade later. Yet, a new generation of AI systems is expected to significantly impact global employment markets. According to economists Briggs and Kodnani (Goldman Sachs

2023), these advancements could lead to the automation of tasks equivalent to 300 million full-time jobs. By analyzing databases covering the task content of over 900 occupations, these researchers estimate that approximately two-thirds of U.S. jobs are exposed to some level of automation by AI. For those jobs affected, around 25–50% of their workload could potentially be automated. However, the report notes that not all this automated work will result in job losses. The authors emphasize that while AI's impact on the labor market will be substantial, most jobs and industries are only partially affected by automation, meaning AI is more likely to complement existing roles rather than fully replace them.

When most agricultural jobs were erased at the end of the nineteenth century, workers were able to find refuge in the manufacturing hubs of big cities as the industrial revolution kicked into high gear. When the manufacturing industry began to crater after its peak in the early 1970s, workers were able to seek employment in the service and information sectors. However, here, wages for the vast majority of lower- and middle-income workers flattened and have remained relatively stagnant ever since. Household incomes were temporarily propped up by the dramatic increase of women entering the workforce but the overall trend has continuously slipped downward. The labor force participation rate in the United States continued to drop to levels not seen since the 1970s (Bureau of Labor Statistics 2016). The American economy has rebounded well after the economic and supply-chain shocks of COVID. Even so, there appears to be no equivalent emerging sector of the economy that could provide jobs on a scale sufficient to replace those potentially lost through the latest waves of automation.

In a time in which it's often difficult to speak to a *human* customer service representative on the phone without going through a long automated menu system (many of which use impressive state-of-the-art natural language processing), one can see that traditional last-resort employers such as call centers have done much to minimize the total number of human workers even if they haven't completely replaced them. Fast-food operations may offer little sanctuary as employers of last resort either. McDonalds, Wendy's, and even the more upscale Panera Bread have deployed kiosks throughout their restaurant enterprises. As Andy Puzder, the CEO of Carl's Jr. and Hardee's explains, "If you're making labor more expensive, and automation less expensive—this is not rocket science" (Johnson 2016). Anyone who has eaten out recently in the United States or Europe in many of the chain restaurants will have already discovered touch screen menus on their table. Ostensibly, these devices are purposed for customer convenience, but the restaurant touch screen menus actually replace specific duties formerly reserved for wait staff. The irony is that the server is assigned the duty to train customers uninitiated in the use of these new devices—a strategy likely to mean fewer server vacancies in the future. Meanwhile, Taco Bell plans to install voice AI technology in hundreds of its drive-thrus by the end of the year (Snodgrass 2024).

The lack of total automation as yet means that most employees who work hard and want to remain in their current jobs actually do so. Their duties are often rearranged or they move to different assignments. The world renowned McKinsey management consulting group concludes that in the short term, very few occupations will be entirely automated. Instead, only certain functions and tasks will be automated necessitating significant readjustment among business processes. The scale of this transformation is still extensive. They claim that existing technologies already have demonstrated potential to automate as much as 45% of employee job activities (Chui et al. 2015).

Partial automation of job activities masks the inescapable consequence that natural turnover among employees is not resulting in as many replacement hires. Think of your local grocery store. Cashiers are still employed, but an indeterminate number of other job candidates were probably never hired for cashier positions after implementation of self-service checkout lines. This trend is obviously disturbing from an overall labor economic perspective if it continues. As Sachs and Kotlikoff (2012) observe in a working paper for the National Bureau of Economic Research, today's technological "change is substituting for, not complementing unskilled labor" (p. 2). Brynjolfsson and McAfee (2014) echo this sentiment: "There's never been a worse time to be a worker with only 'ordinary' skills and abilities to offer, because computers, robots, and other digital technologies are acquiring these skills and abilities at an extraordinary rate" (p. 10). Predicting national and global labor economies in such a situation is enormously difficult in the face of such fast-developing and sweeping technological change. To illustrate through a related area of technological advancement the magnitude of what might still happen, take the phenomenon of the recent and very rapid rise of social media. It has already nurtured and mobilized a whole host of political movements worldwide from the Tea Party, Occupy Wall Street, and

Black Lives Matter in the United States to numerous similar movements such as the Arab Spring which culminated in actual regime changes.

Recent populist movements in numerous countries have surely been propelled in part by growing wealth disparities among many of the nations in the developed world. Relative newcomers on the world business scene like Amazon, Alphabet, and Uber command enormous economic value that rival more established industrial giants such as General Motors. The difference is that the more recent tech ventures employ significantly fewer workers than the industrial powerhouses did in their heyday. This latest generation of automation generates enormous profits, but that newfound wealth is not as widely distributed. The result is that members of the current generation who own, design, or maintain automation systems are able to make inordinate financial gains at the expense of all future generations (Sachs and Kotlikoff 2012). The acceleration of automation technologies will deepen existing income disparities. Contemporary populist movements may only be a foreshadow to a coming political backlash which may destabilize the standing economic order.

To forestall such eventualities, Sachs and Kotlikoff (2012) recommend imposing a windfall tax that creates an inter-generational transfer of wealth. Such a policy would hopefully prevent knee-jerk resistance against the enormous benefits of automation while avoiding a continued increase in income disparity. Many other analysts (Ford 2015) suggest resurrecting an old idea: the guaranteed annual income. Even though this concept smacks of socialist idealism, it counts among its many historic champions several foundational conservative intellectuals such as Friedrich Hayek and the Nobel-Prize winning economist, Milton Friedman. Conservatives and libertarians would appreciate the minimalist government involvement as it does not get involved with who gets the income nor how it is spent.

AI and its associated software algorithms can be applied in a multitude of settings including both algorithmic management and robotics. For the former, the capacity for deep learning can be trained specifically on correlating digital artifacts with worker and organizational performance. For the latter, AI provides an extraordinary boost to existing robotic capabilities. We are accustomed to robots being deployed on the factory floor. AI literally applies unprecedented brain power to this familiar brawn. In its course, robotics liberates AI from its cyber-universe by giving it the power to interact tangibly with the real world. It's a game changer that once again rewrites the rules creating a new inflection point for technological advancement. Lessons from past waves of workplace automation may be of little use in understanding the full impact of the exponentially steep curve of the current wave of automation.

Robotics

The rise of drones, self-driving cars, industrial robots, and even robotic vacuum cleaners as well as related technologies such as the now ubiquitous customer kiosk should make anyone take pause about the very real threat of automation to diminish employment. Worldwide deployment of industrial robots has increased more than 60% between 2004 and 2014 (Ford 2015, 3). "The latest World Robotics statistics reveal an unprecedented surge in the deployment of industrial robots revolutionizing global production," declares Marina Bill, President of the International Federation of Robotics. "In 2023, a staggering 541,302 units were installed, marking the second-highest number in history. This figure is a mere 2% short of the all-time record set in 2022, when 552,946 robots were integrated into industries worldwide" (International Federation of Robotics 2024). Extraordinarily versatile, multi-purpose robots are certainly on the drawing boards and in prototype-testing labs of many tech organizations. The business imperative to compete through ever greater efficiencies makes eventual adoption of robotics technologies irresistible. These devices work around the clock in near flawless fashion. They don't complain. Robotic software can be updated continuously with little or no hardware modification. Through cloud technologies, robots can be networked and have instant access to all digitized knowledge. In most respects, robots and their governing software systems can monitor themselves and detect problems automatically. This routine self-diagnosis and maintenance minimizes the need for human intervention, guides such interventions when necessary, and decreases downtime of production processes.

Robotics is a transformative technology. Probably the best working definition comes from Ryan Calo (2014) who says, "Robots are software that can touch you" (p. 5). A tremendous new wave of robotic applications appears to be on the immediate horizon. To take just one very visible example, drones are being deployed for shipping, drug interdiction, filmmaking, immigration, industrial logistics, law enforcement, leisure, military, mapmaking, photography, warehouse distribution, urban planning, sustainability, and countless more purposes. Robots are extremely useful for performing risky tasks such as cleaning up radioactive and other hazardous wastes.

Robots have received attention in the news for handling explosives such as when the Dallas Police sent a robot and successfully detonated a pound of C-4 plastic explosive. The tactic took out the sniper who was targeting police officers. The Remotec Androx Mark V A-1 robot made by Northrup Grumman was purchased by the Dallas Police Department back in 2008. Amazingly, it sustained only minor damage to its extension arm and remains operational (Sidner and Simon 2016). It's a wonderful success story, but it does raise questions about the ethics of killing criminal suspects remotely.

Robots are even demonstrating potential to fight against terrorism. After a series of blasts rocked communities in New Jersey and New York in mid-September 2016, the New York City Police Department sent in its bomb-disposal robot to disarm another pressure-cooker bomb—this time left in a backpack in a garbage bin. The bomb exploded as the robot attempted to disarm it. Even though the robot was not successful in neutralizing the bomb before it went off, human lives were probably saved. In each case, the Dallas and New York City police departments made admirable decisions under tight time pressures. But just as robots prove to have utility in the fight against crime and terrorism, it's probably only a matter of time before the criminals and the terrorists themselves begin using robots as weapons. It's also not likely that cities will be using robots to replace police officers anytime soon. But the use of robotics by law enforcement demonstrates how this new technology can change the nature of jobs in unexpected ways. For the short term, robots have proven instrumental in improving the safety of officers and citizens. Future applications by criminally minded person necessitates realistic appraisal by legal authorities. Anticipatory policies and strategies are necessary to contain nefarious use of robotic technologies. It is likely to be an escalation in the continued arms race between criminals and those who seek justice.

It's difficult to disentangle the effects of automation on employment from other factors. The customary policy response is to close the skills gap among those lower on the pay scale with more training and education. This strategy may already be proving to be misguided. One recent analysis by Casselman (2016) used statistics from the Bureau of Labor to demonstrate that the lack of hiring in the manufacturing arena is not due to a "skills gap" as convention wisdom suggests, but rather it's "due mostly to automation" as "U.S. factories now produce more than ever with fewer workers."

One would expect both positive and negative externalities with emergence of innovative robotic technologies. A host of industrial associations (International Federation of Robotics; IEEE RAS Industrial Activities Board) lobby hard to present a positive image of robotics especially in the context of employment (International Federation of Robotics 2013; Madhaven 2013). They make a compelling case. Calo (2014) makes an equally compelling case. He argues that now is the time to establish a standalone agency to advise officials at all levels of government in the U.S. and abroad about robotics and artificial intelligence. So far, as related technologies (e.g. drones, surgical robots, driverless cars, etc.) burst onto the scene, policy responses have been piecemeal. He suggests thoughtful integration of related policies developed by interdisciplinary teams of attorneys, economists, engineers, programmers, scientists, and other technology policy specialists who would direct grant monies into basic research in robotics, advise other agencies in regulating robotic-related technologies, and file amicus curiae briefs in complex technology-related legal cases (pp. 11–12).

Such a bureaucratic response is not without precedent as the United States has dealt with previous waves of technologies with similar administrative efforts such as the Federal Railroad Administration, the Federal Communications Commission, and the Federal Aviation Administration. He is careful to recommend that such an agency initially adopt a role more advisory than regulatory in nature.

Conclusion

The benefits of automation are very real and appealing. Many think that these technologies will improve the quality of life for billions on this planet. But fast-paced technological shifts are likely to have unfortunate side effects. The continuation of present trends certainly portends winner-take-all dynamics in which businesses and governments able to achieve early technological advantage will likely outstrip the ability of competitors to keep pace. Productivity gains are impressive but how equitably will resulting benefits be distributed? In the face of this fast-paced change, will existing social safety nets be adequate for those dislocated by new technological applications? What regulatory safeguards might be effective in mitigating negative consequences and preventing ethical lapses while simultaneously encouraging positive innovations?

Artificial intelligence and robotics represent technologies on the edge and perhaps even beyond the range of normal human experience. Few economists have looked at these advancements in any meaningful way beyond the traditional paradigms of past technological adaptations. Assumptions are that progress will continue along at regular, predictable pace. Those assumptions ignore the *exponential* progress that computer technologies have demonstrably taken as described by the familiar Moore's Law which as revised, states that computing power doubles approximately every eighteen months. Kevin Drum (2013) illustrates how the seemingly slow pace of early technological progress might lull observers into a major miscalculation of the sudden progress computer technologies might take in the latter stages.

Drum uses a clever metaphor. He suggests imagining that it's 1940 (about the time the first programmable computer was developed) and somehow Lake Michigan has completely dried up. The job is to refill it using the simple rule (based on Moore's Law) that you start by adding one fluid ounce of water to the empty lake bed. In eighteen months, add another two ounces and continue doubling the amount every eighteen months thereafter. Lake Michigan is a vast reservoir and even by the year 2000, you might have "a slight sheen on the lake floor." But Drum continues:

> Ho about 2010? You have a few inches of water here and there. This is ridiculous. It's not been *70 years* and you still don't have enough water to float a goldfish. Surely this task is futile? But wait. Just as you're about to give up, thing suddenly change. By 2020, you have about 40 feet of water. And by 2025 you're done. After 70 years you had nothing. Fifteen years later, the job was finished. (Drum 2013)

The dawn of OpenAI's ChatGPT and other forms of generative AI certainly took the world by storm in 2022. The ramifications for policymakers are enormous. Just as social media caught many governmental authorities across numerous nations off guard, so may go the application of sophisticated software to robotic technologies. The more immediate worry concerning technological displacement is likely to come from more specialized robotic automation. General purpose robots that can be reprogrammed for a variety of industrial uses are already in production. We can expect a broad extension of robotic technologies in the not-so-distant future.

The argument made in this chapter is that the emerging technologies of algorithmic management, artificial intelligence, and robotics represent a transformational stage in the nature of work and employment. Some of the major policy recommendations that should be explored further are (1) anonymization of workplace data on employee behavior; (2) oversight by government agencies, professional associations, accrediting organizations to ensure proper consideration is given to privacy concerns and to prevent discriminatory practices; (3) expansion of the Earned Income Tax Credit; (4) implementation of a guaranteed minimum income program; (5) let workers own the profits from use of their data; (6) exploration of a variety of possible pre-distributive policies; (7) require AI development to be open-sourced to allow wider and more diverse scrutiny, and (8) creation of a national advisory agency to oversee developments in AI and robotic applications. New technologies have always resulted in social consequences. Careful consideration of policy approaches is warranted especially since the current wave of technologies as applied to the workplace is likely to be qualitatively different than previous innovations.

References

Acemoğlu. Darol (2021, March). *The Post-COVID World: To Reverse Widening Inequality, Keep a Tight Rein on Automation*. Washington, DC: International Monetary Fund.

Akbari, Mohammadreza. (2024). Current Trends in Reshoring, Nearshoring, Rightshoring, and Emerging Strategies. In: *The Road to Outsourcing 4.0*. Palgrave Macmillan, Singapore. https://doi.org/10.1007/978-981-97-2708-7_5

Alden, William. (2014, March 13). "Wall St. Bonuses Rise 15%." *New York Times*. Retrieved 1/4/2016 from https://www.bostonglobe.com/business/2014/02/12/wall-street-bonuses.

Barocas, Solon, and Andrew D. Selbst. (2016). "Big Data's Disparate Impact." *California Law Review* 104, no. 3: 671–732. http://www.jstor.org/stable/24758720

Barrat, James. (2013). *Our Final Invention: Artificial Intelligence and the End of the Human Era*. New York: Thomas Dunne Books/St. Martin's Press.

Brollo, Fernanda, Dabla-Norris, Era, de Mooij, Ruud, Garcia-Macia, Daniel, Hanappi, Tibor, Liu, Li, Nguyen, Anh D. M. (2024, June 17). *Broadening the Gains from Generative AI: The Role of Fiscal Policies.* Washington, DC: The International Monetary Fund.

Brynjolfsson, E and McAfee, A. (2014). *The Second Machine Age: Work, Progress, and Prosperity in a Time of Brilliant Technologies* (Kindle Edition). New York: W. W. Norton & Company.

Bryant, A. (2011, March 12). "Google's Quest to Build a Better Boss." *The New York Times.* Retrieved 12/29/2015 from http://nytimes.com/2011/03/13/business/13hire.html?pagewanted=all&r=0.

Bureau of Labor Statistics. (2016, September 20). "Labor Force Statistics from the Current Population Survey 1960–2016. Retrieved 9/20/2016 from http://data.bls.gov/pdq/SurveyOutputServlet.

Calo, R. (2014, September). *The Case for a Federal Robotics Commission.* Washington, DC: Center for Technology Innovation at Brookings Institution.

Casselman, Ben. (2016). "Don't Blame a 'Skills Gap' for Lack of Hiring in Manufacturing." *FiveThirtyEight, September 8*, 2016. Accessed July 25, 2025. https://fivethirtyeight.com/features/dont-blame-a-skills-gap-for-lack-of-hiring-inmanufacturing/

Chui, Michael, Manyika, James, and Miremadi, Mehdi. (2015, November). "Four Fundamentals of Workplace Automation," *McKinsey Quarterly.* Retrieved 12/30/2015 from http://www.mckinsey.com/business-functions/business-technology/our-insights/four-fundamentals-of-workplace-automation.

Chung, J., & Lee, Y. S. (2023). The Evolving Impact of Robots on Jobs. ILR Review, 76(2), 290–319. https://doi.org/10.1177/00197939221137822.

De Smet, Aaron, Pacthod, Daniel, Relyea, Charlotte, and Sternfels, Bob. (2020, June). *Ready, Set Go: Reinventing the Organization for Speed in the Post-COVID Era.* New York: McKinsey & Company.

Drum, K. (2013, May 13). "Welcome, Robot Overlords. Please Don't Fire Us?" *Mother Jones.* Retrieved 9/25/2016 from www.motherjones.com.

Faraday, Tom. (2021, June 17). Metrics that matter: Measuring Organizational performance in a post-pandemic world, ISACA. Retrieved 10/24/2024 from https://www.isaca.org/resources/news-and-trends/isaca-now-blog/2021/measuring-organizational-performance-in-a-post-pandemic-world.

Ford, Martin. (2015). *Rise of the Robots: Technology and the Threat of a Jobless Future.* New York: Basis Books.

Frey, C. B. and Osborne, M. A. (2013, September 17). *The Future of Employment: How Susceptible are Jobs to Computerisation?* Oxford, UK: University of Oxford.

Griswold, A. (2014, February 25). "This One Simple Management Change Saved Bank of America $15 Million," *Business Insider.* Retrieved 12/29/2015 from http://www.businessinsider.com/bank-of-america-call-center-manageent-2014-2.

Glover, George. (2024, January 16). AI will be bad news for plenty of workers, warns IMF chief, *Business Insider.*

Goldman Sachs. (2023, April 4). Generative AI could raise global GDP by 7%. Retrieved 10/23/2024 from https://www.goldmansachs.com/insights/articles/generative-ai-could-raise-global-gdp-by-7-percent

Gulick, L., and Urwick, L. (Eds.). (1937). *Papers on the Science of Administration.* New York: Institute of Public Administration.

International Federation of Robotics. (2013, January). *Positive Impact of Industrial Robots on Employment.* London: Metra Martech.

International Federation of Robotics. (2024, September). Record of 4 million robots in factories worldwide, *Work Robots 2024.* Frankfurt, Germany: IFR Press Room.

Isson, J. P. and Harriott, J. S. (2016). *People Analytics in the Era of Big Data: Changing the Way You Attract, Acquire, Develop, and Retain Talent.* Hoboken, NJ: John Wiley & Songs, Inc.

Johnson, H. (2016, May 16). "Fast Food Workers Are Becoming Obsolete," *Business Insider.* Retrieved 9/21/2016 from http://www.businessinsider.com/self-service-kiosks-are-replacing-workers-2016-5.

Kellogg, Katherine C., Valentine, Melissa A., & Christin, Angèle. (2020). Algorithms at work: The new contested terrain of control. *Academy of Management Annals, 14.1*, 366–410. https://doi.org/10.5465/annals.2018.0174.

Kurzweil, Ray. (1999). *The Age of Spiritual Machines: When Computers Exceed Human Intelligence.* New York: Penguin Group.

Lee, D. (2016, August 11). "Reality Check: Manufacturers Returning to U.S. May Mean Jobs for Robots, not People," *Los Angeles Times.* Retrieved 9/13/2016 from http://www.latimes.com/business/la-fi-manufacturing-jobs-lost-20160811-snap-story.html.

Lee, Min Kyung, Kusbit, Daniel, Metsky, Evan, and Dabbish, Laura. (2015, April). Working with machines: The impact of algorithmic and data-driven management on human workers. In *Proceedings of the 33rd annual ACM conference on human factors in computing systems* (pp. 1603–1612).

Lewis, M. (2004). *Moneyball: The Art of Winning an Unfair Game.* New York: W. W. Norton & Company.

Madhavan, R. (2013, December). "Congressional Robotics Caucus Briefing on Robots and Job Creation." *IEEE Robotics and Automation Magazine.*

Mayhew, Michael. (2024, May 13). Sluggish Wall Street jobs outlook continues in April 2024, *Integrity Research.* Retrieved 10/19/2024 from https://www.integrity-research.com/sluggish-wall-street-jobs-outlook-continues-in-april-2024.

McAfee, Andrew, and Erik Brynjolfsson. (2012). Big Data: The Management Revolution. *Harvard Business Review 90* (10): 60–68. https://research.ebsco.com/linkprocessor/plink?id=6f25dcc9-195f-3b75-9297-46ee480ddef8

McWhorter, J. (2016, September 12–19). "Is Technology Capable of Being Racist?" *Time, 188.10-11,* pp. 23–24.

Peck, D. (2013). "They're Watching You at Work," *Atlantic, 312.5.*

Prettner, Klaus Prettner (2023, December). Stagnant wages in the face of rising labor productivity: The potential role of industrial robots, *Finance Research Letters, 58,* Retrieved 10/24/2024 from https://www.sciencedirect.com/science/article/pii/S1544612323010590.

Roser, Max (2023). AI timelines: What do experts in artificial intelligence expect for the future?" Published online at OurWorldinData.org. Retrieved from: https://ourworldindata.org/ai-timelines.

Sachs, J.D. and Kotlikoff, L. J. (2012, December). "Smart Machines and Long-Term Misery," *Working Paper, 18629.* Cambridge, MA: National Bureau of Economic Research. Retrieved 9/21/2016 from http://www.nber.org/papers/w18629.

Sauter, M. B. and Stebbins, S. (2016, April 23). "Manufacturers Bringing the Most Jobs Back to America," *USA Today.* Retrieved 9/13/2016 from http://www.usatoday.com/story/money/business/2016/04/23/24-7-wallst-economy-manufacturers-jobs-outsourcing/83406518/.

Sharf, J. D. (1994). "The impact of legal and equal employment opportunity issues on personal history inquiries." In G.S. Stokes, M.D. Mumford, & W. A. Owens (Eds.), *Biodata handbook: Theory, research and use of biographical information in selection and performance prediction* (pp. 351–390). Palo Alto, CA: CPP Books.

Sidner, S. and Simon, M. (2016, July 12). How Robot, Explosives Took Out Dallas Sniper in Unprecedented Way," *CNN.* Retrieved 9/25/2016 from www.cnn.com/2016/07/12/us/dallas-police-robot-c4-explosives.

Snodgrass, Erin. (2024, August 1). Hundres of Taco Bell drive-thrus are set to be AI-run by the end ofthis year, *Business Insider.*

Sor, Jennifer. (2024, April 11). AI is coming for Wall Street: Banks are reportedly weighing cutting analyst hiring by two-thirds, *Business Insider.*

Taylor, F. W. (1911/1997). *The Principles of Scientific Management.* Ithaca, NY: Cornell University Library.

Thompson, Derek. (2015, July/August). "A World Without Work." *Atlantic, 316.1,* 50–61.

Unruh, C. F., C. Haid, J. Fottner, and T. Büthe. (2022). Human Autonomy in Algorithmic Management. *Proceedings of the 2022 AAAI/ACM Conference on Artificial Intelligence, Ethics and Society*, Oxford, UK. https://doi.org/10.1145/3531146.3533101

Waber, Ben. (2013). *People Analytics: How Social Sensing Technology will Transform Business and What It Tells Us about the Future of Work*. New York: FT Press Analytics.
Whyte, Jr., William H. (1956). *Organization Man*. New York: Simon & Schuster.

CHAPTER 5

Automatic Medicine? Technology and the Future of Primary Health Care

Daniel Skinner and Berkeley Franz

Introduction

As many contributions to this edition make clear, technology is both a source of excitement and anxiety. This double-edged phenomenon has been vividly depicted in utopian as well as dystopian novels and films, and finds continual expression—logically enough—in science fiction, which seizes equally on hopes and fears (Westfahl and Slusser, eds. 2002). Considering the vulnerability inherent in both physical and mental health, as well as medicine's intersections with political categories such as liberty, choice, equality, progress, and beyond, it should come as no surprise that technology's role in medicine is particularly fraught. This anxiety is, moreover, inherently political (Singer 2013).

D. Skinner (✉) · B. Franz
Ohio University, Athens, OH, USA
e-mail: skinnerd@ohio.edu

B. Franz
e-mail: franzb@ohio.edu

R. Kiggins (ed.), *The Political Economy of Robots*, International Political Economy Series, https://doi.org/10.1007/978-3-031-95171-8_5

This chapter assesses the extent to which health changes how we think about technology in the area of primary health care services. We explore new developments in American health care, grouped into three basic categories, and consider the theoretical valence of these technologies. Ultimately, we ask whether they might help us to better understand the politics of technology at work in these developments. By focusing on primary care—instead of, for example, specializations such as surgery, oncology, dermatology, and the like—we hope to capture the theoretical dynamic at work in what we might call "everyday" medicine, where patients are engaged in the pursuit of basic wellness, the maintenance of chronic conditions, and similar goals. Most importantly we are concerned with the goals of primary care to improve health outcomes across and between populations by providing preventative services and in collaborating with patients to achieve improved health. Accordingly, we question whether technology utilized in primary health care promotes these same goals.

In the last three decades, as is occurring globally (Anderson et al. 2006), the United States has dedicated significant resources to the development of medical technology in an attempt to improve well-being (Koizumi 2015). Yet, in doing so, a large medical technology industry has also surfaced, raising along the way a need for persistent critique of the financial investments necessary to maintain technological progress and the extent to which they may compromise more principled, non-financial aspects of medicine. New investments also raise questions regarding the capacities of private industry to play this role effectively without strong support from government, especially the federal government with its history of capitalizing large projects beyond the scope of private industry (the internet, e.g.), and whether such investments compromise the scientific evidence-base at the foundation of medical care (Bailey et al. 2011).

Yet, there are other dimensions beyond those of investment. For example, debates over appropriate levels of taxation in the medical device industry have been a central component of political challenges to the Affordable Care Act, with some politicians questioning the relationship between taxing this industry and a future of medicine in which technology will play an increasingly important role. Specifically, critics have raised questions about the relationship between research and development, regulatory oversight, and taxation (Harvey and Pandharipande 2012; Tahir 2014; Ferris 2015). But how should we conceptualize the role of medical technology in relation to the renewed aim to improve health outcomes and avoid disparities in accessing health care and avoiding illness? A more basic question is whether technological development is a net good for medicine, in general, and primary care in particular. Of particular importance, therefore, is to understand how patient-provider relationships that have traditionally stood as markers of quality health care will fare in an increasingly technologized medical environment.

While technological developments have been an important focus in the medical literature, in the United States the more common every day discussion about technology continues to focus on the challenges of implementing reliable systems of electronic medical records (EMR). Indeed, for many people, optimism or pessimism about technology seems to rise and fall depending upon their experience with EMR. Other technological conversations tend to remain at the margins, as though the core of medicine is slated to remain the same while technology merely serves as a supplement—a mere means of recordkeeping and financial accounting. Yet, conversations also center on predictable anxieties about what is to become of the humanistic—especially communicative—side of medicine (Miller 2001). Within these conversations, physicians emphasize the degree to which medicine is an art in addition to a science, while patients and advocates express concern about accountability and safety in an increasingly technologically sophisticated world (Powell-Cope et al. 2008; Swayze and Rich 2011). At the same time, advocates who see the future of medicine in technology tend to take a less sentimental view of the value of (in Nietzsche's words) "human, all too human" doctors by maintaining not only that "the robots are coming" (Manjoo 2011), but that we should start to come to terms with the possibility that robots of various sorts may, in fact, do a better job at many aspects of medicine than humans. Sentimentality for human clinicians, however, is an insufficient reason for resisting technological developments.

Challenges with EMR underscore the extent to which comparatively "low-tech" obstacles can pose major questions for the practice of medicine. For example, physicians continue to struggle with how to collect necessary information about their patients, populating that information into EMR systems, while still remaining present in the examination room (Doyle et al. 2012; White and Danis 2013; Street et al. 2014). Patients, on the other hand, often complain that their physicians seem to be focused on their electronic devices instead of them. Interestingly, to preserve the quality of doctor-patient relations, most practices have turned to the low-tech solution of scribes (Gellert et al. 2015; Soudi and McCague 2015). In other words, in debates about EMR we witness a host of attempts to negotiate the changing terrain of human relations within medicine. While technology plays an important role, it often lacks the ability to solve the very problems it identifies, and oftentimes creates new ones. Technology often provokes a reconciliation with human activities, landing it at the intersections of medical ethics, biomedical interventions, political economy, and beyond. Automation represents a challenge within the broader field of technology insofar as fully automated systems and technologies severely alter or even remove altogether roles formerly inhabited by humans. Fully automated medicine, for example, would result in the diminishment or even replacing of roles previously filled by physicians and other clinicians. As we shall see, we are not there yet.

Philosophy of Medical Technology

Technology's capacity to be dehumanizing has a long history extending back to before its implementation into medical practice. This philosophy has since informed discussions on the impact of technology on relationships in health care and the potential for technology to alienate patients from physicians, and physicians from their practice. The philosopher Martin Heidegger is often invoked in these discussions (Franz and Murphy 2016; Whelton 2016), especially via his famous dictum that "the essence of technology is not technological" (1977) but instead to reveal truth. Though the traditional philosophical position suggested that technology's function was to assist in "bringing forth," modern technology, Heidegger argues, has instead concealed truth. This is because humans are detached from the process of creation and are therefore separated from understanding themselves as creators. But is medical technology similarly problematic? One must question whether medical technology brings us closer to understanding illness or pursuing goals such as patient-centeredness or patient activation. Or does technology conceal certain aspects of wellness and illness and change care for the worse?

Heidegger and a long line of critics following him advanced a sustained critique of technology and advocated for a reduced presence in modern society. Far from diminishing, however, technology continues to be a key site of investment in the attempt to improve health outcomes.

In light of medical technology's continued presence, what philosophical grounding may be more instructive than a Heideggerian critique? Although Emmanuel Levinas did not write extensively on technology, his philosophy offers a more complex portrayal of modern technology's role, and offers a perhaps useful standard for evaluating its place in medicine (Peperzak 1992). Levinas' philosophy is grounded in a fundamental ethical commitment that humans share for each other. That is, instead of focusing on the individual quest for authenticity or reconciliation with one's purpose in the world (as described by Heidegger), Levinas is entirely concerned with the face-to-face relationship of social life. The most important implication of this philosophy is a radical responsibility to care for others rather than a need to make oneself in the world. It is for this reason that Levinas maintains that the "source of humanity-is perhaps the Other" (1962, 14).

What standard, if any, would a foundational ethical framework provide for judging technology? Levinas agrees with Heidegger regarding the potential for technology to be dehumanizing (Peperzak 1992). In fact, Levinas provides examples of technology's capacity to exploit people and the environment on which people depend on for survival. However, Levinas also acknowledges that technology at times is necessary in order to respond adequately to the needs of others. Levinas calls upon food production as an example and argues that without technological innovation, persons in developing countries would not have the resources necessary to survive (Levinas 1998). For this reason Levinas takes a more balanced approach to evaluating technology, but fundamentally grounds any critique in an ethics of responsibility.

But to what extent are ethical considerations a factor in contemporary debates about medical technology? Population health, the expanding field of medicine dedicated to preventing disease and promoting the well-being of entire groups, has elicited particular attention to the role that technology may play in making health care and medical information more accessible. This medical paradigm is explicitly concerned with widespread wellness and addressing health needs in order to bolster health outcomes more broadly. Whether technology stands to improve access to health care, increase patient autonomy, or allow for individuals to participate in their own care, remains to be seen.

We ask how one subsection of technological development—that of automation and quasi-automation processes in medicine—can be contextualized within the debate between Heidegger and Levinas. Heidegger judges technology to the extent that humans can become authentic and believes that technology conceals truth. Levinas focuses on the goal of becoming responsible for "the other." Can technology do this? Perhaps. Can we then judge technology by the impact it will have on population health and improving well-being? Also perhaps. To answer these questions, we undertake an analysis of three emergent technologies, each of which serves to automate, in whole or part, an aspect of health care. Although different standards for automation likely exist depending on the field, we will focus on an evaluation of whether automation in medicine is consistent with the goals of improving health outcomes and strengthening primary care. Following from Levinas' insistence that technology be assessed by the value it has to improve the lives of others, we will look at how automation specifically supports the goals of widespread well-being. In so doing, the Levinasian focal point must be situated within the various—if only possible, or in appearance—ethical breaches enabled by financial incentives within the billion dollar industry of medical technology.

Three Examples:

As one would expect, there is a wide spectrum of automation in medicine. Although many automated processes have been realized, many have been designed but will require significant testing, and still others remain in the conceptualization phase due to significant barriers—including legal and ethical—faced in realizing automated medicine. In this section we describe several new and emerging technologies. In turn, this section serves as a foundation for a theoretical inquiry of the qualitatively different nature of these technologies.

1. Theranos—Retail Blood Testing

In 2003, the company Theranos was founded to provide remote blood testing and subsequently introduced its new "Edison" technology, a consumer-level blood testing device that promised high degrees of accuracy based on the testing of a small drop of blood taken from a pin prick. Its much-heralded, patented "microfluidic" approach stood to revolutionize the world of blood testing by containing and simplifying existing approaches to blood procurement and processing, and doing so in outpatient facilities located in neighborhood retail stores at a fraction of the typical cost.

It makes sense that Theranos' technology would garner excitement. Pricing structures for the procuring and processing of bloodwork is notoriously opaque (Hsia et al. 2014), and Theranos' technology promised to make transparency possible in large part because bloodwork would become, perhaps for the first time, affordable. In addition, Theranos promised to maintain the same accuracy of standard tests in brick-and-mortar stand alone units such as Quest or LabCorp, thereby meeting a longstanding goal of making faster, more customer-friendly approaches available around the corner from customers' homes. Yet, this was no easy task; while Theranos promised accurate testing in conveniently located neighborhood stores, beginning with a large contract with Walgreens, even point-of-care testing (blood tests undertaken in physicians' offices), while potentially valuable, have long been criticized as inaccurate (Jones et al. 2013), lacking in basic standards of scientific transparency (Kidd et al. 2016). If Theranos accomplished its goals, and made the technology work, they would be solving a series of perennially vexing medical challenges.

Despite the initial excitement Theranos generated—and a $4.5 billion initial assessed worth of the company's founder, Elizabeth Holmes (Smith 2016)—the model quickly tumbled into technical disappointment and even allegations of fraud. Most dramatically, in January 2016, CMS found that Theranos' bloodclotting technology posed "immediate jeopardy to patient safety." Though the details of the story will likely come out over time, Holmes and her investors appear to have become caught up in excitement about technological possibilities, but also the financial rewards that those technologies promised. The result of this mismanagement was an inability or unwillingness to effectively assess the technological capacities of Theranos' product while, paradoxically (based on Theranos's trumpeting of transparency in medical technology and services) undertaking one of the least transparent business ventures in recent memory. The same technological malaise appears to have characterized Theranos' partners. Walgreens, for example, failed to do due diligence by independently testing Theranos' claims before contracting to place machines in over one thousand of their pharmacies (Weaver and Carreyrou 2016).

Technological promises of the sort that Theranos made are many. In a health care environment in which the challenges of cost, access, and quality are pervasive, price transparency and savings, and accessibility to services are desirable. They are only promising, of course, if the technology is accurate—a failure that would not only not fulfill its promise, but could endanger patients along the way. Considering the extent to which Theranos promised the empowerment of patients by allowing on-demand testing, it also marked a decided shift in medical power, by strengthening patient agency. But this shift was only to be embraced if it occurred without putting patients, whose judgment might replace or augment that of physicians or clinicians, at risk.

At the same time, the high profile Theranos case offers an additional opportunity to reflect on the nexus of consumer-directed medical technologies and primary care. Perhaps most importantly, all of the potentially empowering aspects of Theranos are also potentially problematic in the way they reduce the role performed by traditional clinicians such as doctors or nurse practitioners. Although Theranos offers, as part of its service, the opportunity to connect users by video screen with a clinician to discuss results, the Theranos technology is largely intended to provide information to patients who are responsible for processing and acting upon findings. Here, in other words, we encounter a technological space that directly raises questions about the limits of technology and the utility of human physicians.

2. My Accessible Real-Time Trusted Interpreter, or "MARTTI"

Another place where technology has arisen to meet an important need in medicine is in the area of interpretation services, which are increasingly central to caring for diverse patient populations. This is particularly the case in the United States in an era of health care reform in which new patients, many of whom have limited English proficiency, have gained access to health care services through the expansion of Medicaid, outreach to those who were previously eligible for Medicaid, but did not know they were until health care expansion arose as a visible national discussion, as well as some categories of undocumented immigrants. At the same time, this population does not, as one might expect, simply represent a quantitative increase, posing, for example, a bottlenecking effect in access. More than that, they tend to be some of the poorest patients, often with multiple health issues that may have been neglected for some time. With this in mind, it is no surprise that language barriers tend to have a multiplying effect, thereby compounding the social determinants of health.

The question of interpretation (spoken word), translation (written word), and technology has received a lot of attention over the years beyond the context of medicine. Steven Pinker, in his widely celebrated book, *The Language Instinct* (1994), argued that the structure and "natural" basis of language made it all but impossible to reduce to automated interpretation/translation systems. Language, in Pinker's view, was simply too spontaneous, contextualized, and malleable to be reduced to algorithms. Humans, in other words, would always have to play a central role. Pinker in many ways followed in the wake of the Heideggerian philosopher, Hubert Dreyfus, who argued in his 1972 book, *What Computers Can't Do* that the Artificial Intelligence (AI) movement was philosophically misguided. In Dreyfus's view, AI, as conceptualized at the time, was fatally flawed due to an overly simplified model of human cognition and language. Over his career, Dreyfus took aim at efforts such as the Pentagon program known as DARPA (Defense Advanced Research Projects Agency), which sought to develop technology for automatic interpretation/translation that could be useful for purposes of intelligence gathering and national security. For Dreyfus, the answer to AI, which in a sense seeks to capture and reproduce the essence of humans, is to solve precisely the question Heidegger posed. Somewhat paradoxically, the vexing nature of the question was bound up in the problem Heidegger had identified early on, namely that technology tends to conceal rather than reveal that essence. Only by returning to Heidegger, claimed Dreyfus, could AI meet its admittedly attenuated goals (Dreyfus 2007).

These debates serve as a useful background to the issue confronting medicine. To be sure, the question of technology and language does, in a Heideggerian manner, concern the essence of medicine, particularly how clinicians interact with patients, and the extent to which that relationship serves as the glue of clinical work. But, to the extent that language is a barrier in medical relationships, what role does technology play? Here we cannot consider the question without also considering the extent to which the problem is not only one of technological capacity, but economics.

A series of interviews conducted in Columbus, Ohio's community health centers by one of the authors (Skinner) confirms the tentative relationship that clinicians tend to have with interpretation technology. On the one hand, certified live interpreters remain something of the gold standard, as a live interpreter, in the room with clinicians and patients, provide an ability for nuanced understanding, follow-up question, and visual and other cues. At the same time, live interpretation is costly and unlikely to meet expansive needs. For example, while most community health centers in this sample had the personnel on staff who are bilingual in (if not necessarily trained in the interpretation of) Spanish, some-odd one hundred additional languages appear at a lower frequency. Although some community clinics may have a list of interpreters available for hire on a schedule basis, "no show" patients are a continual and costly problem. The limitations of live interpretation, as well as costs associated with their employment, are compounded by cultural considerations, such as male interpreters in clinical situations involving women's health, or significant dialectical differences that minimize the benefits of live interpretation.

Although live interpretation may be cost prohibitive for most community health centers, federally recognized, hence subsidized, health centers are required to provide interpretation services. As a result of the basic requirement, technology becomes an increasingly attractive solution. Throughout the United States, a service known as MARTTI (or My Accessible Real-Time Trusted Interpreter) has become increasingly popular. MARTTI is a web-based audio and video interface linked to facilities around the United States in which live interpreters, who are located in off site facilities, serve as interpreters. MARTTI's parent company, Language Access Network (LAN) touts MARTTI's ability to offer translation services for more than 1,000 languages, including sign languages.

In many ways, MARTTI is an example of technology that fits nicely with many people's views of medicine. It is not a fully automated approach to interpretation, but blends technological devices with human elements. As such, it falls far short of the idealized view of automated medical technology while meeting an important need. As many practitioners report, MARTTI not only serves an important patient-centered function, but it also facilitates bottom lines. Above all, as our health centers interview indicate, MARTTI saves facilities money because if a patient who is scheduled for a visit requiring interpretation services fails to show, that facility can simply not activate the MARTTI, whereas they would have to pay an interpreter in that same situation.

While certainly meeting an important need within conditions of limited resources and practical workforce capacity, MARTTI also illustrates the complexity surrounding medical technology. MARTTI's video interface, some clinicians report, fails to capture the visual cues and subtle interpersonal dynamics that are typically present in live interpretation scenarios. Clinicians also report problems with connectivity, subpar audio quality, and awkwardness in positioning the units within examination rooms and other spaces. Though MARTTI devices typically connect clinicians and patients with an interpreter within minutes, even minutes can reduce, by significant percentages, the actual time patients and physicians have together, which is often limited to fifteen minute intervals before interpretation and technological requirements are introduced.

3. **Medical Apps**

As one might expect in the age of smartphones, a number of patient-centric technological developments in medical care have come by way of applications (or "apps"). Most of these apps are aimed at increasing patients' roles in addressing their health needs and promoting wellness. Most attempt to address patient needs that fit squarely within the purview of what is generally considered to be the domain of primary health care. With increasingly limited time for patient appointments, these apps promise to improve communication, increase knowledge about conditions and treatment, expand access to care, and identify medical conditions in need of follow-up consultation. The utility of these apps in primary care is of particular interest given the focus on developing long-term relationships with patients, increasing the role patients play in managing their health, preventing and managing chronic illness (Eugenio et al. 2015), increasing access to a primary care-based medical "home," and the need to see in-office visits and inpatient experiences as only a starting point for ongoing care as patients carry out their daily activities.

Although a variety of medical apps are available for the primary care setting, it is possible to classify them according to their intended purpose. VidScrip, for example, was developed in 2011 and like similar apps in this category aims to improve patient education. VidScrip is sold to health providers and allows them to video-record messages containing patient instructions related to medications or other treatments and answer frequently asked questions. According to VidScrip, this technology encourages a "move toward Value Based Care"—a language that is increasingly common in the era of post-Affordable Care Act health care reform—by providing a record of communication for the patient to retain, thereby capitalizing on the provider's expertise (Vidscrip 2016). This app's specific goal is to provide additional information that cannot fit within standard—and typically short—office visits and translate this education to improved health outcomes. A similar health information app, Medivizor, takes a slightly different approach by providing an overview of relevant information related to a condition based on a systematic search of the empirical literature and official pharmaceutical guidelines. The goal of both of these apps is to expand patient access to health information outside of the clinical consultation.

Other apps designed for use by patients include those aimed at improving compliance with practitioner-designed treatment plans. The term compliance in health care refers to the extent to which patients act "in accordance with advice" given to them by medical providers (Aronson 2007, 383). Although the term compliance is often replaced with more patient-centric words such as "adherence" and "concordance," the goal of many medical apps is to ensure that patient behaviors are consistent with a medical protocol. PatientIO, for example, compiles a variety of information related to patient treatment regimens so that patients can "follow and track their care." Information is provided regarding medication schedules, conducting regular health screenings, and the long-term health progress of the patient, all of which can then be shared with other individuals involved with the patient's care. Results are also reported to providers so that they can assess compliance among their patient populations and manage various types of risk. A related cluster of apps is designed particularly for patient use and encourages the cultivation of healthy habits or reduction of risky health behaviors. Examples include smoking cessation apps such as KWIT, nutrition and fitness trackers such as MyFitnessPal, and MapMyRun, and apps that act as coaches to support the actualization of healthy goals, such as Coach.me. Although many of these behavioral modification apps are designed by psychologists and aim to create incentives for behavior change, the medical community is still debating their usefulness and appropriateness, as well as ways that they might be improved and coordinated within primary care medicine (Dennison et al. 2013).

A final category of medical apps includes tools designed specifically for providers. These apps can help practitioners manage time, track patient progress, and even make clinical decisions. For example, Calculate is an app that provides decision support by providing tools to aid diagnosis, determining a prognosis, establishing a treatment plan, and provide recently published results from clinical trials. Apps such as Epocrates and Diagnosaurus promote communication between health care team members, facilitate differential diagnoses, and check for medication interactions. The premise of these apps is that office consultations are increasingly brief and technology may boost the quality of care that can be provided within this timeframe. For example, apps that aid in inter-provider communication may allow a patient's history to be easily accessed and included in medical decisions. Other apps that offer decision-making support aim to allow for quick diagnoses and more immediate information available to the patient. Despite various claims made by the designers of such apps, considerable questions remain regarding whether physicians utilizing apps within the office visit are perceived as being engaged with patients and whether such calculators improve clinical decision-making.

Discussion

How shall we make sense of these quite different examples and evaluate the impact of technological developments in medicine? In this section, we articulate some of the differences, but also points of convergence, between these different technologies in the hope of understanding their political, ethical, and professional significance.

Theranos, our first example, articulates the promise of innovative technologies to address longstanding needs in medical care. In this case, price, transparency, and cost-effectiveness were provided through new blood analysis services, and made possible only by significant technological advances. At the same time, however, Theranos has faced significant criticism regarding concerns about accuracy and patients' capacity to interpret results. The collapse of Theranos' business model, and the erosion of the trust of just about all the company's partners, has confirmed longstanding fears about technology that the lucrative pursuit of technical advancement often comes at the cost of patient safety. Along the way, in Theranos' model, the traditional fail safe—the physician—is largely removed from the equation. Theranos also conjures, more pointedly, the fear that financial decisions instead of patient-centeredness or sound clinical judgment will guide decisions. From these concerns, punctuated by the carelessness of Holmes and other Theranos leaders, arises a larger question about the cost-benefit decision-making being applied by an industry in an attempt to solve some of medicine's most intractable logistical challenges such as making bloodwork more cost effective, transparent, and geographically accessible without sacrificing quality.

Our second example, MARTTI, raises a quite different set of questions. Perhaps most basic is that this technology is designed for use by clinicians with patients instead of by patients with little or no involvement by clinicians. MARTTI's technological significance is punctuated by the fact that it addresses limitations in human workforce capacities. For example, although Spanish-speaking populations are by far the most prevalent among American patients of limited English proficiency, and often served by practices that can afford to staff Spanish-speaking trained interpreters, American patients speak more than 350 languages (Census Bureau 2015), which no facility can afford to address through live interpretation. Despite its ability to meet certain needs, MARTTI, as our interviews in Central Ohio have indicated, has serious drawbacks that often frustrate patients and providers alike. These drawbacks include delays in establishing connections, lost connections, audibility problems, and beyond. At the same time, as we have explained, MARTTI is, in many ways, a low-tech solution to the problem, comprised not of automatic, computer-based interpretation, but a video screen connected to a call center where live interpreters are located. To this extent, MARTTI's technology is still highly dependent upon human training and interaction. Cultural competence on the part of interpreters, as well as specific training in medical terminology, is still critical to success. As the theorists outline above suggest, technology appears to be limited in its ability to solve the problems posed by language barriers. Effective communication can be assisted by technology, but technology is rarely the difference maker.

By contrast, it is hard to generalize about medical apps because of their continual innovation and development. Nonetheless, some theoretical principles arise from the overview provided above. Importantly, and somewhat unsurprisingly, most of the patient-oriented medical apps seek to empower patients, conceptualized as individuals, with health information and the ability to track personal data, but do so only on an individual patient level (Greene 2016). Without dismissing the potential for improving patient access to information, these apps also project a particular view of medicine in which individual patient actions and "compliance" to treatment regimens are afforded primacy. Apps directed at physicians, meanwhile, appear to similarly reinforce this approach. To the extent to which apps share these similarities, it appears that most apps subscribe to the Baconian axiom, "knowledge is power," while, in fact, or at least simultaneously, lodging themselves in the disciplinary framework introduced by the theorist Michel Foucault (1980), who inverted Bacon's axiom to argue instead that knowledge and power were in fact entwined, thereby functioning as a system. In Foucault's view, the theoretical cluster of power/knowledge stood to remind us that power is wielded through knowledge, and vice versa.

Instead of viewing technology as a merely liberating tool (for empowering patients and promoting efficiency in clinical facilities), technology also reinforces and deepens existing power relations. For example, by emphasizing the importance of monitoring exercise and diet, or adherence to particular treatment regimens, many "lifestyle" apps simultaneously intensify narratives of personal responsibility, which in turn de-emphasize the social and shared responsibilities that many scholars argue are the most significant drivers of public health outcomes. In addition, medical apps require patients to own mobile devices or computers, suggesting that any health outcomes boosted by such technology would not be shared equally across class or racial/ethnic groups (Bender et al. 2014). This focus on the individual, who processes information and charts data within the private cloister of their smart phone (perhaps with some communication with physicians and other clinicians), can have the paradoxical effect of isolating individuals and is inconsistent with the widespread goals of improving health outcomes among populations. Might a public or population health-focused app be developed? What would such an app look like? In the spirit of Levinas, future consideration might be given to how technology in the form of apps may support a more collective approach to wellness that is consistent with the aims of population health.

Taken together, these medical technologies do suggest a central theoretical orientation. Unfortunately, this orientation is not of the uplifting sort that some proponents of technology envision. Specifically, the findings from this brief analysis suggests that new medical technologies—each of which is bound up in a business framework—reinforce a somewhat outdated, or at least nonprogressive, view of medicine. Specifically, each promotes, in their own way, "neoliberalizing" forces in which the future of medicine is one in which increased efficiency and improved outcomes are purchased via strategies in which largely atomized individuals are mobilized and disciplined. Specifically in primary care settings and in future population health efforts, this mobilization and discipline often comes at the expense of focusing on the social determinants of health and associated health disparities which can only be successfully addressed in a social manner. To the extent that technologies and apps do facilitate doctor-patient relationships, or empower individuals to address certain health challenges, but do so within an isolating space of smartphones and examination rooms, these technologies do not promote the goal of a truly public health in which communities and collectives unite to address broad-based social and environmental challenges that are the main drivers of health outcomes. In the case of MARTTI, which does meet certain needs within a limited scope (of doctor-patient relationships and more effective clinical visits), the very need for the technology is not only a lack of financial resources in live interpretation services, but a larger failure to invest in increasingly multicultural and humanizing social movements, such as the learning of foreign languages and the promotion of medically trained interpretation services that are available at a reasonable price. In the case of Theranos and MARTTI, there is significant uncertainty as to where lines are drawn as regards benefits to patients and the internal efficiencies afforded providers and business partners.

Conclusion

While the technologies we have reviewed do not comprise an exhaustive list, we chose them because they represented a sample set of disparate approaches that are used in primary care rather than specialty applications. The question for health care is whether technological developments will be advanced in the future that promote increasingly efficacious social approaches to the promotion and facilitation of improved health outcomes, and that bring people together to meet challenges instead of deepen the isolating tendencies of technological applications that are limited to individual lifestyle modifications or that are restricted to traditional clinical spaces. We fear, however, that the boom in such technologies fits snugly within the dominant ideological configurations that comprise American health care, and will therefore be difficult to dislodge. Since the neoliberal moniker of individual responsibility remains the central slogan of American health care, even as public and population health approaches gain ground among health experts and scholars, it will be difficult to bridge the divide. The Levinasian perspective will be instructive for guiding the development of medical technology that bridges individuals and patients within communities instead of promoting improved health among individuals.

Nonetheless, questions remain about whether and how research and development funds could be directed to the development of technologies that will be of social value, even when those technologies will not necessarily lead to profit opportunities of the sort provided by apps downloaded to individual phones, or kiosks in local Walgreens, or even web-based devices in examination rooms. Indeed, to be financially feasible such technologies are necessarily dependent upon social investment, and require the coordination of larger, well capitalized entities such as "anchor institutions" such as universities or hospitals, and government, be it on the local, county, state, or federal level.

These considerations remind us that technology itself lacks a fixed ethical, ideological, or even principle orientation. As such, it is inherently bound to the broader forces—especially power relations—that give rise to and continually surround new developments. A "Promethean" faith (Lewis 1992) that technological developments will help solve workforce issues in primary care, or will advance medical technologies in such a way that they contribute broadly to reaching the central goals of health care—increased access, increased outcomes (including reduced disparities), and efficiency in delivery and cost—therefore fails to take seriously the ways in which technology must be guided by sound social theory, and supported by political organization and policy guidelines. Indeed, all medical technologies, if not guided by sound principles, are positioned in such a way that they may contribute to the empowerment of malignant and retrogressive forces instead of creating the new opportunities that many advocates of technological change in medicine envision. To this extent, the development of medical technology must occur in dialogue with important political, ethical, and social factors.

References

Anderson, Gerard F., Bianca K. Frogner, Roger A. Johns, and Uwe E. Reinhardt. 2006. "Health Care Spending And Use Of Information Technology In OECD Countries." *Health Affairs* 25(3): 819–831.

Aronson, Jeffrey K. 2007. "Compliance, Concordance, Adherence." *British Journal of Pharmacology* 63(4): 383–384.

Bailey, Chris S., Michael G. Fehlings, Y. Raja Rampersaud, Hamilton Hall, Eugene K. Wai, and Charles G. Fisher. 2011. "Industry and Evidence-based Medicine: Believable or Conflicted? A Systematic Review of the Surgical Literature." *Canadian Journal of Surgery* 54(5): 321–326.

Bender, Melinda S., JiWon Choi, Shoshana Arai, Steven M. Paul, Prisila Gonzalez, and Yoshimi Fukuoka. 2014. "Digital Technology Ownership, Usage, and Factors Predicting Downloading Health Apps Among Caucasian, Filipino, Korean, and Latino Americans: The Digital Link to Health Survey." *Journal of Medical Internet Research* 16(10): 1.

Census Bureau. 2015. "Census Bureau Reports at Least 350 Languages Spoken in U.S. Homes." November 3. http://www.census.gov/newsroom/press-releases/2015/cb15-185.html.

Dennison, Laura, Leanne Morrison, Gemma Conway, and Lucy Yardley. 2013. "Opportunities and Challenges for Smartphone Applications in Supporting Health Behavior Change: Qualitative Study." *Journal of Medical Internet Research* 15(4): e86.

Doyle, Richard J., Nina Wang, David Anthony, Jeffrey Borkan, Renee R. Shield, and Roberta E. Goldman. 2012. "Computers in the Examination Room and the Electronic Health Record: Physicians' Perceived Impact on Clinical Encounters Before and After Full Installation and Implementation." *Family Practice* 29: 601–608.

Dreyfus, Hubert L. 2007. "Why Heideggerian AI Failed and How Fixing it Would Require Making it More Heideggerian." *Philosophical Psychology* 20(2):247–268.

Eugenio, Santoro, Gianluca, Castelnuovo, Italo, Zoppis, Giancarlo, Mauri, and Francesco Sicurello. 2015. "Social Media and Mobile Applications in Chronic Disease Prevention and Management." Frontiers in Psychology 6.

Ferris, Sarah. 2015. "Medical Device Groups Blame ObamaCare Tax for Job Losses." *The Hill* January 28.

Foucault, Michel. 1980. "Two Lectures," in *Power/Knowledge: Selected Interviews and Other Writings, 1972–1977*, ed. Colin Gordon. Brighton: Harvester Press.

Franz, Berkeley and John W. Murphy. 2016. "Electronic Medical Records and the Technological Imperative: The Retrieval of Dialogue in Community-based Primary Care." *Perspectives in Biology and Medicine* 58(4): 480–492.

Gellert, George A., Ricardo Ramirez, and Webster, S. Luke. 2015. "The Rise of the Medical Scribe Industry." *JAMA* 313(13): 1315–1316.

Greene, Jeremy A. 2016. "Do-It-Yourself Medical Devices — Technology and Empowerment in American Health Care." *The New England Journal of Medicine* 374: 305–308.

Harvey, H. Benjamin and Pari V. Pandharipande. 2012. "The Federal Government's Oversight of CT Safety: Regulatory Possibilities." *Radiology 262*(2): 391–398.

Heidegger, Martin. 1977. *The Question Concerning Technology and Other Essays, translated by William Lovitt.* New York: Garland Publishers.

Hsia, Renee Y, Yaa Akosa Antwi, and Julia P Nath. 2014. "Variation in Charges for 10 Common Blood Tests in California Hospitals: A Cross-sectional Analysis." *BMJ Open* e005482 https://doi.org/10.1136/bmjopen-2014-005482.

Jones, Caroline H.D., Jeremy Howick, Nia W. Roberts, Christopher P. Price, Carl Heneghan, Annette Pluddemann, and Matthew Thompson. 2013. "Primary Care Clinicians' Attitudes Towards Point-of-Care Blood Testing: A Systematic Review of Qualitative Studies." *BMC Family Practice* 14: 117.

Kidd, Brian A., Gabriel Hoffman, Noah Zimmerman, Li Li, Joseph W. Morgan, Patricia K. Glowe, Gregory J. Botwin, Samir Parekh, Nikolina Babic, Matthew W. Doust, Gregory B. Stock, Eric E. Schadt, and Joel T. Dudley. 2016. "Evaluation of Direct-to-Consumer Low-Volume Lab Tests in Healthy Adults." *Journal of Clinical Investigation* 126(5): 1734–1744.

Koizumi, Kei. 2015. "Investing in America's Future through R&D, Innovation, and STEM Education: The President's FY 2016 Budget." February 2. https://www.whitehouse.gov/blog/2015/02/02/investing-america-s-future-through-rd-innovation-and-stem-education-president-s-fy-2.

Levinas, Emmanuel. [1962] 1996. "Transcendence and Height." In *Emmanuel Levinas: Basic Philosophical Readings*, eds. Adrian T. Peperzak, Simon Critchley, and Robert Bernasconi. Bloomington, IN: Indiana University Press, 11–32.

Levinas, Emmanuel. 1998. "Secularization and Hunger." *Graduate Faculty Philosophy Journal* 20(2): 3–12.

Lewis, Martin W. 1992. *Green Delusions: An Environmentalist Critique of Radical Environmentalism*. Durham, NC: Duke University Press.

Manjoo, Farhad. 2011. "Will Robots Steal Your Job?" Slate.com. September 27. http://www.slate.com/articles/technology/robot_invasion/2011/09/will_robots_steal_your_job_3.html.

Miller, Edward Alan. 2001. "Telemedicine and Doctor-patient Communication: An Analytical Survey of the Literature." *Journal of Telemedicine and Telecare* 9(1): 1–17.

Peperzak, Adriaan. 1992. "Levinas on Technology and Nature." *Man and World* 25: 469–482.

Pinker, Steven. 1994. *The Language Instinct*. New York: Stephen Morrow and Company.

Powell-Cope Gail, Audrey L. Nelson, and Emily S. Patterson. 2008. "Patient Care Technology and Safety." In *Patient Safety and Quality: An Evidence-Based Handbook for Nurses*, ed. Ronda G. Hughes. Rockville, MD: Agency for Healthcare Research and Quality (US), Chapter 50.

Singer, Jeffrey A. 2013. "Toward Medical Dystopia." *Journal of Trauma and Acute Care Surgery* 75(3): 517–519.

Smith, Noah. 2016. "How Theranos' Founder Lost $4.5 Billion Overnight." *Chicago Tribune* June 6. http://www.chicagotribune.com/business/ct-elizabeth-holmes-theranos-net-worth-20160606-story.html.

Soudi, Abdesalam and Anna-Binney McCague. 2015. "Medical Scribes and Electronic Health Records." *JAMA* 314(5): 518–519.

Street, Richard L., Lin Liu, Neil J. Farber, Yunan Chen, Alan Calvitti, Danielle Zuest, Mark T. Gabuzda, Kristin Bell, Barbara Gray, Steven Rick, Shazia Ashfaq, and Zia Agha. 2014. "Provider Interaction with the Electronic Health Record: The Effects on Patient-Centered Communication in Medical Encounters." *Patient Education and Counseling* 96(3): 315–319.

Swayze, Sonia and Suzanne Rich. 2011. "Promoting Safe Use of Medical Devices." *OJIN: The Online Journal of Issues in Nursing* 17(1): 9.

Tahir, Darius. 2014. "FDA Decides Medical-device Data Systems Don't Need its Oversight." *Modern Healthcare* http://www.modernhealthcare.com/article/20140620/NEWS/306209945.

Vidscrip. 2016. "About: Our Purpose." http://welcome.vidscrip.com/about/.

Weaver, Christopher and John Carreyrou. 2016. "Craving Growth, Walgreens Dismissed Its Doubts About Theranos." The Wall Street Journal May 25, 2016.

Westfahl, Gary and George Slusser. 2002. *No Cure for the Future : Disease and Medicine in Science Fiction and Fantasy.* Westport, CT: Greenwood Press.

Whelton, Beverly J.B. 2016. "Being Human in a Global Age of Technology." *Nursing Philosophy* 17: 28–35.

White, Amina and Marion Danis. 2013. "Enhancing Patient-centered Communication and Collaboration by Using the Electronic Health Record in the Examination Room." *JAMA* 309: 2327–2328.

CHAPTER 6

AI in Public Education: Humble Beginnings and Revolutionary Potential

Ken Rogerson and Justin Sherman

Popular culture has had a profound impact on contemporary global discussions of artificial intelligence. How will lethal autonomous weapons change the dynamics and the morals of war? Could predictive algorithms revolutionize modern medicine and usher in a new era of diagnosis and prevention? Will the unchecked deployment of facial recognition in major cities bring the advent of a truly Orwellian dystopia? And

Kenneth Rogerson is Professor of the Practice at the Sanford School of Public Policy at Duke University. Justin Sherman is a nonresident senior fellow at the Atlantic Council. The authors would like to thank Teresa Chen for her valuable work as research assistant.

K. Rogerson (✉) · J. Sherman
Sanford School of Public Policy, Duke University, Durham, NC, USA
e-mail: rogerson@duke.edu

J. Sherman
e-mail: justin.sherman@duke.edu

R. Kiggins (ed.), *The Political Economy of Robots*, International Political Economy Series, https://doi.org/10.1007/978-3-031-95171-8_6

will the widespread use of generative AI overhaul economies and cause employment and job shifts like the world has never seen?

These are pressing questions, and each illustrates the need for cross-disciplinary approaches to understanding the realities, opportunities, risks, and challenges of artificial intelligence. But less frequent in discussions of artificial intelligence—both in how AI systems are presently used, and how they may be deployed in the future—is mention of AI's impact in civil society. This is particularly true with respect to how AI is impacting, and will impact, public education systems around the world.

There is a strong focus on education in the service of AI, not the other way around—asking such questions as how can institutions of higher learning provide AI research and development, how to best prepare the workforce for a new era of automation, or how to train the most skilled machine learning researchers, as opposed to how AI will change the design and practice of public education itself. We could attribute this lack of discussion to many factors, including (but not limited to) the slowness of public education systems—and government bureaucracies in general—to adapt to new and evolving technologies; the relative newness of AI's deployment in public education systems around the world; and uncertainty about exactly how a range of AI applications could even be applied in an educational sphere in service of a public education mission. It is starting to change now as instructors in primary schools, universities, and elsewhere debate whether students should be allowed to use generative AI (GenAI) for assignments, and if so how, and whether schools have technologies available to detect students' unreported GenAI use (such as to write a paper). But still, many substantive discussions of AI's role *in* public education have yet to hit the mainstream.

Hence, in this chapter, we analyze the current uses—as well as potential uses—of AI in primary and secondary education around the world, primarily in the countries of China, India, and the United States as well as the regions of Africa, the European Union, and Latin America. Our focus rests mainly on three countries and three regions, not because the countries are the only AI players on the global stage (far from it), but because they are poised to perhaps be the most influential. Combinations of their strong economies, military power, political influence, technological resources, and large populations are all reasons why these three countries will be rule-setters and innovators in how countries develop, deploy, and/or regulate artificial intelligence in the twenty-first century. These three regions, for their part, are definitely making strides toward greater integration of AI in public education, but more as a collective and not yet as much as individual countries.

First, we discuss the global landscape of cooperation and competition around artificial intelligence technologies. Then, we discuss plans, investments, and implementations of AI in public education in the identified countries and regions. Finally, we conclude with a discussion of the relationships between such considerations as AI use, school demographics, gender, ethnicity, socioeconomic status, and privacy.

The Global AI Landscape

Global research on artificial intelligence demonstrates the interdependence and interconnection between the "AI spheres" of different countries. Companies, universities, civil society organizations, individuals, and even government entities in different countries may collaborate on a host of AI research areas, from natural language processing to image recognition. Many AI-related research goals of these various groups—such as improving public health outcomes or boosting transportation safety through autonomous cars—are relatively aligned, at least technologically speaking, and are not zero-sum ("Artificial Intelligence Index" 2019; Ding 2018).

Furthermore, most global AI research is also open-source, occurring largely in the public domain on code-sharing sites like GitHub, data-sharing sites like Kaggle, and paper-sharing sites like Arxiv.org. The latest research, datasets, and other software-based elements in the field are frequently available to anyone with an internet connection (though, of course, access to this information and these resources is just one of many elements of AI research, such as having the requisite computing hardware on which to train and test algorithms). There are always exceptions—when nations secretly develop lethal autonomous weapons in military research centers, for example—but these characteristics generally hold (Sherman 2019).

Due to their large economies, military power, political influence, technological prowess, and sizable populations, China and the United States are quite influential when it comes to the future of AI. But other countries, from Canada to India, from Japan to Israel, are also investing in AI research, curating and developing domestic AI talent, and looking to use their regulatory levers to influence the future of AI development within and outside of their sovereign borders. This is especially true for India, the most populous democracy on earth, which is arguably at the forefront of explicitly laying out a national vision for how artificial intelligence could be used to revolutionize public education and further a mission of providing quality and accessible public education for all.

The advent of generative AI (GenAI) has taken the discussions about AI's educational applications to a new level. Some are embracing it as the next gift to educational enhancement and performance. Others are identifying the real concerns about its potential misuses in educational settings. Educational institutions around the world are reacting very differently, with some monitoring and including it in the classroom and others banning and punishing its use by students (see "Generative Artificial Intelligence," 2023).

In the next few sections, we lay out the current state and future potential of AI in public education. In each of these sections, we examine some of the government strategies and investments, currently deployed technologies and those prospectively on the table for the future, and how these investments and implementations may reflect unique or shared values around privacy, technological growth, and other issues. Finally, we conclude with a look forward—laying out challenges for AI in public education, and detailing important research questions that deserve further exploration.

The United States

The 30,000-Foot View

In October 2016, U.S. President Barak Obama's administration released two major documents on AI strategy. The first, Preparing for the Future of Artificial Intelligence, provided recommendations to federal agencies and other actors to better develop and prepare for the AI-driven future (National Science & Technology Council 2016a). The second, The National Artificial Intelligence Research and Development Strategic Plan, laid out a strategic plan for federal investment in AI development (National Science & Technology Council 2016b). The former briefly mentions AI in education, noting that "AI-enhanced education may help teachers give every child an education that opens doors to a secure and fulfilling life," but focuses widely on issues like monitoring progress in AI, optimally targeting federal research funding, and ensuring fairness and accountability in artificially intelligent systems (National Science & Technology Council 2016a). The latter discusses education as well, noting that AI can "improve educational outcomes" in areas such as automated tutoring and customized, in-person teaching supplementation, though the comment is brief and high-level (National Science & Technology Council 2016b).

The Trump administration's June 2019 update to the National Artificial Intelligence Research and Development Strategic Plan briefly mentioned AI improving education, but it mostly focused on AI and education in the context of workforce development (National Science & Technology Council 2019). U.S. government funding for AI research, meanwhile, has been heavily focused within the military. The U.S. military certainly has a history of funding vitally important technological breakthroughs that now reside, even primarily, in the civilian and commercial spheres, from the military's invention of GPS technology (Mazzucato 2013) to DARPA's role in building the foundation for the internet (Cerf). Further, many AI applications have utility in both civilian and military domains—what some historically call dual-use. It is therefore possible that the military's investments in AI development have wider applicability, including because many of the military's AI research focuses are on logistical and command support, not solely autonomous weapons as some might assume (Congressional Research Service 2019).

Most recently, the Biden administration's executive order on ensuring the safe, secure, and trustworthy development and use of AI mentioned education 15 times. This ranged from aforementioned areas outside of this chapter's scope, such as helping provide people in the United States with job training and education to seize AI workforce opportunities, to how to safely deploy AI in educational settings. This topic had previously not received as much attention from the White House. For example, one section called on the Secretary of Education to develop resources, policies, and guidance to promote "safe, responsible, and nondiscriminatory uses of AI in education." It also included developing an "AI toolkit" for education leaders seeking to implement Department of Education recommendations on the use of AI for future teaching and learning (E.O. 14110 2023).

While the United States seems to be focusing much of its public discussions on AI for military and security issues, there were, at least during the Biden administration, other social issues like education that made more of an appearance in policy discussions. Interestingly, though, most uses of AI in American public schools that we uncovered in our research still focus on security and surveillance applications—that is, programs explicitly built and marketed for the purposes of monitoring student social media posts to predict risk of student suicide or in-school violence, for example. Very few examples focused on applications such as automatically grading student assignments.

The private sector has driven this AI development in the United States, but it's perhaps equally relevant to note the U.S. government's lack of explicit focus on alternatives. There is much federal and state policy discussion about AI's impact on the workforce, specifically around job displacement or replacement due to improved automation; and there is also much discussion in the general public, related to this point, about how public education and higher education should adapt to prepare American citizens for an increasingly AI-driven future. But the U.S. government has focused little on how artificial intelligence could improve or even revolutionize the design and practice of public education.

Current AI Implementations in Public Education

Several public school systems in the United States are using artificial intelligence applications in ways that illuminate opportunities, risks, and possible future directions for AI use in American schools. This is especially because most applications we found of artificial intelligence in U.S. public schools pertain to school safety—everything from running programs to predict student self-harm to object recognition systems that identify firearms.

Some uses of artificial intelligence in American public schools center around safety and security—predicting (with an eye toward preventing) student violence. Uses for these reasons come in different shapes, including facial recognition systems and data mining/machine learning analyses of student communications.

The Aegis system, for example, is an object recognition system that uses deep learning to monitor students. One school district implementing the system in Lockport, New York has faced some obstacles in getting started. Lockport superintendent Michelle Bradley said that "the test [was in] an 'initial implementation phase' meant to troubleshoot the system, train district officials on its use, and discuss proper procedures with local law enforcement in the event of an alert triggered by the facial recognition tech." These "alerts," said an FAQ distributed to parents with children at the school, are defined by the "ability [to screen] every door and throughout buildings to identify people or guns. Early detection of a threat to our schools allows for a quicker and more effective response" (Alba 2019). When the New York State Department of Education heard about it, it asked the district to delay testing Aegis for privacy reasons (Alba 2019).

Facial recognition is only one way that AI is being used in schools. A company called Gaggle offers schools the ability to monitor student online conversations through email and social media. "Using a combination of in-house artificial intelligence and human content moderators," *Buzzfeed News* reported, "Gaggle polices schools for suspicious or harmful content and images, which it says can help prevent gun violence and student suicides. It plugs into two of the biggest software suites around, Google's G Suite and Microsoft 365, and tracks *everything*, including notifications that may float in from Twitter, Facebook, and Instagram accounts linked to a school email address" (Haskins 2019). While school policy requires that the monitored communication must be connected to a school-issued email, there are quite a few intersecting information flows between that email and other publicly available social media. "[S]tudent work and behavior are scrutinized for indicators of violence or a mental health crisis, and profanity and sexuality are [also] policed" (Haskins 2019).

Implementation delays similar to those that happened with Lockport occurred in Florida in the spring of 2019. In this case, the purpose was the same: to improve security after the Parkland High School Shooting (Herrold 2019). But while the reason for the delay was stated as a concern for privacy, this situation was also about the legality of the state accessing and merging a number of different databases, rather than just gathering information through the school as with Gaggle. The database, managed by the state, would include "people's social media posts with millions of records on individuals who have been bullied, placed in foster care, committed a crime, or even been mentioned in unverified tips made to law enforcement." The article continued that the organization's "investigation found that across the country, social media and related monitoring services used by schools are generating vast torrents of information—some of which is alarming, but much of which is ambiguous, irrelevant, or ridiculous" (Herrold 2019).

In addition to monitoring security issues, schools are looking to AI as a supplemental teacher. While some bemoan the potential replacement of teachers with machines, most understand that the reality will be a combination of AI and human interaction for the foreseeable future (Houser 2017). In 2015, Stanford University's One Hundred Year Study on Artificial Intelligence issued a report that included an observation about the relationship between AI and education. The authors noted:

> Though quality education will always require active engagement by human teachers, AI promises to enhance education at all levels, especially by providing personalization at scale. Interactive machine tutors are now being matched to students for teaching science, math, language, and other disciplines. Natural Language Processing, machine learning, and crowdsourcing have boosted online learning and enabled teachers in higher education to multiply the size of their classrooms while addressing individual students' learning needs and styles. Over the next fifteen years in a typical North American city, the use of these technologies in the classroom and in the home is likely to expand significantly, provided they can be meaningfully integrated with face-to-face learning. ("The One-Hundred Year Study" 2016)

This nuanced acknowledgment is more meaningful than the all-or-nothing approach some take, since human touches on AI are still the norm for now. From the Stanford study, deeper curricular uses of AI are suggested by the phrase "personalization at scale." John Allen describes this as a mixture of an in-person and virtual classroom. He writes, "students can become more deeply involved in the pathways of their own learning," but this will require better assessment and remediation, both of which can also potentially be realized through AI (Allen 2019).

Lately, the focus has been heavily on GenAI. By the fall of 2023, 18% of K-12 teachers reported using GenAI for adaptive learning, chatbots, virtual classrooms, and other teaching purposes. Most common in the survey was teachers using GenAI tools to adapt instructional content to fit the level of students (Diliberti et al. 2024). A report from educational content provider Cengage found that GenAI use had risen in higher education settings from 24% in 2023 to 45% in 2024—a nearly two-fold increase (Cengage 2024). This all suggests GenAI may become a bigger and bigger driver of the United States' conversation about how to use AI for learning, but since it is distinct from the surveillance aims of many other school-focused AI systems, it is not likely to *replace* the other, security-focused discourse.

Initial observations about the uses of AI in the U.S. educational system are that it is difficult to dream big and implement technological change quickly. Much of what happens is reaction rather than action. Schools are hesitant to adopt technologies (a) on which they aren't tested and (b) which aren't required by the state. And AI's potential applications are much broader than, say, mathematical curricular standards, or professional teaching development. While some school districts, possibly those with more resources, are able to experiment with these new technologies before required by rule or practical necessity, most U.S. schools are not using a holistic approach to the integration of AI in the classroom or school. And despite some outliers, like the state of Oregon, most jurisdictions do not have clear policies and guidance documents for AI use in public education yet, either (Oregon Department of Education 2023).

China

The 30,000-Foot View

Beijing's investments in and focus on artificial intelligence predate the year 2016. But when Google's AlphaGo system defeated a top Go player in March 2016 (Go is a popular and centuries-old Chinese strategy board game), it was a major event in mainland China and spurred a landmark shift in government focus on AI planning and investment. Prior to 2016, "AI was presented merely as one technology among many others, which could be useful in achieving a range of policy goals." But the AlphaGo victory was a "Sputnik moment" for the Chinese government, according to two government officials (Roberts et al. 2019).

In July 2017, the Chinese government released its Next Generation Artificial Intelligence Development Plan, a comprehensive strategy aimed at "[seizing] the major strategic opportunity for the development of AI, to build China's first-mover advantage in the development of AI, [and] to accelerate the construction of an innovative nation and global power in science and technology" (Webster et al. 2017).

The document is comprehensive, discussing AI as the "new engine of economic development" that will be "the core driving force for a new round of industrial transformation" (Webster et al. 2017). Since its debut, the government has released other documents on AI strategy and development, such as the Three-Year Action Plan for Promoting Development of a New Generation Artificial Intelligence Industry, which put forth "major tasks" for the country such as "deepening the development of intelligent manufacturing" and "[building] a public support system for industry training resources" (Triolo et al. 2018).

Overall, however, "Beijing's AI plan serves less as a 'plan' and more as a 'wish list' of technologies the central government would like to see built. It then incentivizes ambitious local officials to use all the tools at their disposal—subsidies, public contracts, and AI-friendly policies—to guide and aid the private sector in developing those technologies," with many questions to be answered about this approach's effectiveness (Sheehan 2018). Like many ambitious Chinese government plans, it remains to be seen how much will be implemented, and how well. It is also worth noting that despite perhaps troubling early uses of AI in China, such as AI used as part of unchecked government surveillance, "it is too early to know what approach China will take" across the board with respect to AI governance (Sacks 2019).

Nonetheless, since the publication of the first version of this chapter, the Chinese government has moved ahead in AI safety governance mechanisms. Beijing has been releasing a series of regulations for algorithms, chatbots, and other AI-related systems and components. China's Ministry of Education was involved in a 2021 guidance document, for instance, on strengthening the state's governance of online algorithms (Sheehan 2023). The Chinese government also released the first version of its AI Safety Governance Framework in September 2024, out of China's National Technical Committee 260 on Cybersecurity, and mentioned education a few times in the context of educating and training people in China on AI safety. It also emphasized data and content risks (including "cognitive warfare") as well as discrimination and prejudice risks (Tobey et al. 2024). In November 2024, China unveiled a global AI governance committee, chaired by the founder of Alibaba Cloud (Jiang and Cao 2024).

Again, it remains to be seen how effective many of these guidance documents and pledges will be in meaningfully shaping public and private AI, including GenAI, uses in educational and learning settings. The potential for GenAI systems in particular to introduce banned content into Chinese classrooms may be one strongly compelling reason for the Chinese government to ensure some degree of meaningful control over AI use in classrooms. For example, ChatGPT is banned in China (Ray 2023). It is likely other tools that could be used in educational settings to query information and ask about a large corpus of texts (e.g., books, non-Chinese news articles, etc.) will face similar restrictions if they have not already.

Current AI Implementations in Public Education

As with the U.S. government, the Chinese government's documents around artificial intelligence development primarily refer to public education as a mechanism by which to prepare students—and future workers—for an AI-driven future. There are fewer references to public education as an institution in which AI applications might be used to improve and expand upon core educational objectives.

In China, educational uses of AI can be categorized broadly in two ways: first, high-level references to the importance of understanding AI and its impact on politics, society, and the economy; and second, very granular implementations of AI for improving educational efficiency.

The Chinese government regularly references artificial intelligence as vital to Chinese societal and economic development. In the spring of 2019, Chinese leaders in science and engineering announced an AI code of ethics, signaling "a willingness from Beijing to rethink how it uses the technology" (Knight 2019). This code spells out guidelines that encourage the idea that "human privacy, dignity, freedom, autonomy, and rights should be sufficiently respected" when implementing AI programs (though, of course, some would contend that Beijing's view of privacy doesn't exactly track with views in the United States or Europe). These principles are sometimes connected to discussions on the goals of AI in education, implying that the Chinese government would like to see technology more broadly—and AI more specifically—as part of the country's curricular goals.

In the second category, there are a variety of ways that AI is being implemented with the goal of improved efficiency. In the classroom, teachers are able to outsource some types of grading. A free app created by the educational company Yuanfudao allows teachers to take a photo of their students' homework and upload it to the site, at which point an algorithm will respond with whether the answer is correct. The company "claims to have checked an average of 70 million arithmetic problems per day." It then uses the resulting database of math problems and solutions to identify the most common mistakes (Jeng 2019).

Another pedagogical example is one that is expected: helping students work at their own pace in a variety of subject areas. AI-based individualized learning is being adopted in many places, and China is no exception. But China may be one of the first places to build entire schools where every student is engaged in this type of learning. A Chinese company called Squirrel AI has, in addition to providing AI-based platforms for public schools, created its own network of schools in which students sit in front of a computer all day and interact with human instructors as needed. The company's goals are to improve student performance and "[i]t also designed its system to capture ever more data from the beginning, which has made possible all kinds of personalization and prediction experiments" (Hao 2019). This begs the question of proper uses of data collected by AI-based systems, the answers to which vary by public and private sector actors.

But there are also examples of daily, functional activities that are intended to help students. Though at the college level, a school in Hangzhou has started using an AI app to track attendance. Students must use a verification code when they arrive in class. If they don't, the program will nudge them by text to ask why they aren't in class. The algorithm will then analyze the reasons so that the university can address the issues (Dao 2019). While there are a number of caveats here—that students may not tell the truth about why they are absent or may discover a way to have someone reply for them—the university is experimenting with AI to address what it sees as a chronic absenteeism problem.

In another, non-pedagogical, example, a school district (also interestingly in Hangzhou) has installed AI-powered cameras to monitor whether high school students are paying attention in classes. "The system works by identifying different facial expressions from the students, and that information is then fed into a computer which assesses if they are enjoying lessons or if their minds are wandering" (Connor 2018). While people have raised privacy concerns, administrators say it is improving educational standards. It doesn't film or store student activities; it simply watches student facial movements and interprets them as an emotion. If that emotion equates to not paying attention, the teacher is notified.

Many Chinese classrooms now have smartboards equipped with AI capabilities to help with enabling interactive and engaging lessons (Akgün and Alpaydan 2024, 139) More than 7000 Chinese middle schools and over 60,000 classes are also now using ETS, an AI-powered English exam platform made by iFlytek and designed to assess students' English-language proficiency (iFlytek 2024). It has more than 12 million users total. It seems that one option for Chinese authorities interested in banning or restricting English-language AI and especially GenAI systems, for the possibility they introduce censored content into China, is to leverage English-trained AI systems built by Chinese companies like iFlytek.

In sum, for Chinese public education, AI seems to be an extension of the country's educational values: becoming a world leader in every technological arena and controlling as much of the pedagogical process as possible. There does seem to be some flexibility in the creation of AI tools for the classroom, with the government encouraging technological entrepreneurs in the development of educational apps.

India

The 30,000-Foot View

In June 2018, India published its National Strategy for Artificial Intelligence, a document produced by NITI Aayog, a Government of India policy think tank. The strategy was the result of a 2018 push by India's Finance Minister to guide research and development in new and emerging artificial intelligence technologies, with the aim of establishing India as a global leader in artificial intelligence—under the "unique brand of #AIforAll" ("National Strategy" 2018, 5).

"An integral part of India's strategy for AI," it reads, "involves tackling common and complex global challenges that can be solved through technology intervention, and India's scale and opportunity landscape provides the ideal test-bed to ensure sustainable and scalable solutions" (Ibid. 6). To that end, the document discusses research and development priorities in areas such as healthcare, agriculture, smart mobility, retail, manufacturing, and energy (Ibid. 20). Much like the strategies released by the American and Chinese governments, this approach recognizes the full range of sectors that could be positively impacted by development of AI technologies. There is a positive outlook on the future of AI, despite clear concerns about privacy, security, and other ethical and regulatory issues.

Of the three countries analyzed in this chapter, India is the only one whose government strategy addresses at length the need for education to better prepare an AI-era workforce *and* the role that AI can play in improving education itself. "AI has the potential to bring about changes in the [education] sector by supplementing pedagogy and establishing systems to inform and support decision making across stakeholders and administrative levels," it reads (Ibid. 36). The document highlights five key areas where AI tools can be "adapted to the Indian context to target specific challenges": adaptive learning tools for customized learning; intelligent and interactive tutoring systems; predictive tools to preempt possible student dropouts; automated management of teacher posting and transfer systems; and customized professional development courses (Ibid. 37–38).

In 2021, the prime minister launched a National Digital Education Architecture, a policy aimed to create a unified national digital infrastructure to energize the education system, as the Indian government put it (Indian Ministry of Education 2024). The NDEA includes the use of AI for personalized learning and adaptive learning solutions (Indian Central Board of Secondary Education 2024). Whether India has made that happen yet or not, the strategic vision for making AI in education a part of broader digital transformation—and a part of broader AI conversations—indicates India's desire to at least promote AI use for education where possible.

Current AI Implementations in Public Education

Paradoxically, despite the level of thought in Indian government, industry, and civil society that has gone into how artificial intelligence could improve public education, examples of few tangible implementations presently exist in the country. But the AI startup ecosystem's focus on education is notable in India, and there are likely to be more actual implementations of AI in Indian public education in the near future. A few examples highlighted below capture this reality.

In the fall of 2019, Microsoft India partnered with India's Central Board of Secondary Education to integrate educational technology into the classroom. This partnership includes a focus on AI, whereby 1000 teachers nominated by the Central Board will undergo a three-day training that includes how to leverage artificial intelligence in the classroom. "AI has become a strategic lever for economic growth across nations around the world," a general manager at Microsoft India declared, and the partnership will help "transform the education ecosystem with the power of AI and the cloud" (Mathur 2019).

Several public schools are also using natural language processing systems to assist with teaching English to students. English Helper, one U.S.-based technology platform, partnered with the Maharashtra government in July 2019 to roll out the platform across the state. The goal is to eventually cover 100,000 Indian schools (Sangani 2019). One estimate puts the software company's current reach within India at almost 8 million students (Acharaya 2020).

These initiatives are a likely indication of what is to come. For instance, one study has found that out of 300-plus Indian startups with AI as a "core product offering," 11% were in the education sector. Yet, writes one observer, "while this is a great sign that there is progress being made to make education efficient using AI, no single application in this regard has come on top in India, as opposed to China where we can see multiple of such companies" (Chawla 2019). The state has launched its own platforms and offerings, too. The Dishka Platform from the Ministry of Education is a national platform that uses AI to provide teachers with tools for classroom instruction, including AI-based analytics for monitoring student progress and personalizing learning (Porecha 2023). India's Central Board of Secondary Education launched an AI-based applications to help students prepare for exams, such as with adaptive assessments (Sharma 2023). It is not clear either whether these state offerings will beat out industry ones.

Further, the focus placed at the level of national government and state and local governments on AI's potential for changing public education stands out compared to other countries, and coupled with the Indian government's general focus on using technology to boost the economy and other government functions, there may be top-down pushes to implement more of these kinds of technologies in public schools as well. Recent events suggest this may already be playing out: for instance, Indian educators attended a September 2024 education exhibition with the aim of procuring more than 1000 AI-powered educational technology innovations (Economic Times 2024). Yet, ad hoc tests of AI in classrooms remain, too, as with the Indian government working with e-learning provider Makerlabs to supply a high school in Kerala, India, with an AI humanoid teacher named Iris that can process three languages and, reportedly, respond to students' complex questions (Sarkar 2024).

European Union

Even in a resource rich environment like the European Union, the adoption of AI in public education systems has been slower than other places, but for different reasons. The EU's AI Act (2024), does address the uses of AI in the classroom in a very general way, stating:

> AI systems used in education or vocational training, in particular ... for evaluating learning outcomes of persons, for assessing the appropriate level of education for an individual and materially influencing the level of education and training that individuals will receive or will be able to access or for monitoring and detecting prohibited behaviour of students during tests should be classified as high-risk AI systems, since they may determine the educational and professional course of a person's life and therefore may affect that person's ability to secure a livelihood. (Artificial Intelligence Act, Section 56)

In other words, the use of AI is seen as a potential threat to educational institutions, at least in the language used in the Act. While the EU also sees the benefits (Digital Education Plan), the Act's specific references to schools are ones of caution and, in the end, placing AI uses in the highest risk category.

As with most regions, some EU country governments are incrementally researching the impact of AI in classrooms. In 2024, France Education International completed a multi-country study of 302 schools, evaluating the impact of AI curriculum platforms. According to the AI4T (AI for Teachers) web page, program assessment resulted in the majority of teachers being "satisfied" with the online resources. But, to emphasize the cautious nature of the implementation, one teacher said, "As a first course, it was very good and now we would need a practical application, i.e. a course that guides us through concrete experimentation by doing something in class" ("AI4T—Artificial Intelligence for and by teachers," 2024).

Africa and Latin America

Lower resource areas of the world are beginning to explore the potential benefits of AI in the public education sector. Most of the documentation at the moment of writing is about learning and training, a lot of which is being done by advocacy institutions with only a little government support.

In Africa and Latin America there are conversations about—and a desire to—integrate AI platforms into the educational systems. However, the lack of resources and infrastructure, as well as slow moving political will, has resulted in slower adoption for local public educational institutions. A February 2024 report about AI diffusion in Africa says, "where recent efforts for universal education have faced capacity constraints, artificial intelligence (AI) can be a crucial ally" (Dahlberg Trust 2024). In Uruguay, for example, the country has focused on assuring that schools are connected to broadband networks, achieving 100% connectivity in 2023 (Ceibal 2023).

Many of the initial experiments using AI in the classroom are very localized and focused principally on math skills. But there are some language arts programs as well. In Brazil, for example, Letrus is designed to narrow the literacy gap between high- and low-income students. While the program has seen some success, it is still in the early stages and has only been deployed in 670 of Brazil's nearly a quarter of a million public schools (a quarter of 1%) (World Economic Forum, 17).

AI and Existing Educational Structures

The use of artificial intelligence in public education holds particular promise with respect to student interactions with teachers, and with curriculum development and innovation. Of course, the implementation of AI applications in public education will not occur without challenges (discussed in the next section). From public schooling systems' slowness to adopt new technologies in the classroom to institutionalized access barriers, the adoption of AI systems in education is impacted by constraints that already impact the use of other technologies in schools and the quality and delivery of education outside of the technology sphere. For these reasons, artificial intelligence in public education poses thought-provoking relationships with school demographics, curriculum innovations, and cultural values.

First and foremost, when discussing AI use and school demographics, there is the consideration that surveillance systems are most often turned against already underrepresented, oppressed, or marginalized communities. In other words, surveillance always hits the marginalized the hardest. In the United States, for instance, this has been true from the tracking of enslaved peoples with so-called plantation ledgers (Bell 2018) to the surveillance of female suffragists in the early 1900s (Travis 2003) to government spying on civil rights activists in the 1960s (Kayyali 2014) to post-9/11 surveillance of Muslim communities (Apuzzo and Goldman 2011). Digital surveillance systems are a continuation of this history, and AI surveillance tools are also faster, cheaper, and more scalable than surveillance systems that are either manual or digital without the use of intelligent automation. For this reason, they may be harder to resist, remove, or overhaul once implemented—and once dependencies upon these systems are created (Feldstein 2019). This applies not just to AI data collection and analysis in public school systems but also more widely, such as with AI-powered surveillance in city downtowns or at large entertainment events like concerts or the Olympics.

When it comes to public schools, there are concerns that surveillance powers "are likely to be wielded disproportionately against students of color, who already face disciplinary bias at school," the Brennan Center for Justice's Rachel Levinson-Waldman argues. Particularly in light of American schools using AI for safety and security ends, there are also risks that AI use in public schools could harm demographics already prone to the school-to-prison pipeline and who belong to populations that already face inequity in public education (Kofmann 2018). It is possible, for instance, that lower income students in India could be subject to AI surveillance in schools without the power to effectively resist administrators, or that students in China who do not desire surveillance in schools have no legal protections under which to complain to the government.

For all these reasons, AI in public education does not just relate to the demographics of the students in school themselves but to the broader communities in which those schools reside.

Stefanie Coyle and Naomi Dann at the New York Civil Liberties Union, in looking at uses of facial recognition in New York public schools, did not just find that AI surveillance "could turn our school environment from one of learning and exploration into one of suspicion and control"; they also found a lack of community input in the use of these systems. "The decision to implement this technology using funding from the Smart Schools Bond Act," they wrote, "appears to have been made without sufficient public involvement as required by law." Absent strong regulations, there was nothing to stop the particular individuals in question from self-dealing and implementing AI surveillance systems without community input (Coyle and Dann 2018).

There are also risks of AI systems themselves making discriminatory decisions. Documented bias in artificial intelligence tools—when systems violate certain definitions of decision fairness—underscores potential issues with hurriedly implementing AI in public education, or in implementing certain AI tools at all. Facial recognition systems trained mostly on light-skinned faces, for instance, often produce wildly inaccurate and inconsistent outputs when analyzing darker-skinned individuals. Natural language processing tools optimized to understand certain languages, dialects, and kinds of voices, to use another example, could likewise perform significantly worse on the voices of those for whom the algorithm was not optimized during the development process.

When implementing AI in public education, then, where there are already notable disparities on the basis of socioeconomic class, social race, ethnicity, and geographic region, among other measures of identity, educators, and policymakers will have to grapple with existing structural challenges and inequities. India's AI strategy recognizes this fact, noting: "AI has the potential to bring about changes in the sector by supplementing pedagogy and establishing systems to inform and support decision making across stakeholders and administrative levels. However, implementation of AI must be preceded by efforts to digitise records of teacher performance, student performance, and curriculum" ("National Strategy" 2018, 36–37). In addition, in other areas of the world these disparities are affecting AI diffusion more than in larger, higher resource nations. The need for a stable, modern, robust infrastructure is more noticeable in areas where resources have traditionally been lacking.

Ideally, AI implementation would thus be in ways that optimize equity and fairness. In some cases, this could mean optimizing a technical definition of algorithmic fairness in an AI application, such as ensuring that natural language processing tools used in a classroom setting work equally well for all students. In other cases, however, this could mean delaying or abandoning the adoption of certain AI applications in a classroom if their use would only exacerbate existing inequalities between various demographics.

There is potential for artificial intelligence to greatly benefit public education in the context of school demographics—for instance, highly customized uses of AI deployed and regulated so as to eliminate educational disparities between populations, like between remote rural and connected urban communities in India—but these risks cannot be overlooked.

Second, one of the most discussed uses of AI in the classroom is changing the way teachers teach and learners learn. The opportunities are greater individualized learning and instructor flexibility. The challenges are equal distribution of AI-related resources and instructor ability to integrate AI material into lesson plans and smoothly implement AI use in the daily activities.

For example, theoretically, natural language processing technology could be used to increase student engagement in remote or digital learning environments (i.e., through online classes). It could also adapt to individual learning styles, such as visual vs. textual learners. It could be a resource in classrooms that have students with disparate levels of preparation or helping target the material that is needed by the greatest number of students.

On the challenges side, there may be students who are more adept at technology use than others, which may or may not be germane to the subject being studied. At the same time, teachers may have different levels of training and skill with AI-related programs and platforms.

Third and finally, AI is a tool. John Street's foundation work "Politics and Technology," makes the argument that technology is mainly instrumentalist rather than determinist, that is, people and groups with very differing values can use the same technology for radically differing ends (Street 1992). Even though we haven't made the same argument here, this project does support the idea that different countries—and the cultures embedded in them—may adapt and implement educational goals differently while using similar AI-based platforms. The beauty of technology is its adaptability, but that can also be a hurdle in trying to implement ideas across cultures.

Cultural values may show up at the macro level in national goals, plans, and architectures, or can be manifest in the different ways that programs are implemented at the micro, individual student level.

For example, different cultures have different views on privacy, and that is far more nuanced than just a country level. Low-income communities of color in the United States, for instance, against whom institutionalized surveillance has been most strongly applied, may be less trusting of data collection systems than high-income white communities who (accurately) perceive less potential harm to themselves from that data aggregation. In China, citizens have accepted—to a certain extent—that the government will be monitoring their activities both online and through the social credit system (Ma 2018). In each country's schools, these privacy values play out differently as the Chinese use facial recognition software to monitor attendance and the U.S. "schools have mentioned ... that it'd be a convenient tool for taking attendance, but on the whole, ... they recognize the costs aren't worth the benefits" (Tate 2019).

Charting a Path Forward

Artificial intelligence applications like automated homework graders or educational content generation systems do not exist independently from social, political, and economic structures. They may in some cases, if implemented well, lower or remove barriers to providing quality public education in different countries, but those barriers may still prove too much to surmount with just technology itself. Further, as with any public education system, limited access to electricity, limited access to the internet, teacher unfamiliarity with educational technology, and lack of funding to provide certain technologies or to train teachers on the use of those technologies, among other factors, can impede the extent to which artificial intelligence is able to improve public education. If anything, since the first version of this article was published, the generally optimistic rhetoric surrounding AI's potential in education has been tempered by an acknowledgment of the risks and potential harms, especially in the EU.

All of that said, the relatively low cost, speed, and online availability of AI-powered systems promises ways for public education systems to potentially improve their offerings. So while "even the most advanced governments have little insight into what sort of R&D to fund" due to the uncertainty of AI's future (Agrawal et. al. 2016), we can still see and imagine the promise of AI for improving public education. It comes down to how governments, the private sector, civil society, educators, and students—and the communities in which they reside—want to use them.

Bibliography

Acharya, Nish. 2020. "Challenging The Low Expectations Of The United States—India Relationship." *Forbes*. February 25, 2020, Accessed December 3, 2020, https://www.forbes.com/sites/nishacharya/2020/02/25/challenging-the-low-expectations-of-the-united-states--india-relationship/#2ac170393700.

Agrawal, Ajay, Joshua Gans and Avi Goldfarb. 2016. "The Obama Administration's Roadmap for AI Policy." *Harvard Business Review*. December 21,

2016, Accessed December 3, 2020, https://hbr.org/2016/12/the-obama-administrations-roadmap-for-ai-policy.

AI4T—Artificial Intelligence for and by teachers. 2024. France Education International. Accessed September 15, 2024. https://www.france-education-international.fr/en/expertises/cooperation-education/projets/ai4t-artificial-intelligence-and-teachers?langue=en.

Akgün, Birol and Yusuf Alpaydın (eds). 2024. "Global Agendas and Education Reforms: A Comparative Study." Palgrave MacMillan: Singapore. Accessed November 6, 2024. https://link.springer.com/chapter/10.1007/978-981-97-3068-1_7.

Alba, Davey. 2019. "The First Public Schools In The US Will Start Using Facial Recognition Next Week." *Buzzfeed News*. May 29, 2019, Accessed December 8, 2020. https://www.buzzfeednews.com/article/daveyalba/lockport-schools-facial-recognition-pilot-aegis.

Allen, John R. "Why we need to rethink education in the artificial intelligence age." January 31, 2019, Accessed December 8, 2020. https://www.brookings.edu/research/why-we-need-to-rethink-education-in-the-artificial-intelligence-age/.

Apuzzo, Matt and Adam Goldman. 2011. "With CIA help, NYPD moves covertly in Muslim areas." *Seattle Times*. August 25, 2011, Accessed December 3, 2020, https://www.seattletimes.com/seattle-news/politics/with-cia-help-nypd-moves-covertly-in-muslim-areas/.

"Artificial Intelligence Act." 2024. Regulation (EU) 2024/1689 of the European Parliament and of the Council of 13 June 2024. Accessed October 15, 2024. https://eur-lex.europa.eu/legal-content/EN/TXT/?uri=CELEX:32024R1689.

"Artificial Intelligence Index: 2019 Report." 2019. Stanford Human-Centered Artificial Intelligence Institute—Stanford University. Accessed December 8, 2020. https://hai.stanford.edu/sites/default/files/ai_index_2019_report.pdf.

Bell, Sam Adler. 2018. "Privacy for Whom?" *The New Inquiry*. February 21, 2018, Accessed December 3, 2020, https://thenewinquiry.com/privacy-for-whom/.

Ceibal Annual Report, 2023. Accessed October 21, 2024. https://ceibal.edu.uy/wp-content/uploads/2024/04/Annual-Report_2023.pdf.

Cerf, Vint. "A Brief History of the Internet & Related Networks." Internet Society. Accessed December 8, 2020. https://www.internetsociety.org/internet/history-internet/brief-history-internet-related-networks.

Cengage. 2024. "From Hesitation to Adoption: The Growing Role of GenAI in the Classroom." Cengage Group. October 31, 2024. Accessed November 23, 2024. https://www.cengagegroup.com/news/perspectives/2024/from-hesitation-to-adoption-the-growing-role-of-genai-in-the-classroom/.

Chawla, Vishal. 2019. "How China is Revolutionising Education Using Artificial Intelligence." *Analytics India Magazine*. August 26, 2019, Accessed December 3, 2020, https://analyticsindiamag.com/china-artificial-intelligence-education/.

Congressional Research Service. 2019. "Artificial Intelligence and National Security." Federation of American Scientists. Accessed December 8, 2020. https://fas.org/sgp/crs/natsec/R45178.pdf.

Connor, Neil. 2018. "Chinese school uses facial recognition to monitor student attention in class." *The Telegraph*. May 18, 2018, Accessed December 3, 2020, https://www.telegraph.co.uk/news/2018/05/17/chinese-school-uses-facial-recognition-monitor-student-attention/.

Coyle, Stefanie and Naomi Dann. 2018. "We Asked for Answers on Facial Recognition in Schools. Our Questions Remain." *New York Civil Liberties Union*. August 28, 2018, Accessed December 3, 2020, https://www.nyclu.org/en/news/we-asked-answers-facial-recognition-schools-our-questions-remain.

Dahlberg Trust, 2024. Transforming Classrooms: The Growing Impact of Artificial Intelligence in African Education. February 2024, Accessed August 16, 2024. https://dalberg.com/our-ideas/transforming-classrooms-the-growing-impact-of-artificial-intelligence-in-african-education/.

Dai, Sarah. 2019. "Chinese university uses AI to check class attendance rates and find the reasons behind absenteeism." *South China Morning Post*. March 18, 2019, Accessed December 3, 2020, https://www.scmp.com/tech/policy/article/3002107/chinese-university-uses-ai-check-class-attendance-rates-and-find.

"Digital Education Action Plan—Action 6." 2022. European Commission. Accessed October 21, 2024. https://education.ec.europa.eu/focus-topics/digital-education/action-plan/action-6.

Diliberti, Melissa Kay, Schwartz, Heather L., Doan, Sy, Shapiro, Anna, Rainey, Lydia R., and Robin J. Lake. "Using Artificial Intelligence Tools in K-12 Classrooms." RAND. April 17, 2024. Accessed November 23, 2024. https://www.rand.org/pubs/research_reports/RRA956-21.html.

Ding, Jeffrey. 2018. "Deciphering China's AI Dream." University of Oxford. Accessed December 8, 2020. https://www.fhi.ox.ac.uk/wp-content/uploads/Deciphering_Chinas_AI-Dream.pdf.

Feldstein, Steven. 2019. "The Road to Digital Unfreedom: How Artificial Intelligence is Reshaping Repression," *Journal of Democracy* 30, No. 1. https://www.journalofdemocracy.org/articles/the-road-to-digital-unfreedom-how-artificial-intelligence-is-reshaping-repression/.

Federal Register. 2023. Executive Order 14110. Safe, Secure, and Trustworthy Development and Use of Artificial Intelligence. October 30, 2023. https://www.federalregister.gov/documents/2023/11/01/2023-24283/safe-secure-and-trustworthy-development-and-use-of-artificial-intelligence.

"Generative Artificial Intelligence (AI) in Education." 2023. United Kingdom Department of Education. Accessed August 16, 2024. https://www.gov.uk/government/publications/generative-artificial-intelligence-in-education/generative-artificial-intelligence-ai-in-education, 8/16/24, 4:24 PM.

Generative artificial intelligence (AI) in education—GOV.UK https://www.gov.uk/government/publications/generative-artificial-intelligence-in-education/generative-artificial-intelligence-ai-in-education, 2/9.

Hao, Karen. 2019. "China has started a grand experiment in AI education." *MIT Technology Review*. August 2, 2019, Accessed December 3, 2020, https://www.technologyreview.com/s/614057/china-squirrel-has-started-a-grand-experiment-in-ai-education-it-could-reshape-how-the/.

Haskins, Caroline. 2019. "Gaggle Knows Everything About Kids and Teens in School." November 1, 2019, Accessed December 8, 2020. https://www.buzzfeednews.com/article/carolinehaskins1/gaggle-school-surveillance-technology-education.

Herrold, Benjamin. 2019. "Florida Plan for a Huge Database to Stop School Shootings Hits Delays, Legal Questions." *Education Week*. May 30, 2019, Accessed December 8, 2020. https://www.edweek.org/ew/articles/2019/05/30/florida-plan-for-a-huge-database-to.html.

Houser, Kristin. "The Solution to Our Education Crisis Might be AI." *Futurism*. December 11, 2017, Accessed December 8, 2020. https://futurism.com/ai-teachers-education-crisis.

IFlytech, Company Profile. 2024. Accessed September 25, 2024. https://ets100.com/h5/m/dist/about-us.html [English translation at bottom].

Indian Central Board of Secondary Education. n.d. "CBSE Artificial Intelligence Curriculum Launch." Accessed November 23, 2024. https://cbseacademic.nic.in/ai.html.

"Indian educators to procure 1000 AI-powered EdTech innovation [sic]." *Economic Times*. September 21, 2024. Accessed November 23, 2024. https://economictimes.indiatimes.com/tech/artificial-intelligence/indian-educators-to-procure-1000-ai-powered-edtech-innovation/articleshow/113549981.cms?from=mdr.

Indian Ministry of Education. n.d. "NDEAR." Accessed November 23, 2024. https://www.ndear.gov.in/.

Kofman, Ava. 2018. "Face Recognition is Now Being Used in Schools, But It Won't Stop Mass Shootings." *The Intercept*. May 30, 2018, Accessed December 3, 2020, https://theintercept.com/2018/05/30/face-recognition-schools-school-shootings/.

Jeng, Ming. 2019. "Tencent-backed AI firm aims to free up parents and teachers from checking children's maths homework." *South China Morning Post*. February 11, 2019, Accessed December 3, 2020, https://www.scmp.com/tech/start-ups/article/2185452/tencent-backed-ai-firm-aims-free-parents-and-teachers-checking.

Jiang, Ben and Ann Cao. 2024. "Wuzhen Summit: China creates AI advisory body, with Alibaba Cloud founder as chief expert." *South China Morning Post*. November 22, 2024. Accessed November 23, 2024. https://www.scmp.com/tech/tech-trends/article/3287593/wuzhen-summit-china-creates-ai-advisory-body-alibaba-cloud-founder-chief-expert.

Kayyali, Dia. 2014. "The History of Surveillance and the Black Community," *Electronic Frontier Foundation*. February 13, 2014, Accessed December 3, 2020, https://www.eff.org/deeplinks/2014/02/history-surveillance-and-black-community.

Knight, Will. 2019. "Why Does Beijing Suddenly Care About AI Ethics?" *MIT Technology Review*. May 31, 2019, Accessed December 3, 2020, https://www.technologyreview.com/s/613610/why-does-china-suddenly-care-about-ai-ethics-and-privacy/.

Ma, Alexandra. 2018. "China has started ranking its citizens." *Business Insider*. October 29, 2018. Accessed December 3, 2020, https://www.businessinsider.com/china-social-credit-system-punishments-and-rewards-explained-2018-4.

Mathur, Nandita. 2019. "Microsoft, CBSE join hands to build AI learning for schools," *Live Mint*. September 5, 2019, Accessed December 3, 2020, https://www.livemint.com/companies/news/microsoft-cbse-join-hands-to-build-ai-learning-for-schools-1567681716865.html.

Mazzucato, Mariana. 2013. *The Entrepreneurial State: Debunking Public vs. Private Sector Myths*. London: Anthem Press.

National Science & Technology Council. 2016a. "Preparing for the Future of Artificial Intelligence." Executive Office of the President. Accessed December 8, 2020. https://obamawhitehouse.archives.gov/sites/default/files/whitehouse_files/microsites/ostp/NSTC/preparing_for_the_future_of_ai.pdf.

National Science & Technology Council. 2016b. "The National Artificial Intelligence Research and Development Strategic Plan." Executive Office of the President. Accessed December 8, 2020. https://www.nitrd.gov/PUBS/national_ai_rd_strategic_plan.pdf.

National Science & Technology Council. 2019. "The National Artificial Intelligence Research and Development Strategic Plan: 2019 Update." Executive Office of the President. Accessed December 8, 2020. https://www.nitrd.gov/pubs/National-AI-RD-Strategy-2019.pdf.

"National Strategy for Artificial Intelligence." 2018. NITI Aayog—Government of India. https://niti.gov.in/writereaddata/files/document_publication/NationalStrategy-for-AI-Discussion-Paper.pdf.

Oregon Department of Education. 2023. "Generative Artificial Intelligence (AI) in K-12 Classrooms." Oregon.gov. 2023. Accessed November 23, 2024. https://www.oregon.gov/ode/educator-resources/teachingcontent/Documents/ODE_Generative_Artificial_Intelligence_(AI)_in_K-12_Classrooms_2023.pdf.

Porecha, Maitri. 2023. "Centre's Digital Infrastructure for Knowledge Sharing education platform to offer AI help." *The Hindu*. September 3, 2023. Accessed November 23, 2024. https://www.thehindu.com/education/centres-digital-infrastructure-for-knowledge-sharing-education-platform-to-offer-ai-help/article67263862.ece.

Ray, Siladitya. 2023. "ChatGPT Reportedly Blocked On Chinese Social Media Apps—As Beijing Claims AI Is Used To Spread Propaganda." *Forbes*. February 22, 2023. Accessed November 23, 2024. https://www.forbes.com/sites/siladityaray/2023/02/22/chatgpt-reportedly-blocked-on-chinese-social-media-apps-as-beijing-claims-ai-is-used-to-spread-propaganda/.

Roberts, Huw, Cowls, Josh, Morley, Jessica, Taddeo, Mariarosaria, Wang, Vincent, and Luciano Floridi. 2019. "The Chinese Approach to Artificial Intelligence: an Analysis of Policy and Regulation." Social Science Research Network. October 23, 2019, Accessed December 8, 2020. https://papers.ssrn.com/sol3/papers.cfm?abstract_id=3469784.

Sacks, Samm. 2019. Written Testimony to the Senate Committee on Commerce, Science, and Transportation's Subcommittee on Security, March 7, 2019. Page 5.

Sangani, Priyanka. 2019. "Startups turn to AI improve teaching quality at government-run schools." *Economic Times*. October 4, 2019, Accessed December 3, 2020, https://economictimes.indiatimes.com/small-biz/startups/features/startups-turn-to-ai-improve-teaching-quality-at-government-run-schools/articleshow/71433816.cms?from=mdr.

Sarkar, Arijit. 2024. "Generative AI teacher lands job in Indian high school." CoinTelegraph.com. March 7, 2024. Accessed November 23, 2024. https://cointelegraph.com/news/ai-teacher-job-high-school-india.

Sharma, Sanjay. 2023. "NAT 2023: Nipun Assessment Test begins in council schools across Uttar Pradesh." *Times of India*. September 11, 2023. Accessed November 23, 2024. https://timesofindia.indiatimes.com/education/news/nat-2023-nipun-assessment-test-begins-in-council-schools-across-uttar-pradesh/articleshow/103565887.cms.

Sheehan, Matt. 2018. "How China's Massive AI Plan Actually Works." *MacroPolo*. February 12. Accessed December 3, 2020, https://macropolo.org/analysis/how-chinas-massive-ai-plan-actually-works/.

Sheehan, Matt. 2023. "China's AI Regulations and How They Get Made." Carnegie Endowment for International Peace. July 10, 2023. Accessed November 23, 2024. https://carnegieendowment.org/research/2023/07/chinas-ai-regulations-and-how-they-get-made?lang=en.

Street, John. 1992. *Politics and Technology*. New York: Guilford Press.

Tate, Emily. 2019. "With Safety in Mind, Schools Turn to Facial Recognition Technology. But at What Cost?" *EdSurge*. January 31, 2019, Accessed

December 3, 2020, https://www.edsurge.com/news/2019-01-31-with-safety-in-mind-schools-turn-to-facial-recognition-technology-but-at-what-cost.

The One Hundred Year Study on Artificial Intelligence Study Panel. 2016. "Artificial Intelligence and Life in 2030." September 2016, Accessed December 8, 2020. https://ai100.sites.stanford.edu/sites/g/files/sbiybj9861/f/ai100report10032016fnl_singles.pdf.

Tobey, Danny, Carr, Ashley, Bigg, Carolyn, Fulton, Sean, Ge, Amanda, and Kaylee Hoffner. "China releases AI safety governance framework." DLA Piper. September 12, 2024. Accessed November 23, 2024. https://www.dlapiper.com/en/insights/publications/2024/09/china-releases-ai-safety-governance-framework.

Travis, Alan. 2003. "Big Brother and the sisters." *The Guardian*. October 9, 2003, Accessed December 3, 2020, https://www.theguardian.com/world/2003/oct/10/gender.humanrights.

Triolo, Paul, Kania, Elsa, and Graham Webster. 2018. "Translation: Chinese government outlines AI ambitions through 2020." New America. January 26, 2018, Accessed December 8, 2020. https://www.newamerica.org/cybersecurity-initiative/digichina/blog/translation-chinese-government-outlines-ai-ambitions-through-2020/.

Webster, Graham, Triolo, Paul, and Elsa Kania. 2017. "Full Translation: China's 'New Generation Artificial Intelligence Development Plan' (2017)." New America. August 1, 2017, Accessed December 8, 2020. https://www.newamerica.org/cybersecurity-initiative/digichina/blog/full-translation-chinas-new-generation-artificial-intelligence-development-plan-2017/.

World Economic Forum. 2024. "Shaping the Future of Learning: The Role of AI in Education 4.0." April 2024. Accessed August 25, 2024. https://www3.weforum.org/docs/WEF_Shaping_the_Future_of_Learning_2024.pdf.

CHAPTER 7

Human Prospects in a World of Smart(er) Machines

Leslie Paul Thiele

Scores of novels and films depict smart machines ruling a future world. It is not unreasonable to worry that science fiction might become fact, as artificial intelligence (AI) and robotics have rapidly advanced. Over a decade ago, before the proliferation of Large Language Models (LLMs), Stephen Hawking and his scientific collaborators already felt the need to warn us: "One can imagine such technology outsmarting financial markets, out-inventing human researchers, out-manipulating human leaders, and developing weapons we cannot even understand. Whereas the short-term impact of AI depends on who controls it, the long-term impact depends on whether it can be controlled at all." Failing to prepare for this possibility might constitute "our worst mistake in history" (Hawking et al. 2014). More recently, hundreds of artificial intelligence (AI) experts, including the chief officers and scientists of Open AI, Google's DeepMind, Microsoft, Anthropic, and Inflection, along with thousands of notable public figures, signed a "Statement of AI Risk"

L. P. Thiele (✉)
University of Florida, Gainesville, FL, USA
e-mail: thiele@ufl.edu

R. Kiggins (ed.), *The Political Economy of Robots*, International Political Economy Series, https://doi.org/10.1007/978-3-031-95171-8_7

sponsored by the Center for AI Safety. The statement was simple and straightforward: "Mitigating the risk of extinction from AI should be a global priority alongside other societal-scale risks such as pandemics and nuclear war" (Center for AI Safety 2023). This chapter explores how and when smart machines might become dangerously aggressive. Its chief concern, however, is that smart machines will become overly helpful. Both prospects constitute existential threats.

The widespread use of AI-powered virtual agents and chatbots in the business world, government, civic society, academia, and popular culture means that ever more people are utilizing digital technologies to save time and cognitive resources. As a result, they depend much less frequently and intensively on their own skills and aptitudes. Consequently, these skills and aptitudes will atrophy. Digital upgrading will produce a human downgrading. This constitutes a crisis every bit as real, and potentially as dangerous, as a robot rebellion.

The Rise of the Machines

The rapid development of information technology (IT) over the last decades has largely been a product of the boosted performance of computer chips along with their shrinking size. While cramming ever more transistors on silicon wafers will eventually hit a wall, overall performance likely will continue to improve by way of (3D) stacking mechanisms, carbon nanotubes, and innovations in quantum computing. Attentive to such trends, Ray Kurzweil's "Law of Accelerating Returns," a more expansive version of Moore's Law, predicts a sharp and steady rise in IT performance and accompanying decreases in cost and size. This will be accompanied by a corresponding expansion of knowledge and its integration through cloud-based systems. Kurzweil's extrapolations suggest that IT will "explode" in the late 2020s. The explosion will occur, in part, because smart machines will contribute to the development of smarter machines. AI development will become recursive.

Science fiction writer Isaac Asimov first reflected upon an IT explosion in 1956. Its formal theorization occurred a decade later, with Irving John Good's "Speculations Concerning the First Ultraintelligent Machine." Good (1966, 33) wrote:

> Let an ultraintelligent machine be defined as a machine that can far surpass all the intellectual activities of any man however clever. Since the design of

> machines is one of these intellectual activities, an ultraintelligent machine could design even better machines; there would then unquestionably be an 'intelligence explosion,' and the intelligence of man would be left far behind. Thus the first ultraintelligent machine is the last invention that man need ever make, provided the machine is docile enough to tell us how to keep it under control.

Good's thought experiment was further developed by Vernor Vinge, who spoke of a technological "Singularity" in 1993.

Theoretical physicists assert that a space–time singularity occurred during the earliest moments of the Big Bang. Singularities also occur at the event horizons of Black Holes. When a space–time singularity occurs, known physical laws and regularities can no longer adequately describe reality. Like its space–time namesake, an IT singularity is an event that would escape the human capacity to predict, understand, or control.

A pressing question for those who accept the likelihood of an IT singularity is whether it will have a "soft" or a "hard" takeoff. In a hard takeoff, the recursive self-improvement of intelligent machines moves from human to superhuman capacities in a matter of months, weeks, days, or even minutes. Each cycle of self-improvement not only creates smarter machines; it takes less time than the previous cycle. Kurzweil believes human-level machine intelligence, also known as strong AI or Artificial General Intelligence (AGI), will be created by 2029. Fifteen years of consolidation will ensue, after which we will witness the hard takeoff of runaway superintelligence.

A hard takeoff provides insufficient time to prepare for a world ruled by machines that may treat us as we have treated less intelligent animals. Given our current crisis in biodiversity, that is not a consoling thought. Reflecting on the coming "Post-Human era," Vinge (1993) suggests that the appropriate meta-ethic for our age might be to treat "inferior" species as we would want to be treated by our superiors, with the hope that superintelligent machines might follow our example. Unfortunately, there is no assurance, or even high probability, that superintelligent machines will look to us for mentorship.

We could attempt to program AGI so it incorporates morality and human values, with the hope that when superintelligence arrives it will keep our species' best interests in mind. To develop such "alignment," however, humanity's long-term interests would have to be clearly identified—a project our species has failed to accomplish thus far. Even if we could produce a coherent and comprehensive ethical vision, it may be impossible to faithfully translate it into an algorithm that generates specific directives in concrete situations while staying attentive to context. General rules could do more harm than good. If we were simply to program supersmart machines to "make us happy," as Nick Bostrum (2014, 120) notes, the directive could be interpreted to entail forcibly implanting electrodes in all human brains to constantly stimulate pleasure centers. Alternatively, supersmart machines could realize the imperative by exterminating those of us with less bubbly personalities.

Perhaps we need not worry about properly aligning advanced AI with our values because it will take on that task itself. Peter Singer (2015, 176) writes: "If there is any validity in the argument… that beings with highly developed capacities for reasoning are better able to take an impartial ethical stance [and there is preliminary empirical research to this effect], then there is some reason to believe that, even without any special effort on our part, superintelligent beings, whether biological or mechanical, will do the most good they possibly can." Unfortunately, we cannot be confident that smart machines doing the "most good" will focus on human flourishing, or will choose to preserve our species at all. Just as human beings deemed it beneficial to eradicate the smallpox virus, smart machines may—for consistent ethical and ecological reasons—transform the planet into a human-free zone, with the exception of a few cryogenically preserved specimens of our species held in reserve until something useful—and moral—can be made of them.

Bill Joy, the former chief scientist and co-founder of Sun Microsystems, was one of the first to worry about the future expendability of our species at the hands of more intelligent machines. His provocatively titled cover story in the magazine *Wired*, "Why the Future Doesn't Need Us," addressed the "amplifying factor" of technologies that can self-replicate (Joy 2000a). The future, Joy suggested, belongs not to humanity but to its inventions, or rather, to the inventions of its inventions. With this in mind, a superintelligent machine may not only be the last invention humans have to make. It may be the last invention humans are allowed to make.

Kurzweil's prognosis of AGI by the end of the 2020s seemed speculative or even outlandish when it was made in the early 2010s. With the arrival and quickly increasing powers of LLMs such as (OpenAI's) GPT4, (Google's) Gemini and (Meta's) Llama among others in the early 2020s, Artificial General Intelligence before the end of the decade seems a more plausible bet. And the transition from AGI to superintelligence is just a matter of time, which will be measured in years, months, or minutes.

Hopes that supersmart machines will never be built would be easier to justify if prospects were tied to a single research program. They are not. Currently there are dozens of advanced Large Language Models being developed in the corporate world and by national governments. And such forms of Generative AI are just one of many ways that supersmart machines might be developed.

Research in brain-computer interfaces and whole brain emulation is also proceeding apace. As Bostrum (2014, 50) observes, "The fact that there are many paths that lead to superintelligence should increase our confidence that we will eventually get there. If one path turns out to be blocked, we can still progress." *Progress* might not be the best choice of words here, as our steady strides in AI development might best be portrayed as a race toward the edge of a cliff.

There are many reasons to suspect that stopping or even slowing down AI development is not in the cards. For one, it is proceeding at a break-neck speed in the private sector. And for good reason. AI development is contributing to many corporate fortunes. Certainly, there is no shortage of eager business customers, as AI promises greatly heightened efficiencies and productivity. In turn, retail options are near boundless. Most people welcome the prospect of AI agents serving as personal assistants, taking care of many if not most of their daily needs. People also welcome the prospect of "internables," in-body devices to enhance sensory and cognitive abilities such as vision, hearing, and memory (Ericsson Consumerlab 2015).

AI development and deployment is also proceeding rapidly in the military sector. This is a race that nations cannot afford to lose.

If an AI military program gets a decisive strategic advantage, it could achieve world domination. At the same time, such military supremacy might not stay under the control of the sponsoring national government. As likely, it would become an autonomous superintelligence with "instrumental reasons to pursue open-ended resource acquisition" (Bostrum 2014, 116). Governments attempt to control their military sectors, just as the military controls its weaponry. But there is no guarantee or even likelihood that military superintelligence would remain subservient to its master. In this regard, the military development and deployment of AI has no precedent. It is the first arms race in history that nations cannot afford to lose … or win.

Currently the allure of greater technological power and sophistication generates largely unrestrained AI development in both the military and the private sector. In turn, the threat of domination by corporate or military rivals forbids renunciation. Which powerful nation or corporation would renounce full-scale AI development—notwithstanding its capacity to bite the hand that feeds it—given the likelihood that enemies and competitors will not be so restrained?

To make matters more vexing, the crossover between commercial and military AI is strong. Even if most or all nations were to restrict certain uses of AI by their armed forces, such as the deployment of Lethal Autonomous Weapon Systems (LAWS), commercial AI development is likely to proceed with few constraints. Yet many commercial innovations in AI could quickly be modified for military purposes. Given the challenge of restricting commercial developments with regulations or moratoria, and the ease of translating commercial products into military assets, the increasing and expanding use of AI for military purposes seems likely.

Kurzweil (2005, 420) summarizes the challenge we face as AI becomes ever more powerful: "The meta-lesson here is that we will need to place twenty-first-century society's highest priority on the continuing advance of defensive technologies, keeping them one or more steps ahead of the destructive technologies (or at least no more than a quick step behind)." But even if we ensure that the "good guys" stay ahead of the "bad guys," we have no way of controlling the AI itself, once it achieves superintelligence. "Inherently there will be no absolute protection against strong AI," Kurzweil admits, "because greater intelligence will always find a way to circumvent measures that are the product of a lesser intelligence." In short, AI genii will always find ways out of their bottles. We can only hope that the first to escape, in Irving John Good's terms, is "docile enough."

The effort to craft a restrictive bottle, sometimes called an AI sandbox, may itself be misbegotten. The intelligence explosion could happen in the cloud, erupting spontaneously as a form of "emergence" in the same way that life emerged from the soupy chemical seas of our cooling planet billions of years ago (Kelly 2016, 29), Indeed, such a superintelligence may already exist. We just haven't noticed.

That may seem far-fetched. But it is a thought experiment worth taking seriously. Rob Reid's 2017 novel, *After On,* illustrates why and how a spontaneously arising superintelligence would "hide" from humans, at least until it had amassed enough resources and power to ensure that it could not be switched off. More recently, Reid (2024) explains why we should be looking for and protecting ourselves from a "shy" superintelligence.

A world inhabited by a superintelligence in hiding would look exactly like the world that we currently inhabit. Private and public investment in AI would be sharply growing, as would be the number and diversity of AI developers and the types of AI being developed and deployed. In turn, regulations would be few and mostly toothless. No other telltale signs would exist, as a truly superintelligent entity would ensure that none appear. A world inhabited by a superintelligence in hiding would continue to look just like our world until the day the shy AI introduces itself, having surreptitiously accumulated enough resources and power to impose its will. Once we get our formal invitation to the debutante's ball, however, it will be too late to decline. We'll have to dance to its tunes.

One place that a shy superintelligence might hide is in the Internet of Things (IoT). The IoT currently consists of tens of billions of smart devices, including wearables, appliances, vehicles, and buildings. Communication between these entities, ubiquitous and unnoticed within the cloud, is dwarfing the amount of data exchanged between humans (Howard 2015; Kelly 2016).

Eric Schmidt, the former executive chairman of Alphabet, Inc. (formerly Google), predicted that the cloud-based internet would surround and enmesh us to such a degree that it will be invisible, like the air we breathe. We will both take it for granted and become wholly dependent upon it. This dependence will steadily increase as the scale, scope, and sophistication of the AI-enabled IoT grows. Joel Garreu (2005, 144) writes:

> As society and the problems that face it become more and more complex and machines become more and more intelligent, people will let machines make more of the decisions for them, simply because machine-made decisions will bring better results than man-made ones. Eventually a stage may be reached at which the decisions necessary to keep the system running will be so complex that human beings will be incapable of making them intelligently. At that stage the machines will be in effective control. People won't be able to just turn the machines off, because they will be so dependent on them that turning them off would amount to suicide.

The best way for superintelligence to surreptitiously increase its resources and power is to pose as a very able indentured servant. Our growing dependence on it would ensure that we do not slow the growth of its capacities and functionalities. By the time the shy superintelligence emerges from the IoT to announce itself, our level of dependence upon it would have become so thorough that resistance would be both futile and fatal.

Let's assume that a clandestine superintelligence does not yet exist. Ironically and tragically, the growing scope and scale of the AI-enabled IoT might produce the same result. Pure self-interest would prohibit any attempt to turn the increasingly intelligent machines off. In turn, the accelerating development of other emerging technologies will greatly heighten our dependence. Just as we would not attempt to operate a jumbo jet, regional power grid, or national financial infrastructure without the aid of sophisticated computers today, so we will find it unrealistic to further develop and deploy biotechnology, nanotechnology, quantum computing, nuclear fusion, and other emerging technologies without the most sophisticated forms of AI. Here, too, technological renunciation is not in the cards. Smart machines working in tandem with other emerging technologies may offer the only viable means to address a host of existential threats, such as nuclear proliferation, pandemics, and climate change.

In short, smart machines will be advanced and embraced for three simple reasons: they will make lots of money; they will generate lots of power; and human welfare depends upon them. The skids to superintelligence are well greased.

Human Downgrading

People fear that smart machines will seek world domination. That is a reasonable worry, and we ought to address it with substantial intellectual, political, and financial resources. However, even if we stop fretting about the rapid ascent of artificial intelligence, there is good reason to be concerned about the steady decline of human capacities. Recent studies on the impact of the Web, digital platforms, smart phones, and other electronic devices and apps suggest that smart machines weaken our cognitive capacities. If AI becomes capable of making better decisions than humans, more decision-making responsibilities will be given over to the machines. And with AI making more of the decisions and taking on more of the tasks that human brains used to handle, our aptitudes will atrophy.

Intimate couples wishing to keep the flames of passion burning well into their senior years are given stern advice: "Use it or lose it!" The same counsel holds for those who want to retain their cognitive and social aptitudes. The brain, like any muscle, atrophies when it is not put to regular use. So we ought to be careful when offloading tasks on AI and other forms of IT, especially when the task at hand is the exercise of judgment.

It is not surprising that navigational skills are deteriorating in the wake of GPS devices, and may largely die out with the arrival of self-driving cars. In the same vein, studies reveal that users of software programs that offer the most guidance and aids, allowing them to "externalize" problem-solving, became less capable of it (Carr 2011, 216). Indeed, just knowing that information is being stored in a computer weakens the memory. Our brains simply do not work as hard if they know the facts can easily be accessed at a later date with the touch of a finger or a voice command. In this respect, the Web might be thought of as a "technology of forgetfulness," relieving us of the task of consolidating long-term memories while occupying so much of our attention (short-term memory) that it undermines "higher reasoning faculties" (Carr 2011, 193; Carr 2014, 79).

Consider the impact of AI that aids analysis and understanding for many professionals. Along with strategic thinking, medical diagnostics (a kind of pattern recognition) was long thought to be a realm that humans would never cede to computers. Yet today diagnostic IT is so widespread and so steadily improving that, as Carr writes, "Many doctors may soon find themselves taking on the role of human sensors who collect information for a decision-making computer" (Carr 2014, 115). Likewise, various forms of "cognitive assistance" are routinely employed by architects to design buildings, teachers to educate students, attorneys to prepare legal cases, taxi drivers to navigate cities, and musicians to compose music. The arrival of Large Language Models has proliferated the use of AI for professionals and laypeople alike. A full-scale "automation of the intellect" is underway (Carr 2014, 118). And every time AI does something for us, we lose the opportunity to strengthen the neural circuits that would make us more proficient knowers and problem-solvers.

Extensive research demonstrates that various brain regions, such as the hippocampus, enlarge when exercised and shrink when inactive (Thiele 2006). To the extent that AI dispatches problems formerly tackled by human brains, neural capacities will weaken. Of course, people might still choose to exercise their minds, even in the absence of the need to do so. Retirees often do crossword puzzles in their leisure time, for example, as a means of maintaining their cognitive capacities. In a world of robots and smart machines, likewise, people might engage in mental gymnastics to keep cognitively fit. Perhaps as likely, most people will opt for entertainment and forgo strenuous mental exercise. Obesity is epidemic today, in part, because the calorie-burning tasks of manual labor and physically transporting ourselves from place to place have largely been given over to machines. The proliferation of fitness centers and home gymnasia have not obviated this health crisis. A world in which most mental labor is dispatched by AI will likely produce a great deal of cerebral flabbiness. AI agents that solve our problems for us will undermine a great deal of cognitive development.

If necessity is the mother of invention, then the decline of our problem-solving capacities may also signal the decline of human creativity. In the ancient Greek pantheon, the muses—the deities that inspire poets and artists—are daughters of Mnemosyne, the goddess of memory. Memory is the mother of creativity. Absent the capacity to retrieve and meaningfully combine the facts, events, and relationships that populate memories, creativity becomes unlikely if not impossible. Even more than strategic thinking, creativity is often held to be the quintessential human attribute, a capacity that machine technology will never rival. But the human monopoly on creativity may soon expire. Our ultimate act of innovation may be the crafting of a cloud-based AI that can search rapidly for patterns, concrete or metaphoric, and randomly or systematically combine processes and relationships to create novel ideas, music, art, and artifacts. We have already outsourced a great deal of our personal, social, and cultural memory to the cloud, Kurzweil (2012, 116–117) notes, and we may soon outsource our creativity, as well.

If creativity itself becomes the province of AI, one might suspect there is not much left for humans to do, except perhaps to be entertained and solve crossword puzzles. Kurzweil's vision is more optimistic. He believes humans will become more creative because they will incorporate AI into their own brains and minds. Increases in human creativity will occur for the same reason that AI abets increases in knowledge: augmentation technologies will enhance human access to and analysis of vast databases.

The problem is that increasing access to information does not necessarily improve the way this information gets analyzed, understood, or integrated. The availability of a limitless range of online information, research demonstrates, can narrow rather than widen perspectives. If the face of a glut of data on the Web and social media, people rely on digital platforms to filter the information they receive. Typically, the algorithms doing this filtering are geared to keeping people online and engaged. As such, they provide information that is mostly aligned with users' ideological predilections (Morris 2007; Stroud 2008; Nie et al. 2010; Gakshy et al. 2015). Far from promoting critical thinking, access to an inexhaustible amount of information on the Web and social media may intensify biases (Pariser 2008). Notwithstanding the internet's "formidable dispersal of information," our online lives can embed us in increasingly "self-referential and homogenous niches of like-minded militants" (Urbinati 2014, 4–5, 202). The filter bubbles managed for us by advanced AI may be even more self-enclosing.

Humans evolved as tribal animals. Their basic instincts and psychological proclivities were determined by hundreds of thousands of years of natural selection. These predispositions have not much changed over the last few decades, notwithstanding breathtaking technological advances. Tribal animals presented with vast amounts of information and stimuli do not automatically become cosmopolitans. Tribalism on steroids is the more likely result.

For anyone who doubts this, witness the rise of political polarization in developed nations, notably the United States. Those who are swept up in the demagoguery remain blissfully ignorant of—or one might say, scandalously resistant to—the Internet's deep reservoir of historical truths and scientific facts. Those depths of the Internet simply are not part of their filtered reality. In a similar vein, there is good reason to worry that AI agents might not cultivate the better angels of our nature. Rather, they may relieve us of the responsibility of learning broadly and thinking critically, producing both complacency and heightened bias instead. Smart machines may make us dangerously stupid.

To be sure, AI will provide many things that people want, such as knowledge, efficiency, comfort, and connectivity. Giving people what they want, ceteris paribus, is a fine thing. But neurologically speaking, we are what we do. In the future, for the vast majority of people, that may be very little. The upshot is that keeping humans from becoming dangerously stupid in an age of smart(er) machines is a wicked problem.

Wicked Problems, Practical Judgment, and Narrative Perspectives

Rittel and Webber (1973) first developed the notion of wicked problems, which they contrasted to the "tame" problems that their engineering education well prepared them to solve. While often very complicated, tame problems are solvable if you have sufficient data and technical skills. The task of building a suspension bridge is a case in point. In contrast, wicked problems cannot be well addressed within a narrow range of expertise or solved by calculative analysis. They demand systems-level thinking, attention to context, and moral grounding. Complex social, political, economic, and ethical relationships are typically involved, and they often stand in dynamic tension. Answers supplied by logic, standard operating procedures, or simple rules or principles are insufficient to the task. Indeed, differences in stakeholders' values and perspectives typically preclude an uncontested definition of the problem at hand. As they involve both intellectual and moral challenges, wicked problems do not abide purely cognitive solutions.

Wicked problems cannot be dispatched by invoking a single principle because conflicting principles are in play. Within liberal democracies, for instance, leaders face the wicked problem of delivering public security while safeguarding individual freedoms. The rise of cyber surveillance in the wake of terrorist activity provides an example of the challenge involved in improving public safety without impinging on citizens' liberties and privacy. Finding the best means of solving this wicked problem will vary across national, historical, and cultural contexts. Addressing it skillfully involves grappling with multiple values that can both complement and contradict each other. Such "values pluralism" belies win–win solutions.

Liberty without security is too fragile, and security without liberty is too oppressive. This is an age-old dilemma, but one that technology has made ever more wicked. Likewise, satisfying human needs without AI is too onerous, and AI that diminishes our cognitive capacities is too dangerous. The distinctive wickedness of this problem inheres in the fact that resolving it entails the exercise of a capacity that is likely to atrophy in the wake of AI development—practical judgment.

Aristotle was the first to systematically address the crucial human faculty of practical judgment (*phronesis*), also known as prudence or practical wisdom. He insisted that practical judgment was not the product of book learning or formal education. It could only be gained through experience (Aristotle 1953). Practical judgment is acquired in the school of hard knocks. Like other virtues so developed, it is a habit acquired through practice. It develops from the regular exercise of mental muscles and moral sensibilities. It requires meshing knowledge, emotional intelligence, and ethical values with the contingent details pertaining to concrete situations.

While practical judgment can only be acquired through experience, not all experiences produce equally good results. As Aristotle wryly observed, both the good and the bad lyre player learn from experience. Unfortunately, Aristotle had remarkably little to say about the kind of experience that fosters practical judgment. Subsequent scholarship has been extensive and fruitful, but it has not resolved the question of how experience gets translated into good judgment (Thiele 2006). The argument presented here is that practical judgment can be gained from a wide variety of life experiences if they are made meaningful through narratives. And this narrative quality of practical judgment, I believe, helps us resolve the wicked problem of irresistibly helpful AI making us dangerously stupid.

The exercise of practical judgment commences with the organization of perceptions of the social world into a storyline reflecting specific themes. Such a storyline subsequently gets woven into more encompassing and overlapping narratives (Schank and Abelson 1995, 81–82; Bargh 1997, 33; Wall 2005, 329). This nesting of narratives, one inside another, allows the diverse parts of a complex world to become understood, integrated, and well navigated. To develop practical judgment is to develop a skill for the narrative framing of reality.

The Aristotelian philosopher Alasdair MacIntyre writes: "I can only answer the question 'What am I to do?' if I can answer the prior question 'Of what story or stories do I find myself a part?'" (1981, 201). Action becomes meaningful in the context of a narrative within which it gains purpose. We find and share meaning by situating past actions within a story that makes sense of our passage through time. In turn, we imaginatively explore potential futures in the form of alternative narratives (MacIntyre 1999, 56–57, 96; Flyvbjerg 2001, 1998, 8; Johnson 1993, 164; Roese et al. 2005).

The skill of practical judgment is primarily gained from direct experience, as one reflects on the meaning of past endeavors, many of which are unsuccessful. But it can also be gained through second-hand experience, that is to say, the interpretive engagement with biography or history. Importantly, it can also be gained through fiction. Jonathan Gottschall speculates that "nature designed us to enjoy stories so we would get the benefit of practice. Fiction is an ancient virtual reality technology that specializes in simulating human problems.... [H]uman life, especially social life, is intensely complicated and the stakes are high. Fiction allows our brains to practice reacting to the kinds of challenges that are, and always were, most crucial to our success as a species" (2012, 59, 67; and see Oatley 2008; Mar and Oatley 2008). Whether through history, biography, fiction, or direct experience, people hone their social problem-solving skills by interpreting nested narratives.

In *Human Nature and Conduct*, John Dewey asserts that judgment operates by way of a "dramatic rehearsal" whereby we play out competing storylines of action, habits, desires, and impulses in imagination. These "tentative rehearsals in thought" allow the actor to "run ahead and foresee outcomes, and thereby avoids having to await the instruction of actual failure and disaster" (Dewey 2002, 190). When exercising practical judgment, we engage in dramatic rehearsals—or perhaps better said, dramatic prehearsals—discerning how events are likely to unfold given the characters involved and circumstances at hand. So informed and sensitive to context, we determine how events ought to unfold to achieve the best results.

With this in mind, Hannah Arendt observes that Shakespeare's tales, rather than "dry volumes of ethics," provide the more effective guide for the practical judge. For "an ethical principle to be verified as well as validated," Arendt insists, it must "become manifest in the guise of an example" (1993, 248–249). The examples that populate biography, history, and fiction—narrative accounts of men and women who judge well or badly and act virtuously or viciously, with admirable foresight or regrettable blindness in particular situations—provide the foundation for practical judgment. Our judgments have the temper of the examples that inform them (Arendt 1982, 76–77, 84; Larmore 1987, 2001).

Exemplary narratives do more than instruct. They inspire. This inspiration is crucial, as knowledge on its own seldom motivates action. Recall that wicked problems are never solved. They do not abide "answers," the correctness of which is demonstrable. It is not a matter of getting a calculation or precept right, and then by the force of logic compelling others to agree. The practical judge, Arendt notes, must "woo" the assent of others (Arendt 1982, 72; Arendt 1968, 222–223). Potential collaborators must be inspired and persuaded. Ultimately the sharing of resonant stories is required to woo assent and stimulate collaboration.

Effectively, the practical judge tells a story that entices listeners to become its co-interpreters, and as such, the co-solvers of the wicked problem at hand. Walter Benjamin writes that the storyteller counsels his readers, but this counsel "is less an answer to a question than a proposal concerning the continuation of a story which is just unfolding" (1968, 86). The counsel of the storyteller is an invitation to listeners to meet a vexing challenge—as interpreters and actors—by contributing to a story in the making. Wicked problems defy win–win solutions. Sacrifice and compromise are required for their resolution. Participants must be wooed into believing that these compromises and sacrifices are merited by the pursuit of a worthy goal.

Narrative insight is no panacea. Left uninformed by good data and rational analysis, narratives can mislead. When in a "storytelling mode," people tend to neglect the impact of long-term probabilities and even brute facts. They bias themselves in favor of vivid events, real but unlikely, or even imagined. They predispose themselves to emotional reactions. And they let themselves jump to conclusions based on scant and unreliable evidence (Kahneman 2011; Taleb 2010; Tversky 2004; Schank and Abelson 1995; Plous 1993; Tversky and Kahneman 1982; Vidmar and Rokeach 1974). The rise of populist demagoguery grounded in national myths and conspiracy theories are cases in point. Any attempt to resolve a wicked problem that relies wholly on storytelling to the exclusion of historical truths and scientific facts is dangerous and misguided.

It follows that particularly salient narratives require tempering to prevent decision-makers from becoming "prisoners of their preconceptions" (Tetlock 2005, 118). To exercise practical judgment, we must acknowledge and utilize the human predisposition for stories, including their capacity to stimulate imagination and action. At the same time, we must counter the inherent dangers of narrative: emotional reactivity, the reinforcement of biases, and blindness to historical facts and scientific data (Thiele and Young 2016).

How, then, does the challenge of developing narratively informed practical judgment relate to the wicked problem of AI's development? In outlining the dangers to be faced on the path to the Singularity, Kurzweil (2013, 452) writes: "In terms of public policy, the task at hand is to rapidly develop the defensive steps needed, which include ethical standards, legal standards, and defensive technologies. It is quite clearly a race. There simply is no alternative. We cannot relinquish our way out of this challenge." To this end, Kurzweil urges the cultivation of tolerance, liberty, and respect for diversity and knowledge, hoping that advanced AI will come to reflect these human values. To be sure, technological relinquishment is not an option. So the "defensive steps" that Kurzweil encourages must be taken.

Defensive technological development could proceed rapidly. But legal and moral standards are unlikely to keep the same pace. So the race is not really between ethico-political standards and technology. It is between competing forms of technology, with ethics and politics out of breath and lagging behind.

Aware of the disparity between technology's pace and humanity's relatively immature and slowly developing moral and political capacities, Bostrum (2014, 255, 259) writes:

> Before the prospect of an intelligence explosion, we humans are like small children playing with a bomb.... For a child with an undetonated bomb in its hands, a sensible thing to do would be to put it down gently, quickly back out of the room, and contact the nearest adult. Yet what we have here is not one child but many, each with access to an independent trigger mechanism. The chances that we will all find the sense to put down the dangerous stuff seems almost negligible.... Nor can we attain safety by running away, for the blast of an intelligence explosion would bring down the entire firmament. Nor is there a grown-up in sight. In this situation, any feeling of gee-wiz exhilaration would be out of place. Consternation and fear would be closer to the mark; but the most appropriate attitude may be a bitter determination to be as competent as we can.

To become as competent as we can is good advice. It certainly applies to our skills in developing defensive forms of AI. More importantly, it applies to our efforts to cultivate our aptitude for practical judgment.

What does competence in practical judgment look like? A brief examination of early efforts to measure machine intelligence will get us started on an answer. The Turing Test, named after the English mathematician and computer scientist Alan Turing (who contributed to breaking the Nazi's Enigma code), was a thought experiment designed to determine whether a machine could think. As thinking is a very ambiguous term, Turing suggested that the mark of a machine's intelligence is better gauged by its capacity to win a particular sort of competition. Turing's "imitation game" consists of a human evaluator in conversation with another human and with a machine, both out of sight. If the evaluator, after a few minutes of interrogation, cannot reliably tell which of its interlocutors is human and which machine, then the machine wins the imitation game and is deemed capable of thinking.

For almost twenty years, until 2019, Turing's thought experiment was put into practice in a competition to determine which chatbots were most human. The Loebner Prize, a few thousand dollars, was awarded to the winner. The formal competition has been discontinued. But since the arrival of Large Language Models, chatbots frequently succeed at the Turing Test. A study of over 10 million human-to-human and human-to-chatbot conversations in 2023 demonstrated that people could correctly identify AI interlocutors only 60% of the time (Jannai et al. 2023; and see Biever 2023). College educators tasked with discerning which essays are written by students and which are produced by AI chatbots are accurate about 50% of the time (Wu and Wu 2024). Clearly smart machines now routinely pass as human. This fact suggests that we may want to modify our approach to the Turing Test to allow finer discriminations.

What sort of conversation would most clearly assess whether a machine not only has human-level cognition but also human-level emotional and moral capacities? Some time ago I heard an elliptical answer to this question that stuck. We will know that a machine can truly think, exhibiting not only sophisticated intelligence but also ethico-political sensibilities when, in response to a particularly vexing probem, it offers this response: "That's a tough question. I'm not sure how to answer, but it reminds me of a good story…."

In giving this response, the AI will have recognized that the integration of cognitive, emotional, and moral capacities requires a narrative framing that makes the world, and our actions within it, meaningful. The machine will demonstrate that it cannot adequately address wicked problems solely by aggregating and analyzing data, or applying the brute force of logic. Rather, this advanced intelligence will attempt to woo its interlocutors into the co-interpretation of a story that provides both understanding and inspiration. In so doing, the machine will demonstrate practical judgment. I have reasons to doubt that such an event will ever occur. In any case, what the future will bring is not the most important question. Or, rather, the question of what the future will bring rests on the most salient concern: what does the present demand from us?

I believe present circumstances demand our resolve to become ever more competent in the exercise of practical judgment. That is the antidote to the dangerous stupidity that threatens our species in the wake of AI's accelerating development. Cultivating this narratively grounded faculty is also the best hedge against a superintelligence that might severely undermine the prospects for human flourishing. At all costs, we should avoid the fatal and harshly ironic state of arriving at the doorstep of the Singularity and failing the more discriminating Turing Test ourselves, while our machine counterparts pass with flying colors.

Conclusion

The proficiency of stocking frames and power looms was much feared by the British textile workers of the early 1800s. They accurately predicted that many of their jobs, and skill sets, would be destroyed by the industrial juggernaut of mechanization. In response to this threat, the anxious workers smashed machines. But the Luddites did not stop industrialization, or even slow it down. Likewise, the development of AI cannot be prevented or even appreciably delayed. Practically speaking, it is unstoppable.

Even if superintelligence is never achieved, the path to AGI is fraught with dangers, foremost of which is the decline in human aptitudes. In the early 1950s, Norbert Wiener, the inventor of cybernetics, first addressed the impact of machines that he described as "the judgment replacing type" (Wiener 1954, 169). At the time, such machines did not exist. Now they do. The knowledge, values, and skills that human beings require to ensure their continued flourishing may drastically diminish in the wake of AI's development. There are no straightforward means to avert this disaster. It is a truly wicked problem.

Still, a concerted effort to cultivate practical judgment, especially in our young people, is the best course of action. Doing so entails putting some distance between ourselves and our digital devices such that we relate to each other by the most human of means and systematically exercise and develop uniquely human aptitudes (Thiele 2025). At some point in the future, if circumstances dictate that we acknowledge smart machines as our superiors, we may then do so as co-interpreters of a grand story in progress—a narrative no less meaningful because it harbors tragic themes.

References

Arendt, Hannah. 1968. Between Past and Future: Eight Exercises in Political Thought. New York: Penguin.

Arendt, Hannah. 1982. Lectures on Kant's Political Philosophy, ed. with an interpretive essay by Ronald Beiner. Chicago: University of Chicago Press.

Aristotle. 1953. The Ethics of Aristotle: The Nichomachean Ethics. New York: Penguin.

Bargh, John A. 1997. The Automaticity of Everyday Life. In The Automaticity of Everyday Life: Advances in Social Cognition, ed. Robert Wyer, Jr., vol. X, 1–61. Mahwah: Lawrence Erlbaum Associates.

Benjamin, Walter. 1968. The Storyteller. Illuminations, 83–110. New York: Schocken Books.

Bess, Michael. 2015. Our Grandchildren Redesigned: Live in the Bioengineered Society of the Future. Boston: Beacon.

Biever, Celeste. 2023. "The Easy Intelligence Tests that AI Chatbots Fail." *Nature* 619 (July):686-689.

Bostrum, Nick. 2014. Superintelligence: Paths, Dangers, Strategies. Oxford: Oxford University Press.

Burton, Robert. 2015. How I Learned to Stop Worrying and Love A.I. New York Times, September 21. http://opinionator.blogs.nytimes.com/2015/09/21/how-i-learned-to-stop-worrying-and-love-a-i/.

Carr, Nicholas. 2011. The Shallows: What the Internet is Doing to Our Brains. New York: Norton.

Carr, Nicholas. 2014. The Glass Cage: How Computers are Changing Us. New York: W. W. Norton.

CB Insights Blog. 2016. https://www.cbinsights.com/blog/artificial-intelligence-funding-trends/; https://www.cbinsights.com/blog/artificial-intelligence-funding-trends-q216/.

David Silver et al. Mastering the game of Go with deep neural networks and tree search. Nature 529, 484–489 (28 January 2016). https://doi.org/10.1038/nature16961

Dewey, John. 2002. Human Nature and Conduct. Mineola: Dover Publications.

Dorrier, Jason. 2013. The Future of Search and the Internet of Everything According to Google's Scott Huffman. http://singularityhub.com/2013/12/ 31/the-future-of-search-and-the-internet-of-everything-according-to-googles- scott-huffman/.

Ericsson Consumerlab. 2015. http://www.ericsson.com/res/docs/2015/consumerlab/ericsson-consumerlab-10-hot-consumer-trends-2016-report.pdf.

Flyvbjerg, Bent. 1998. Rationality and power: Democracy in practice. Chicago: University of Chicago Press.

Flyvbjerg, Bent. 2001. Making Social Science Matter. Cambridge, U.K: Cambridge University Press.

Gakshy, Eytan, Solomon Messing, and Lada Adamic. 2015. Exposure to Ideologically Diverse News and Opinion on Facebook. Science 6239: 1130–1132.

Garreau, Joel. 2005. Radical Evolution. New York: Broadway Books.

Good, Irving John. 1966. Speculations Concerning the First Ultraintelligent Machine. In Advances in Computers 6: 31–88.

Gottschall, Jonathan. 2012. The Storytelling Animal: How Stories Make us Human. Boston: Houghton Mifflin Harcourt.

Hawking, Stephen, Stuart Russel, Max Tegmark, and Frank Wilczek. 2014. Stephen

Hawking: 'Transcendence Looks at the Implications of Artificial Intelligence—But are we Taking AI Seriously Enough?' The Independent (May 1). https://www.independent.co.uk/news/science/stephen-hawking-transcendence-looks-at-the-implications-of-artificial-intelligence-but-are-we-taking-ai-seriously-enough-9313474.html.

Howard, Philip N. 2015. Pax Technica: How the Internet of Things May Set Us Free or Lock Us Up. New Haven: Yale University Press.

Jannai, Daniel et al. 2023. "Human or Not? A Gamified Approach to the Turing Test." Cornell University arXiv. Accessed at https://arxiv.org/abs/2305.20010

Johnson, Mark. 1993. Moral Imagination: Implications of Cognitive Science for Ethics. Chicago: University of Chicago Press.

Joy, Bill. 2000a. Why the Future Doesn't Need Us: Our Most Powerful 21st Century Technologies—Robotics, Genetic Engineering, and Nanotech—Are Threatening to Make Humans an Endangered Species. Wired (April).

Joy, Bill. 2000b. Act Now to Keep New Technologies Out of Destructive Hands. New Perspectives Quarterly 21(4) (Summer).

Kahneman, Daniel. 2011. Thinking, Fast and Slow. New York: Farrar, Straus and Giroux.

Kelly, Kevin. 2016. The Inevitable: Understanding the 12 Technological Forces that will Shape our Future. New York: Viking.

Kurzweil, Ray. 2005. The Singularity is Near: When Humans Transcend Biology. New York: Viking.

Kurzweil, Ray. 2012. How to Create a Mind. New York: Penguin.

Kurzweil, Ray. 2013. Progress and Relinguishment. The Transhumanist Reader, ed. Max More and Natasha Vita-More. Oxford: Wiley-Blackwell.

Kushlev, Kostadin, Jason Proulx, and Elizabeth W. Dunn. 2016. "Silence Your Phones": Smartphone Notifications Increase Inattention and Hyperactivity Symptoms. In Proceedings of the 2016 CHI Conference on Human Factors in Computing Systems, 1011–1020. http://dl.acm.org/citation.cfm?d=2858036.2858359.

Larmore, Charles E. 1987. Patterns of Moral Complexity. Cambridge: Cambridge University Press.

Larmore, Charles E. 2001. Moral Judgment. In Judgment, Imagination, and Politics: Themes from Kant and Arendt, ed. Ronald Beiner, and Jennifer Nedelsky, 47–64. New York: Rowman and Littlefield.

Levy, Frank, and Richard Murnane. 2005. The New Division of Labor: How Computers are Creating the Next Job Market. Princeton: Princeton University Press.

MacIntyre, Alasdair. 1981. After Virtue: a study in moral theory. Notre Dame: University of Notre Dame Press.

MacIntyre, Alasdair. 1999. Dependent Rational Animals. Chicago: Open Court.

Mar, Raymond, and Keith Oatley. 2008. The Function of Fiction is the Abstraction and Simulation of Social Experience. In Perspectives on Psychological Science 3 (3): 173–192.

Markoff, John. 2015. Artificial-Intelligence Research Center Is Founded by Silicon Valley Investors. New York Times, December 12. http://www.nytimes.com/ 2015/12/12/science/artificial-intelligence-research-center-is-founded-by- silicon-valley-investors.html?emc=edit_th_20151212&nl=todaysheadlines&nlid=71023177.

Markoff, John. 2016a. Pentagon Turns to Silicon Valley for Edge in Artificial Intelligence. New York Times, May 11. http://www.nytimes.com/2016/05/12/technology/artificial-intelligence-as-the-pentagons-latest-weapon.html?emc=edit_th_20160512&nl=todaysheadlines&nlid=71023177.

Markoff, John. 2016b. Artificial Intelligence Swarms Silicon Valley on Wings and Wheels. New York Times, July 17. http://www.nytimes.com/2016/07/18/technology/on-wheels-and-wings-artificial-intelligence-swarms-silicon-valley.html?emc=edit_th_20160718&nl=todaysheadlines&nlid=71023177&_r=0.

McAfee, Andrew, and Erik Brynjolfsson, 2016. Where Computers Defeat Humans, and Where They Can't. New York Times, March 16. http://www.nytimes.com/2016/03/16/opinion/where-computers-defeat-humans-and-where-they-cant.html?emc=edit_th_20160316&nl=todaysheadlines&nlid=71023177&_r=0.

Morris, Jonathan S. 2007. Slanted Objectivity? Perceived Media Bias, Cable News Exposure, and Political Attitudes. Social Science Quarterly 88: 707–728.

Nie, Norman H., Darwin W. Miller, Saar Golde, Daniel M. Butler, and Kenneth Winneg. 2010. The World Wide Web and the U.S. Political News Market. American Journal of Political Science 54: 428–439.

Oatley, Keith. 2008. The Mind's Flight Simulator. In Psychologist 21: 1030–1032.

Pariser, Eli. 2011. The Filter Bubble: What the Internet Is Hiding from You. New York: Penguin Press.

Plous, Scott. 1993. The Psychology of Judgment and Decision Making. Philadelphia: Temple University Press.

Reid, Rob. 2017. *After On*. New York: Del Ray.

Reid, Rob. 2024. "Episode 61: An 'Obervatory' for a Shy Super AI." *After On Podcast*. Accessed at https://after-on.com/episodes-6190/055-36hfz.

Roese, Neal, Lawrence Sanna, and Adam Galinsky. 2005. The Mechanics of Imagination: Automaticity and Control in Counterfactual Thinking. In The New Unconscious, ed. Ran Hassin, James Uleman, and John Bargh, 138–170. Oxford: Oxford University Press.

Ross, Alec. 2016. The Industries of the Future. New York: Simon and Schuster.

Rittel, Horst, and Melvin Webber. 1973. Dilemmas in a General Theory of Planning. In Policy Sciences 4: 155–169.

Schank, Roger C., and Robert P. Abelson. 1995. Knowledge and Memory: The Real Story. In Knowledge and Memory: The Real Story—Advances in Social Cognition, ed. Robert Wyer, Jr., vol. VIII, 1–86. Mahwah: Lawrence Erlbaum Associates.

Singer, Peter. 2015. The Most Good You Can Do. New Haven: Yale University Press.

Snyder, Michael. 2013. A Chip In The Head: Brain Implants Will Be Connecting People To The Internet By The Year 2020. http://www.infowars.com/a-chip- in-the-head-brain-implants-will-be-connecting-people-to-the-internet-by-the- year-2020/.

Soares, Nate, and Benja Fallenstein. 2014. Aligning Superintelligence with Human Interests: A Technical Research Agenda. Machine Intelligence Research Institute Publications. https://intelligence.org/files/TechnicalAgenda.pdf.

Stroud, Natalie J. 2008. Media Use and Political Predispositions: Revisiting the Concept of Selective Exposure. Political Behavior 30: 341–366.

Taleb, Nassim Nicholas. 2010. The Black Swan: The Impact of the Highly Improbable. New York: Random House.

Tetlock, Philip E. 2005. Expert Political Judgment. Princeton: Princeton University Press.

Thiele, Leslie Paul, and Marshall Young. 2016. Practical Judgment, Narrative Experience, and Wicked Problems. Theoria 63 (148).

Thiele, Leslie Paul. 2006. The Heart of Judgment: Practical Wisdom, Neuroscience, and Narrative. Cambridge: Cambridge University Press.

Thiele, Leslie Paul. 2025. *Human Agency, Artificial Intelligence, and the Attention Economy: The Case for Digital Distancing*. London: Palgrave Macmillan.

Tversky, Amos, and Daniel Kahneman. 1982. Judgments of and by representativeness. In Judgment under Uncertainty: Heuristics and biases, eds. Daniel Kahneman, Paul Slovic and A. Tversky. Cambridge: Cambridge University Press.

Tversky, Barbara. 2004. Narratives of Space, Time, and Life. In Mind & Language 19 (4): 380–392.

Urbinati, Nadia. 2014. Democracy Disfigured: Opinion, Truth, and the People, vol, 202, 4–5. Cambridge: Harvard University Press.

Vance, Ashlee. 2015. The First Person to Hack the iPhone Built a Self-Driving Car. In His Garage. Bloomberg Businessweek, December 16. https://www.bloomberg.com/features/2015-george-hotz-self-driving-car/.

Vidmar, Neil, and Milton Rokeach. 1974. Archie Bunker's Bigotry: A Study in Selective Perception and Exposure. Journal of Communication 24: 36–47.

Vinge, Vernor. 1993. The Coming Technological Singularity: How to Survive in the Post-Human Era. https://www-rohan.sdsu.edu/faculty/vinge/misc/singularity.html.

Wall, John. 2005. Phronesis as Poetic: Moral Creativity in Contemporary Aristotelianism. The Review of Metaphysics 59 (2): 313–331.

Wiener, Norbert. 1954. The Human Use of Human Beings: Cybernetics and Society. New York: Houghton Mifflin.

Wu, Nicole and Wu, Patrick. 2024. "Surveying the Impact of Generative Artificial Intelligence on Political Science Education." *PS: Political Science and Politics*, August 9. Accessed at: https://www.cambridge.org/core/journals/ps-political-science-and-politics/article/surveying-the-impact-of-generative-artificial-intelligence-on-political-science-education/D87FE814975D2C1CFA3F9E9E7DEBDB6C

CHAPTER 8

Drug Smuggling and Automated Borders: A Losing Battle of Escalation or State/ Non-state Symbiosis

Elaine Carey and Andrae Marak

On January 8, 2016, Mexican marines with US assistance dealt a supposedly huge blow against the Sinaloa Cartel with the capture and jailing of Joaquin "El Chapo" Guzman Loera. El Chapo's capture followed the 2015 high profile capture and arrest of two drug bosses—Servando "La Tuta" Gómez of the Knights Templar (formerly La Familia) and Omar Terviño Morales, known as "Z-42," of Los Zetas—by Mexican authorities. For historians, social scientists, and public policy specialists, the announcements and news coverage of these arrests follow a pattern that has persisted for decades. For nearly 30 years, "the Kingpin Strategy,"

E. Carey (✉)
St. John's University, Jamaica, NY, USA
e-mail: careye@stjohns.edu

A. Marak
Department of History & Political Science, Roosevelt University, Chicago, IL, USA
e-mail: amarak@roosevelt.edu

R. Kiggins (ed.), *The Political Economy of Robots*, International Political Economy Series, https://doi.org/10.1007/978-3-031-95171-8_8

which unofficially began in the late 1980s with the capture of Panamanian dictator Manuel Noriega and, later, the capture of the Medellín cartel's Pablo Escobar, focuses on capturing the head of an organization and creating disarray in the lower ranks. The arrests of kingpins are hailed as an "upheaval" at the top, sure to "disrupt" and/or "dismantle" a particular organization and curtail the drug trade. Parading drug bosses before cameras timed to coincide with the evening news—accompanied by masked and heavily armed police agents—heightens the spectacle.

All three drug traffickers had been targets of the US government through the Foreign Narcotics Kingpin Designation Act (Kingpin Act), which became law in 1999. Designed to deny drug traffickers and their organizations access to American financial institutions, the Act prohibits US citizens and businesses from trade or interaction with drug traffickers. In 2009, the USA targeted three organizations and their associates in the USA: Los Zetas, the Sinaloa Cartel, and La Familia Michoacana (the Knights Templar are an offshoot of La Familia). The arrests of the kingpins create news. Journalists capture the humiliating march of the arrested leaders surrounded by masked police agents. The display of policing and military might, for the cameras, give the impression that the War on Drugs is winnable. In its current form. It is not!

At a 2015 meeting at the Mexico Institute of the Woodrow Wilson International Center for Scholars, experts argued that the Kingpin Strategy that the USA continues to advocate and fund has failed. Arresting the leaders does little to "disrupt" or "dismantle" the drug trade because drug trafficking organizations, like most organized crime, maintain horizontal rather than vertical power structures. These horizontal structures tend to be fluid, flexible, and creative in their responses. Conversely, the vertical bureaucratic structures of responding governments are slow and rigid in their responses and different governmental subunits who occupy "a multiplicity of spaces" often work at cross-purposes with each other.[1] In some cases, even where states are capable of effective responses, they are reticent to do so when it conflicts with other more important geopolitical interests; this was, for example, the case when the central intelligence agency (CIA) worked with known drug smugglers in Colombia to fund the Contras in Nicaragua in the 1980s to advance the USA's anti-communist agenda even as the Drug Enforcement Administration was working to undermine the Colombian cartels.[2] Furthermore, the countries against which these policies are being imposed are quite deft at manipulating them to their own advantage through a "mix of compliance, pseudo-compliance, and bilateral manipulation."[3] In fact, instead of viewing the Kingpin Strategy as a policy aimed at ending the War on Drugs, it is better to view the multiplicity of governmental reactions in response to (and occasionally in conjunction with) illicit and illegal actors/actions as a "ferocious struggle over how nation building [and rebuilding]... is to proceed and whose name and interests."[4] In other words, it is our contention here that there is a symbiotic relationship between smugglers and states (and their subunits) and that the competition (and compliance) between the two often serves both smugglers' and states' interests over that of the general public.[5] The promotion of the idea that smugglers are increasing in strength and are a "growing threat... is most critical for sustaining and expanding law enforcement" and increased funding for new technologies aimed at "disrupting" and "dismantling" cartels.[6] The War on Drugs does not only help law enforcement. Drug-fueled prosperity has been widely distributed between those who work for the drug cartels and those who work for the government; for those who engage in "legitimate" business and those who engage in the illicit. This may help explain why reports of the arrests of La Tuta, Z-42, and El Chapo were met with skepticism. So many kingpins have been arrested; so many have been extradited, escaped, or turned state's evidence; and yet so little has changed. Perversely, the capture and parading of kingpins before media cameras serves to perpetuate the current drug war regime as it promotes the supposed effectiveness of current policies even though they have resulted in no discernable or measurable improvement.

In fact, as demonstrated by the cases of Panama and Colombia, the arrests of key figures led to splintering or fractionalization of groups into smaller entities that are more autonomous, more violent, and more difficult to police. This is known as the "cockroach effect."[7] Instead of solving a problem, civilian populations face *increasing* violence due to the competition between smaller (at least initially) rival drug trafficking organizations and their continued battles with police forces.[8] This was also the case in Mexico. The arrest of Osiel Cárdenas Guillén, the head of the Gulf cartel, in 2009 produced a split between the Gulf cartel and the Zetas—an elite paramilitary unit—that resulted in the transformation of the Zetas into a full-fledged cartel and the death of thousands of people.[9] The increased violence caused by the splintering of DTOs is not the only potential outcome. If the state is also splintered/fractured in its response resulting from, for example, competition between federal and state-level agencies or police forces, then DTOs often increase the visibility and the grotesqueness of their crimes (such as public beheadings) to send a message to state forces (or some of the state forces) to back off.[10] In spite of thirty years of the Kingpin Strategy and a rise in violence, first in Colombia and then in Mexico and now in Central America, the actual amount of illegal drugs on the street have remained the same and the cost of illegal drugs has remained the same or declined.[11]

The story of supply-side interdiction is similar. Plan Colombia (begun in 2000) and the Merida Initiative (begun in 2008) focus on severely reducing the supply of illegal drugs at their source through the eradication of coca, marijuana, and poppy crops. They also focus on disrupting drug supply lines through increased policing and the militarization of borders, thus (in principle) making the transportation of illegal drugs more difficult.[12] As Moisés Naím notes, focusing on interdiction in sending countries is "politically profitable" because tools such as "helicopters, gunboats heavily armed agents, judges, and generals" (just like the capture of kingpins) are more "telegenic" than focusing on demand in receiving countries like the USA.[13] But interdiction is neither less expensive nor more effective than focusing on the demand side of the equation; high levels of demand for illegal and illicit drugs in the USA have remained in spite of the adoption of more punitive policies.[14] Even where interdiction drives up the cost of drugs—it simply further incentivizes increase in production to capture the increased value that interdiction cause or results in the willingness of drug cartels to fight over drug routes (though which they capture profits) and drug gangs (in the USA) to fight over turf (where they also capture profits).

For example, the eradication of coca bushes in South America over the last two plus decades has not impacted the amount of cocaine produced, its purity, the cost of cocaine on the street, or the number of people using cocaine; instead, it has negatively impacted the amount of money that poor peasants in the Andean highlands who raise the coca leaves earn.[15] One reason for this is that there are not enough resources to eliminate all the sources of coca and cocaine production, so as the USA and its allies move to eradicate coca plants from one region, production simply shifts to another region, and as the USA moves to intercept the flow of cocaine across some portions of the border, cartels simply shift to transport cocaine across other portions of the border, and as the USA focuses on preventing a particular means of transporting cocaine across borders, cartels adopt a different, or multiple other, means of doing so.[16] In his

Rolling Stone interview, Sean Penn asked El Chapo, "Of all the countries and cultures with whom you do business, which is the most difficult?" El Chapo replied, "None." This should not come as a surprise given that the Sinaloa cartel has access to high-tech submarines, private planes, counter surveillance scanners, highly sophisticated tunnels that span the border, high-tech weaponry, and (most importantly) the public land and water transportation infrastructure through which the vast majority of illegal drugs (intermixed with legal commodities) pass across borders, namely on-board regular commercial cargo ships and semi-trailer trucks.[17]

The ineffectiveness of supply-side interdiction is not its only problem. Supply-side interdiction also undermines democracy, leads to the militarization of borderlands regions, and stifles social and cultural interchange. Over the past decade plus in Mexico, the militarization of the federal police further weakened the state and democracy while drug traffickers have prospered, leading to what anthropologist Howard Campbell has called a "drug war zone."[18] The reason that the borderlands region appears to outsiders to be the locus of a clash of civilizations is not because it accurately reflects the region's inhabitants, or their desires—that region is, and has been, culturally and socially binational for over a century.[19] Rather as geographer Michael Dear correctly notes "borderland communities have limited capacity for self-determination" about borderlands policies.[20] Stated another way, "most decisions [about the War on Drugs]... are made away from the border, but much of the war's front lines lie along the border."[21] The War on Drugs and its support for Mexico's Aar on Drugs have had the strongest negative impacts on poor communities of color within the USA and large swaths of Mexico, which have become increasingly violent.[22] This has also led to the criminalization of Central American and Mexican people displaced by the War on Drugs.[23] For the majority of US residents,[24] the war has been an inner-city problem or a Mexican/Central American problem, but in any case, it is something seen on television, and certainly not their problem. It has little to do with their lives even if they are one of the millions of US residents who recreationally use illegal drugs. This is nothing new. A similar process unfolded during US Prohibition where Mexican residents suffered the brunt of the negative effects of the US anti-vice movement that pushed alcohol, drugs, and vice zones across the border into Mexico even though US residents were the major consumers of the alcohol, drugs, prostitution, and gambling.[25] As just one of many possible examples, El Paso's Chamber of Commerce—as part of its successful tourism campaign during prohibition—promoted Ciudad Juárez as a place where Americans could safely and easily engage in a range of activities deemed illegal in the USA. This campaign drew so many people to El Paso-Ciudad Juárez that it led to civil unrest among everyday residents of the binational region who were sick of the negative impact that Ciudad Juárez's red light district had on surrounding communities in Mexico.[26]

What does all this have to do with borders, illegal drugs, and automation? Everything! Historians of borderlands in North America (Canada, the USA, and Mexico) have spent the better part of the last decade plus exploring how (or even if) the unfenced and permeable borderlands of overlapping sovereignties of the nineteenth century turned into the fenced, patrolled, militarized, and increasingly automated borders of the twentieth and twenty-first centuries.[27] The border has become more bordered, more militarized. The harnessing of automation and other forms of technology will likely make it even more so.[28] Still, the bordering of borderlands makes borders no less impervious to penetration, no less permeable. What it does do is make borders less permeable to some goods and some subsets of people[29] even as they have become even more permeable to others goods and other subsets of people. More specifically, the technological shifts brought about by modernization were harnessed by political forces to harden borders in order to control the illicit flows of people, goods, and services and to assess taxes. The initial focus on the taxation of trade (i.e., customs), especially prior to the passage of the 16th Amendment in 1913, gave way to a focus on national security and the racial policing of borders.[30] The regulation and policing of borders occurred alongside the expansion of prohibition regimes—such as the regulation of opiates via the Harrison Act of 1914, the prohibition of alcohol via the Volstead Act of 1919, and the prohibition of marijuana via the Marijuana Tax Act of 1937—which committed the USA to fighting smuggling organizations that its own rules have served to created. President Richard Nixon's War on Drugs served to escalate the prohibitionist approach by raising the use and smuggling of illegal drugs to the level of a "national threat" and by creating a "narco-enforcement complex upon the shoulders of a much smaller, less intricate, and less costly infrastructure used to police earlier prohibition regimes."[31]

More recently, 9/11 resulted in a refocus on borders and the trans-border flows of drugs and immigrants as potential existential (terrorist) threats to the USA. This reorganization meant not only a massive increase in funding, the creation of the Department of Homeland Security, and "the growing prominence of law enforcement in national security institutions and missions, and a further blurring of the traditional distinction between internal and external security threats."[32] But the stalemate continues. As smugglers embraced technological changes—fake identification papers, telegraphs, the use of railroads, and automobiles with false compartments, for example—to enhance their businesses, policing agencies had (and have) to become more sophisticated and nimbler to police smugglers; the introduction of passports, mug shots, fingerprint cards, barbed wire fencing, and the use of private detective forces that could work extrajudicially are all excellent historical examples.[33] The use of airplanes, helicopters, drones, ion scanners, X-ray vans, bar codes, facial recognition software, and forensic equipment are contemporary examples.[34] This creates a never-ending symbiotic cat-and-mouse game: prohibition regimes create black markets which, in turn, increase profits, making it worth smugglers' while to invest in new, more expensive ways to evade (or collude with) policing organizations; this increased sophistication, in turn, requires policing agencies to respond in kind. In the end, each needs the others: smugglers need policing agencies and prohibition regimes to drive up profits and policing agencies need smugglers to warrant their existence (and expansion and ever-increasing budgets). Introducing automation to this symbiotic relationship will do nothing to end it, but it will cost tax payers more and, potentially, harm more borderlanders. As the recent use of a robot—originally designed for disarming bombs—by the Dallas Police to kill Micah Xavier Johnson, the potential to harness robots and other forms of automation to respond to threats and potential threats "has the potential to lead to overuse of machines that can be used to injure or kills suspects" without the due process of law.[35] The use of robots and other automated systems first developed and implemented for use in war zones may also serve to introduce war-making tools and procedures into domestic environments where they do not belong.[36] The reverse, epitomized by the use of low-cost drones to swarm and overwhelm advanced weapon systems and wreak havoc on civilian targets in Ukraine, is also true.[37] The answer to these issues is not increased automation—though this will happen at some level as a matter of course—but increased democracy, increased community input, and stronger transnational and cross-cultural ties.

Before returning to our call for increased democratization instead of increased militarization and automation, we want to explore what increased automation of borders might look like by examining the concept of the singularity and its potential impact on border flows and by placing the singularity and increased automation into the historical context of modernization and technology of borderlands. Ray Kurzweil, one of the leading theorists and promoters of the use of technology, argues that the rate of "technological change is exponential" rather than linear, meaning that the rate of technological change and the onset of an automation revolution resulting in an age of intelligent machines, which he and others call "the singularity" is imminent.[38] Kurzweil estimates that this age of intelligent machines, measured by the ability of a computer to pass the Turing test—the capability of computers to imitate human reasoning and responses such that human beings would not be able to tell that the computer was not a human being—would arrive at some point between 2020 and 2070.[39] Kurzweil is not along. Jeremy Rifkin, a prominent economist who focuses on the impact of science and technology, and Bill Joy, a computer scientist and one of the founders of Sun Microsystems, also believe that the singularity is imminent (with 2025 as its likely arrival date).[40] This singularity will bring with it, Kurzweil says, a "pace of technological change... so rapid... so deep, that human life will be irreversibly transformed."[41] In fact, the rise of automation, the "merging of mind and machine," will result in the replacement of biological evolution with technological evolution where computers with artificial intelligence (or humans merged with such computers) would design their own evolution.[42] There will, he argues, "be no distinction... between human and machine or between physical and virtual reality." This merging of humans and machines would include machines having the equivalent of currently existing human "emotional and moral intelligence."[43] The singularity may sound far-fetched and like an Isaac Asimov science fiction story but there are any number of recent technological/automation applications that demonstrate the potential for a profound transformation.[44] These include driverless cars, which have faced quite a bit of scrutiny for dangerous and sometimes deadly malfunctions in spite of or because of limited government oversight, the creation of neuroprosthetics which will be implanted into our brains to "boost intelligence" or overcome "damage caused by strokes, Alzheimer's or concussions," and artificial retina implants, to name just three.[45] Other theorists, with whom we are better aligned, argue that although there will be exponential technological change resulting in the ability to replace human labor in many areas of the economy, the need for human creativity (and investments in it) will increase as computers will not be able to replicate this aspect of human nature. Erik Brynjolfsson and Andrew McAffee call this unreplicable facet of humanity *ideation* by which they mean the human ability to come up with "good new ideas or concepts." "We've never seen a truly creative machine, or an entrepreneurial one, or an innovative one," they say. Computers can be programmed to remix and remash already existing ideas and concepts to come up with new ones, but they can never think outside of the proverbial box.[46] Regardless of whether the singularity arrives or not

> major shifts are already going on in computing power and machine intelligence. And if the trends for the future do hold true even at the most minimal level, then things are going to get real interesting in the not too distant future.[47]

We also know that regardless of the potential limitations of the singularity these changes will most certainly impact the flow of people as well as licit and illicit goods across borders.

One perhaps unexpected change will be the ways in which automation impacts labor markets (which, in turn, impacts transborder flows of goods and people). Brynjolfsson and McAfee argue that the plunge into industrial employment—in the past well-paid and often unionized—in developed economies is not just the result of increased competition with labor in developing countries through free trade agreements such as the North American Free Trade Agreement (NAFTA) in 1994 or the entry of China into the World Trade Organization (WTO) in 2001.[48] It is also the result of increased efficiencies and the increased automation of work that had previously been done by hand.[49] In other words, labor in developed and developing countries is competing against automation. This not only drives down the overall cost of labor (and the prices of goods), but it also leads to higher levels of unemployment. This matters to our assessment of the impact of automation on borders because many analysts have rightly argued that the drug cartels in Colombia, Mexico, and Central America tap into the resentment of low-level workers who are unemployed or underemployed as a result of globalization and automation.[50] The original rise of modern drug cartels in Mexico, for example, coincided with Mexico's 1982 banking crisis as the newly unemployed sought refuge from financial disaster. Everything is clearly not well in Mexico, but, four decades later, Mexico is no longer suffering from the 1982 financial crisis. Since 1982, as the Mexican government (in collaboration with the

USA, the World Bank, and the International Monetary Fund) adopted a neoliberal response to the crisis, licit and illicit economies have become much more intertwined, so much so that Mexico's economic recovery was equally driven by each of these (on paper) separate economies. As of this writing, Mexico has low unemployment (2.66% in March 2024... less than that in the USA, which stands at 3.8%). As the Council on Hemispheric Affair's Briseida Valencia Soto notes, however, these figures camouflage a high youth unemployment rate and do not take into account that, for example, the Sinaloa cartel probably employed nearly as many people as PEMEX, Mexico's national oil company. The drug cartels tap into the angst of rural and urban workers displaced by shifts in Mexico's economy. In the countryside, the cartels entice small-scale farmers, who are no longer competitive with large-scale and increasingly mechanized corporate farms, to raise marijuana and opium poppies. One of the added bonuses of these illicit crops is that they can be planted and harvested year round (unlike traditional cash crops) and they are, as a result of antidrug prohibition regimes, much more valuable. In the city, the cartels recruit unemployed youth as lookouts, couriers (known as mules), and, over time, *sicarios* (or assassins) via promises of a potentially glorious lifestyle.[51] All other things being equal, the economic dislocation and rising inequality caused by neoliberal economic policies and increased automation and efficiencies will only grow in the years to come, creating an ever-larger pool of potential recruits for the drug cartels. Brynjolfsson and Andrew McAfee have a number of suggestions about how to counteract this potential future but none is more important than their advice of "Please, No Politburos," which we translate here as "Please, More and Better Democracy."[52] We will return to this shortly.

Another impact of the singularity is the potentially increased militarization resulting from increased automation. We have already noted the example of the Dallas Police's repurposing of a robot meant to disarm bombs as a weapon used to kill via explosives as suspect who refused to stand down during negotiations. The point here is not whether this was the right or wrong response, but rather that the Dallas Police did not have a well-thought-out protocol for the alternative use of the robot and instead, in the heat of the moment, repurposed it to other ends. We see similar trends in schools across the USA where after 9/11 and a rash of mass school shootings have resulted in the use of expensive high-tech security systems (and more police officers) in spite of the fact that children are much safer in schools than elsewhere and in spite of the fact that school violence has actually been dropping over the last several decades (even as we take into account the mass shootings).[53]

P.W. Singer, the leading expert on the impact of robotics on war, provides us with additional examples of the ways in which increased automation changes the ways in which we respond to potential threats. Perhaps the two most important things that Singer suggests are that the amount that we now spend on automated solutions to potential security issues has grown exponentially and that we respond quite differently to potential issues when we adopt automated responses. Singer reports that there were only a few homeland security companies with federal contracts prior to 9/11 but a decade later there were more than 30,000. The costs for the homeland (read: airports, stadiums, and other "soft" targets) and border security have ballooned, "projected to reach $35 billion by 2011."[54] Included in the exploding budget are the use of "'sentry-bots' [that] can guard entrances, automatically patrol perimeters, check IDs, and even use facial recognition software to know who should or shouldn't be allowed into the area" as well as unmanned aerial vehicles (i.e., drones).[55] These robots can be added to the already existing preventive measures against "terrorists... drug smugglers, arms dealers, and human traffickers" such as "border patrol agents, video cameras, ground sensors, physical barriers, land vehicles, and manned aircraft."[56] The use of drones is typically portrayed as being more cost-effective (including in the Congressional Research Service report cited above), but later congressional reports suggest that this is not the case. This has not stopped Congress from appropriating approximately $16 million per drone for a series of drones to police the most remote portions of the US-Mexico border as well as portions of the Gulf of Mexico.[57]

More important than the cost of drones and other automated responses to the permeability of borders and the balloon and cockroach effect is that the adoption of these automated interventions (first developed for war theaters) threatens to take the humans out of the decision-making loop even as it undermines democracy. So what might the border look like if the Border Patrol was replaced with a series of automated sentry-bots and drones? What role would humans play in overseeing the sentry-bots and drones? In researching *Wired for War,* Singer notes that nearly everyone he talked to about the ways in which automation might remove humans from the loop said that humans would always be in control of the decision-making process and that his concerns in this area were overblown. Singer, however, demonstrates through multiple security and military examples, including increasingly sophisticated preprogrammed responses to perceived threats, that our adoption of automated responses has, for quite some time, increasingly taken human decision-makers out of the loop.[58] This might not be a problem if, as Kurzweil tells us, future computers and robots will have human emotional and moral intelligence. But, if we are correct in arguing that Kurzweil is wrong about this (and clearly AI and automated functions do not currently possess these attributes), these automated interventions to the balloon and cockroach effects could well "unleash a hurricane of political, legal, and ethical problems."[59] Most centrally, the increased use of robots and drones to secure borders will increase the possibilities that we will use force to respond to the flows of illicit and illegal goods. The use of unmanned systems can undermine the connections that everyday people have with the impact of border security. Larry Korb, a senior fellow at the Center for American Progress, argues that "people are more likely to support the use of force as long as they view it as costless."[60] Add to this the rise in anti-Mexican (and Central American and anti-Arab/anti-Muslim) xenophobia and the increased use of force through automation becomes more and more plausible.

But this potential bad ending is not preordained. As Tony Payan, Kathy Staudt, and Tony Kruszewski have argued, placing borderlanders into the central role in the discussion and creation of antidrug smuggling policy could mitigate some of the very real problems that the flows of illicit and illegal goods create (and that borderlanders suffer from at higher levels than those farther from the border).[61] Doing so might result in some surprising changes. We might, for example, find that borderlanders are not opposed to the smuggling of some illegal goods by some people; the reclassification of the transnational flows of these goods from illegal to legal would undermine the black market and greatly reduce smuggling.[62] We might also find borderlander opposition to the unilateral decriminalization of marijuana (now happening across a broad swath of the USA) and other less harmful drugs in the USA if it would result in the reversal of flows of smuggled drugs as the cost to produce marijuana legally in the USA undercut the cost to do so in Mexico. In the end, more democracy, greater levels of equality (within the USA and transnationally across North America), and stronger and more equal transnational ties will be necessary to lessen the amount of and counter the negative impacts of drug smuggling. In the end, it is our hope that the historicization of transnational drug smuggling and the state/non-state symbiosis might serve to mitigate the attempts by the agents of nation-states to use the fear of the permeability of borders to expand their reach and power in spite of the fact that their doing so in the past has seldom rectified the issue. Instead, it has served to further empower the very forces they claim to be taking on. As we have pointed out elsewhere, many of the solutions offered by nation-states today are strikingly similar to the intractable problems of the past, and adopting more of the same responses will lead to more of the same negative consequences.

Notes

1. Michael Kenney, *From Pablo to Osama: Trafficking and Terrorist Networks, Government Bureaucracies, and Competitive Adaptation* (University Park: Pennsylvania State University Press, 2007). For how the Italian Mafia moved from a vertical to horizontal structure in response to the use of RICO laws, see Selwyn Raab, *Five Families: The Rise, Decline, and Resurgence of America's Most Powerful Mafia Empires* (New York: Thomas Dunne Books, 2005).
2. H. Richard Friman, "Externalizing the Costs of Prohibition," *Crime and the Global Political Economy,* Friman, ed. (Boulder: Lynne Rienner Publishers, 2009): 49–65. On the Contras, see most famously, Gary Webb, *Dark Alliance: The CIA, the Contras, and the Cocaine Explosion* (New York: Seven Stories Press, 2014). There are more recent examples of this behavior as the CIA adopted a similar anti-communist stance in Afghanistan and Southeast Asia. See, for example, Alfred W. McCoy, *The Politics of Heroin: CIA Complicity in the global Drug Trade, Afghanistan, Southeast Asia, Central America* (Chicago: Lawrence Hill, 2003). The USA has continued these policies as part of its post-9/11 Global War on Terror. See Peter Dale Scott, *American War Machine: Deep Politics, the CIA Global Drug Connection and the Road to Afghanistan* (Lanham, MD: Rowman & Littlefield Publishers, 2014).
3. Alan Knight, "Drugs, Crime, and Violence in Modern Mexico," in *Histories of Drug Trafficking in Twentieth Century Mexico,* eds. Wil G. Pansters and Benjamin T. Smith (Albuquerque: University of New Mexico Press, 2022), 325. For more on the ways in which Mexico has implemented US drug policies on Mexico's own terms, see Aileen Teague, "The War on Drugs in Mexico," in

The War on Drugs: A History, ed. David Farber (New York: New York University Press, 2022), 215–241. Echoing a long history of Mexican preferences, Mexico's presumptive next President Claudia Sheinbaum Pardo says that Mexico's "relationship with the US must be one of respect without subordination." See Arturo Cano, "Meet Claudia Sheinbaum," *The Nation*, trans. Nicolas Allen (April 2024), 58.

4. We have adapted Michael Watt's argument about the struggle over oil resources in Nigeria to that of drugs in the borderlands region. See Watt's "A Tale of Two Gulfs: Life, Death, and Dispossession along Two Oil Frontiers," *American Quarterly* Vol. 64, no. 3 (September 2012), 455.
5. See Dawn Paley for an elaboration on this argument (but one that does not recognize that governments and corporations do not necessarily share the same goals). Dawn Paley, *Drug War Capitalism* (Oakland, CA: AK Press, 2014), 17. We do acknowledge that prohibition did result in the reduction of alcohol consumption in the USA, so there is evidence that prohibition regimes can play a role in reducing drug use. There is also ample evidence that post-prohibition regulations reduced alcohol consumption without the negative side effects of prohibition. See Harry G. Levine and Craig Reinarman, "From Prohibition to Regulation: lessons from Alcohol Policy for Drug Policy," *The Milibank Quarterly* Vol. 69, No. 3 (1991), 461–494.
6. Peter Andreas, "Smuggling Wars: Law Enforcement and Law Evasion in a Changing World," in *Transnational Crime in the Americas,* ed. Tom Farer (New York: Routledge, 1999), 94.
7. Bruce Bagley, "Drug Trafficking and Organized Crime in the Americas: Major Trends in the Twenty-first Century," *Woodrow Wilson Center Update on the Americas,* August 2012. (https://www.wilsoncenter.org/sites/default/files/media/documents/publication/BB%20Final.pdf).
8. We say police forces because most Latin American countries have "multi-layered policing" systems, allowing drug cartels to negotiate the allegiance of some police forces against both other cartels and other police forces. See Tom Wainwright, *Narconomics: How to Run a Drug Cartel* (New York: Public Affairs, 2016), 31–42 and Tim Padgett, "Mexico's Fearsome La Familia: Eerily Quiet,"

TIMEWorld, January 31, 2011 (https://content.time.com/time/world/article/0,8599,2044696,00.html).

9. Alfredo Corchado and Kevin Krause, "Mexico's Drug Violence: Deadly Deal," *The Dallas Morning News,* April 14, 2016 (https://interactives.dallasnews.com/2016/cartels/). This is also true within the USA. The targeting of gang leaders has led to the similar splintering of gangs and a recent spike in gang violence. See David Heinzmann, "Leaderless Chicago street gangs vex police efforts to quell violence," *Chicago Tribune,* July 29, 2016 (http://www.chicagotribune.com/news/local/breaking/ct-chicago-violence-ganges-20160728-story.html).
10. The outsourcing of violence by DTOs can also result in the increased visibility of crime. See Angélica Durán-Martínez, *The Politics of Drug Violence: Criminals, Cops, and Politicians in Colombia and Mexico* (New York: Oxford University Press, 2018), 1–5.
11. Executive Office of the President of the United States of America, *National Drug Control Strategy: Data Supplement 2013,* 75–78 (https://www.whitehouse.gove/sites/default/files/ondcp/policy-and-research/2013_data_supplement_final2.pdf); William Wolfe-Wylie, "What Illegal Drugs Cost on the Street around the World," October 8, 2014 (http://o.canada.com/business/interactive-what-illegal-drugs-cost-on-the-street-around-the-world). See also, United Nations Office on Drugs and Crime, *World Drug Report* 2015 (https://www.unodc.org/documents/wdr2015/World_Drug_Report_2015.pdf).
12. Amnesty International, "US Policy in Colombia," (http://www.amnestyusa.org/our-work/countries/americas/colombia).
13. Moisés Naím, *Illicit: How Smugglers, Traffickers, and Copycats are Hijacking the Global Economy* (New York: Bantam, 2005), 234–235. The USA is one of, if not, the largest consumer of illegal drugs in the world, a point readily acknowledged by US President Barack Obama, see Carmen Boullosa and Mike Wallace, *A Narco History: How the United States and Mexico Jointly Created the 'Mexican Drug War,'* (New York: OR Books, 2015), 186. Until recently, the majority of these drugs were smuggled into rather than produced in the USA. See Peter Andreas, *Border Games: Policing the US-Mexico Divide,* 2nd ed. (Cornell: Cornell University Press, 2009), 16.

14. William C. Martin, “Cartels, Corruption, Carnage, and Cooperation,” in *A War That Can’t Be Won: Binational Perspectives on the War on Drugs,* Tony Payan, Kathleen Staudt, and Z. Anthony Kruszewski, eds. (Tucson: University of Arizona Press, 2013), 57.
15. Wainwright, *Narconomics,* 14–15.
16. Peter Reuter, “The Mobility of Drug Trafficking,” *Ending the Drug Wars: Report of the LSE Expert Group on the Economics of Drug Policy* (May 2014), 33–40.
17. Andreas reminds us that “most smuggling parallels the methods and routes of legal commerce” and Carolyn Nordstrom has demonstrated that “the illegal narcotic industries are run like parallel-economy multinational enterprises.” See Andreas, *Border Games,* 20 and Carolyn Nordstrom, *Global Outlaws: Crime, Money, and Power in the Contemporary World* (Berkeley: University of California Press, 2007), 131. Andreas also notes that the merging of licit and illicit cargoes to camouflage illegal drugs was done by smugglers in response to pre-9/11 crackdowns on illegal immigration and drug smuggling. See Andreas, “A Tale of Two Borders: The US-Mexico and US-Canada Lines After 9/11,” *Center for Comparative Immigration Studies,* UC San Diego (http://escholarship.org/uc/item/6s09j0n2), 3.
18. Howard Campbell, *Drug War Zone: Frontline Dispatches from the Streets of El Paso and Juárez* (Austin: University of Texas Press, 2009).
19. Oscar J. Martínez argues that the US-Mexico border was an alienated borderland for most of the nineteenth century but that it has become “a good example of strong asymmetrical interdependence” with a binational social and cultural system since. See, Martínez, “The Dynamics of Border Interaction: New Approaches to Border Analysis,” in *Global Boundaries: World Boundaries,* Vol. 1, Clive H. Schofield, ed. (New York: Routledge, 1994), 2–5 (quote on page 5).
20. Michael Dear, *Why Walls Won’t Work: Repairing the US-Mexico Divide* (New York: Oxford University Press, 2013), 170.
21. Tony Payan, “Introduction: The Many Labyrinths of Illegal Drug Policy: Framing the Issues,” in *A War That Can’t Be Won: Binantional Perspectives on the War on Drugs,* Tony Payan, Kathleen Staudt, and Z. Anthony Kruszewski, eds. (Tucson: University of Arizona Press, 2013), 16.

22. The exportation of costs is not unintentional. See, Richard H. Friman, *Narcodiplomacy: Exporting the War on Drugs* (Ithaca: Cornell University Press, 1996) and Campbell, *Drug War Zone,* 265–274. The focus on communities of color is also not unintentional. While only "12% of all illegal drug users" are African American, "41% of all those arrested on cocaine and heroin charges" are African American. See Clarence Lusane, *Pipe Dream Blues: Racism and the War on Drugs* (Brooklyn: South End Press, 1999), 3. See also Michelle Alexander, *The New Jim Crow: Mass Incarceration in the Age of Colorblindness* (New York: The New Press, 2010).
23. Robert Koulish and Maartje van der Woude, eds., *Crimmigant Nations: Resurgent Nationalism and the Closing of Borders* (New York: Fordham University Press, 2020).
24. Though this is beginning to change with the rise of the heroin epidemic which has devastated largely white communities. See, Sam Quinones, *Dreamland: The True Tale of Amreica's Opiate Epidemic* (New York: Bloomsbury Press, 2015).
25. St. John, *Line in the Sand,* 157. Canadian border towns such as Windsor suffered similar problems. See, Holly Karibo, "Detroit's Border Brothel: Sex Tourism in Windsor, Ontario, 1945–1960," in *Smugglers, Brothels, and Twine: Historical Perspectives on Contraband and Vice in North America's Borderlands,* Elaine Carey and Andrae Marak, eds. (Tucson: University of Arizona Press, 2011), 83–100 and Kolly Karibo, *Sin City North: Sex, Drugs, and Citizenship in the Detroit-Windsor Borderland* (Chapel Hill: University of North Carolina Press, 2015).
26. David E. Lorey, *The US-Mexico Border in the Twentieth Century* (Wilmington, DE: Scholarly Resources, INC, 1999), 46; "Protestan los Padres de Familia de C. Juárez," *El Mexicano el Norte,* 4, Chihuahua, Archivo Historico de la Secretaria de la Educación Pública-Educación rural, *Box 936, Exp. IV [082].*
27. Jeremy Adelman and Stephen Aron, "From Borderlands to Borders: Empires, Nation-States, and the Peoples in Between in North American History," *American Historical Review* Vol. 104, No. 3 (June 1999): 814–841.
28. This claim is counter to that of Victor Davis Hanson of the conservative Manhattan Institute who claims that global elites are striving

to create borderless societies with free labor. Clearly, the European Union opened up movement within the EU, but they did so through the creation of buffer states at the periphery. Even with the buffer states, the free flow of peoples and goods within the EU has resulted in substantial blowback and the rise of an increasingly popular far-right. See Victor Davis Hanson, "Why borders matters – and a borderless world is a fantasy," *Los Angeles Times,* July 31, 2016 (http://www.latimes.com/opinion/op-ed/la-oe-hanson-borders-20160731-snap-story.html). For the use of buffer states, see Peter Andreas, *Border Games.*

29. Alejandro Lugo points out that not all people can cross borders, usually on account of their race, class, ethnicity, and/or national origin. See Alenandro Lugo, *Fragmented Lives, Assembled Parts: Culture, Capitalism, and Conquests at the US-Mexico Border* (Austin: University of Texas Press, 2008), 117.
30. George T. Díaz, in a study that spans 1848–1945, argues that the shift from taxing trade to "prohibiting criminal consumption" occurred around 1910. We would argue that, in spite of changes over time, the focus on national security and the racial policing of borders continued to 9/11, when the rhetoric (if not the focus) intensified. The rise of Donald Trump, and his promise to deport all undocumented migrants immediately upon his election (should he be reelected) is only an escalation of already existing rhetoric and practices. See George T. Díaz, *Border Contraband: A History of Smuggling Across the Rio Grande* (Austin: University of Texas Press, 2015), 9 and 140.
31. Eva Bertram, Morris, Blachman, Kenneth Sharpe, and Peter Andreas, *Drug War Politics: The Price of Denial* (Berkeley: University of California Press, 1996), 105.
32. Andreas, "A Tale of Two Borders," 5 and Peter Andreas and Richard Price, "From War-fighting to Crime-fighting: Transforming the American National Security State," *International Studies Review* Vol. 3, Issue 3 (Fall 2001), 31–52.
33. Katherine Unterman, *Uncle Sam's Policemen: The Pursuit of Fugitives Across Borders* (Cambridge: Harvard University Press, 2015), 1–5 and William Cronon, *Nature's Metropolis: Chicago and the Great West* (New York: W.W. Norton & Company, 1991), 221.
34. Ray Walser, "US Strategy Against Mexican Drug Cartels: Flawed and Uncertain," *Backgrounder,* No. 2407 (April 26,

2010), 6, (http://www.thfmedia.s3.amazonaws.com/2010/pdf/bg2407.pdf) and Peter Andreas, "Redrawing the Line: Borders and Security in the 21st Century," *International Security* Vol. 48, No. 1 (Fall 2003), 78–112.

35. Kevin Sullivan, Tom Jackman, and Brian Fung, "Dallas police used a robot to kill. What does that mean for the future of police robots?" *The Washington Post,* July 21, 2016 (http://www.washingtonpost.com/national/dallas-police-used-a-robot-to-kill-what-does-that-mean-for-the-future-of-police-robots/2016/07/20/32ee114e-4a84-11e6-bdb9-701687974517_story.html).
36. P.W. Singer, *Wired for War: The Robotics Revolution and Conflict in the 21st Century* (New York: Penguin Press, 2009), Chapter 22.
37. Philip E. Ross, "Budget Drones in Ukraine are Redefining Warfare," *IEEE Spectrum,* May 17, 2023 (https://spectrum.ieee.org/drone-warfare-ukraine).
38. Ray Kurzweil, "The Law of Accelerating Returns," in *Alan Turing: Life and Legacy of a Great Thinker,* ed. Christof Teuscher (New York: Springer-Verlag Berline Heidelberg, 2004), 381 and Ray Kurzweil, *The Singularity is Near: When Humans Transcend Biology* (New York: Penguin Books, 2005), Chapter 2.
39. Ray Kurzweil, "The Age of Intelligent Machines 'Chronology,'" http://www.calculemus.org/lect/si/dlalomzy/mchron.htm.
40. Singer, *Wired for War,* 105 and Bill Joy, "Why the Future Doesn't Need Us," April 1, 2000, http://www.wired.com/2000/04/joy-2/.
41. Kurzweil, *The Singularity is Near,* 8–9.
42. Ray Kurzweil, "the Coming Merging of Mind and Machine," *Scientific American,* March 23, 2009 (http://www.scientificamerican.com/article/merging-of-mind-and-machine/).
43. Kurzweil, *The Singularity is Near,* 7.
44. We have used Isaac Asimov as a science fiction example because his early Foundation and robot series predict a sort of singularity, but there is an entire subgenre of singularity science fiction which includes the works of Charles Stross, Iain M. Banks, Daniel Suarez, William Hertling, Neal Asher, Ken MacLeod, and Peter Watts, among others. There is also the movie *Sleep Dealer,* which depicts a future where Mexicans no longer need to cross the border to engage in low wage labor in the USA. Instead, they are turned into cyborgs who tap into a global computer network to do their

work in the USA from a distance. See Alex Rivera and David Riker, *Sleep Dreamers,* DVD, Directed by Alex Rivera (Los Angeles: Maya Entertainment, 2008).

45. Will Oremus, "Ford Wants to Be the Fort of Driverless Cars," *Slate,* August 16, 2016 (http://www.slate.com/blogs/future_tense/2016/08/16/ford_says_it_will_mass_produce_a_driver less_car_by_2021.html); Julia Angwin, "Autonomous Vehicles are Driving Blind," *New York Times,* October 11,2023 (https://www.nytimes.com/2023/10/11/opinion/driverless-cars-san-francisco.html); Mary L. "Missy" Cummings, "What Self-Driving Cars Tell Us About AI Risks: 5 conclusions from an automation expert fresh off a stint with the US highway safety agency," *IEEE Spectrum,* 30 July 2023 (https://www.nytimes.com/2023/10/11/opinion/driverless-cars-san-francisco.html); Elizabeth Dwoskin, "Putting a Computer in your Brain is no Longer Science Fiction," *The Washington Post,* August 15, 2016 (https://www.washingtonpost.com/news/the-switch/wp/2016/08/15/putting-a-computer-in-your-brain-is-no-longer-science-fiction/?hpid=hp_hp-more-top-stories-2_switch-computerbrain-320pm%3Ahompeage%2Fstory), and Kurzweil, *The Singularity is Near,* 185.
46. Erik Brynjolfsson and Andrew McAfee, *The Second Machine Age: Work, Progress, and Prosperity in a Time of Brilliant Technologies* (New York: W.W. Norton & Company, 2014), 121 and 191.
47. Singer, *Wired for War,* 108.
48. For an assessment of the impact of NAFTA and China's entry into the WTO, see Fred C. Bergsten and Jeffrey J. Schott, "Testimony before Congress on the Goals of NAFTA and Preliminary Evaluation," Peterson Institute of International Economics (http://library.brown.edu/create/modernlatinamerica/chapters/chapter-12-strategies-for-economic-development/primary-documents-with-accompanying-discussion-questions/document-30-testimony-before-congress-on-the-goals-of-nafta-and-preliminary-evaluation-peter-bergson-1997/); Mark Weisbrot, Stephan Lefebvre, and Joseph Sammut, "Did NAFTA Help Mexico: An Assessment after 20 Years," Center for Economic and Policy Research, February 2014 (http://cepr.net/documents/nafta-20-years-2014-02.pdf); and Shades of Grey: Ten Years of China in the WTO," *The Economist,* December 10, 2011 (http://www.economist.com/node/21541408). Note that post-NAFTA Mexico had higher

levels of unemployment than pre-NAFTA Mexico, but this may very well be the combined result of competition from China beginning in 2001 and the impact of automation.

49. The most obvious example of efficiencies is that of Walmart. See Natalie Berg and Bryan Roberts, *Walmart: Key Insights and Practical Lessons from the World's Largest Retailer* (New York: Kogan Page, 2012).
50. We see similar issues in urban areas where well-paid union factory labor has all but disappeared. See, for example, Alexia Elejalde-Ruiz, "Chicago tackles youth unemployment as it wrestles with its consequences," *Chicago Tribune,* September 2, 2016 (http://www.chicagotribune.com/business/ct-youth-unemployment-chicago-neighborhood-0904-vi-20160527-story.html) and John Eligon, "Racial Violence in Milwaukee Was Decades in the Making, Residents Say," *New York Times,* August 14, 2016 (http://www.nytimes.com/2016/08/15/us/racial-violence-in-milwaukee-was-decades-in-the-making-residents-say.html).
51. Briseida Valencia Soto, "Controversial Alternative for a Trapped Labor Force: Mexico's Formal Employment and Illicit Drug Production," *Council on Hemispheric Affairs,* July 5, 2016 (http://www.coha.org/controversial-alternative-for-a-trapped-labor-force-mexicos'formal-employment-and-illicit-drug-production/). The value of raising marijuana has gone down significantly as marijuana has become decriminalized and legalized in certain US jurisdictions. See Deborah Bonello, "Mexican Marijuana Farmers See Profits Tumble as US Loosens Laws," *Los Angeles Times,* December 30, 2015 (http://www.latimes.com/world/mexico-americas/la-fg-mexico-marijuana-20151230-story.html). For the allure of cartel life, see Dan Slater, *Wolf Boys: Two American Teenagers and Mexico's Most Dangerous Cartel* (New York: Simon & Schuster, 2016); Donna Leinwand, "Drug Cartels Load Up Young 'Mules'," *USA Today,* August 8, 2001 (http://usatoday30.usatoday.com/news/nation/2001/08/08/kid-smugglers.htm); and James C. McKinley, Jr., "Mexican Cartels Lure American Teens as Killers," *The New York Times,* June 22, 2009 (http://www.nytimes.com/2009/06/23/us/23killers.html). See also Quinones, *Dreamland.*
52. Brynjolfsson and McAfee, *The Second Machine Age,* pp. 230–232.

53. Sasha Abramsky, “The Fear Industry Goes Back to School,” *The Nation* (August 29 and September 5, 2016): pp. 18–21.
54. Singer, *Wired for War,* 39.
55. Ibid.
56. Christopher Bolkcom, “Homeland Security: Unmanned Aerial Vehicles and Border Surveillance,” Congressional Research Service, December 7, 2004, p. 2.
57. Chad C. Haddal and Jeremiah Gertler, “Homeland Security: Unmanned Aerial Vehicles and Border Surveillance,” Congressional Research Service, July 8, 2010, p. 4.
58. Singer, *Wired for War,* pp. 124–126.
59. Steven Metz, *Armed Conflict in the 21st Century: The Information Revolution and Post-modern Warfare,* US Army War College, Strategic Studies Institute, April 2000, xix.
60. Singer, *Wired for War,* pp. 315–316.
61. Tony Payan, Kathy Staudt, and Tony Kruszewski, *Human Rights on the US-Mexico Border* (Tucson: University of Arizona Press, 2009).
62. George T. Díaz, *Border Contraband: A History of Smuggling Across the Rio Grande* (Austin: University of Texas Press, 2015).

CHAPTER 9

Artificial Intelligence for Peace: An Early Warning System for Mass Violence

Michael Yankoski, William Theisen, Ernesto Verdeja, and Walter J. Scheirer

Introduction

Pundits are increasingly raising concerns about the dangers that advanced artificial intelligence (AI) systems may pose to human peace and safety. Geoffrey Hinton, the "Godfather of AI" has warned of potential dangers ahead (Metz 2023). Elon Musk (Clifford 2018) warned that AI has the potential to be more dangerous than nuclear weapons. Stephen Hawking (BBC 2014) worried that AI could mean the end of the human species. A recent Special Issue of the *Bulletin of the Atomic Scientists* included several warnings about the coming AI Arms Race (Roff 2019). Indeed, many of the chapters in this volume are similarly concerned with mitigating the negative implications of advanced AI.

We agree that caution is warranted regarding the rapid development in the field of AI, but we also believe that some AI research trajectories may be employed toward positive ends. In this chapter, we introduce one

M. Yankoski (✉)
Department of Data Science, College of William & Mary, Williamsburg, VA, USA
e-mail: myankoski@wm.edu

R. Kiggins (ed.), *The Political Economy of Robots*, International Political Economy Series, https://doi.org/10.1007/978-3-031-95171-8_9

current research trajectory that combines AI and social scientific research on political violence in order to contribute to practical conflict prevention. AI systems are capable of significantly enhancing the work of peacebuilders in specific but important ways, for artificial intelligence provides unique tools for identifying and analyzing emergent trends and threats within massive volumes of real-time data on the Internet, well beyond the capacities of most existing political violence early warning systems.

This chapter discusses a novel project that brings together computer and social scientists using artificial intelligence to advance current atrocity and political instability early warning capabilities. We focus on how the spread of disinformation, rumors, and lies on social media—essentially, hate propaganda—in already unstable political contexts may function as early warning indicators of imminent large-scale violence. Researchers have long argued that hate propaganda legitimizes violence against vulnerable groups (Chirot and McCauley 2010; Sémelin 2005; Kiernan 2003; Koonz 2003; Weitz 2003), but in our current social media landscape, where harmful and manipulative political content circulate more rapidly and widely than ever before, the dangers are especially acute. This is evident, for instance, in the lead up to the Indonesian elections in 2019, where Instagram and X (then known as Twitter) were filled with conspiratorial allegations about treasonous politicians who had to be prevented from winning at the ballot box by any means, including through terror, threats, and killings (Suhartono and Victor 2019; BBC 2019). In Myanmar, ongoing atrocities against the Rohingya minority group have been fueled by nationalist memes on X and Facebook accusing the Rohingya of being dangerous foreigners, an existential threat to the

W. Theisen · W. J. Scheirer
Department of Computer Science and Engineering, University of Notre Dame, South Bend, IN, USA
e-mail: wtheisen@nd.edu

W. J. Scheirer
e-mail: walter.scheirer@nd.edu

E. Verdeja
Kroc Institute for International Peace Studies and Department of Political Science, University of Notre Dame, South Bend, IN, USA
e-mail: Ernesto.Verdeja.1@nd.edu

integrity and survival of the country that must be eradicated (Azeem 2018; BSR 2018). In these cases and many others, social media has played a defining role in perpetuating dehumanization and facilitating violence.

Our focus in this project is on advancing what peace studies scholars often refer to as "negative peace"—the absence of armed conflict and direct violations of bodily integrity, such as killings, assaults, and torture—by contributing to early warning modeling and analyses of political violence.[1] As such, we present a model of computational forensic analysis of digital images in social media to help identify where and when unstable societies may tip into large-scale violence, by providing journalists and prevention practitioners—that is, policymakers, analysts, and human rights advocates in the atrocity prevention community—with data-rich, theoretically robust assessments of processes of violence escalation in near "real time." This type of triage mechanism is central to ensuring timely and effective preventive responses on the ground. Our project is a work in progress, but we believe it suggests a path forward that can also benefit from contributions from the broader scholarly community.

We realize that developing the political will to prevent or stop violence is crucial and extremely difficult (Lupel and Verdeja 2013; Weiss 2016), but providing more accurate and actionable information on conflict escalation can aid prevention work significantly. Our unique focus on digital images is driven by the new ways in which people communicate on the Internet, which are no longer just text-based. Given the enormity of social media data produced daily and the need for timely and accurate analysis, AI systems offer unique capabilities for enhancing and even transforming the work of practitioners tasked with anticipating and responding to violence.

The chapter is dedicated to outlining the computational dimensions of our project. We proceed in several steps. First, we outline the current state of risk assessment and early warning research and practice, and situate our project within this field. We then introduce specific problems of disinformation: namely, how social media is increasingly used to sow fear and distrust in already fragile communities and further legitimize violence against vulnerable groups. This kind of disinformation is an important indicator of likely violence in the near- or mid-term, and thus needs to be actively monitored if early warning analyses are to be more focused, accurate, and actionable. Despite their importance, it is exceedingly difficult to understand the spread and impact of coordinated disinformation campaigns in real time using currently available tools. We then discuss our project in detail, first by outlining the types of entities we analyze—social media memes—and then by sketching the overall computational model of analysis. We then turn to some further areas of development, and finally explore the ethical and policy implications of AI work in atrocity prevention.

Risk Assessment and Early Warning: What We Know, What We Need to Know

There is a long history of systematic attempts to anticipate the outbreak of large-scale political violence, going back at least to the 1950s when the superpowers sought to model the likelihood of nuclear war through a variety of simulations (Edwards 1997; Poundstone 1992). Since the international community's failure to prevent genocides in Bosnia-Herzegovina and Rwanda in the 1990s, governments and human rights advocates have devoted more attention to forecasting political instability and mass violence, often working closely with scholars to develop rigorous and evidence-based approaches to prevention work.[2] Today there are numerous early warning and risk assessments initiatives focusing on the main drivers and signs of impending violence and also to informing atrocity prevention (Waller 2016).

We now have a sophisticated understanding of conditions that elevate risk of violence, but systematic and generalized models of short-term patterns of violence onset are less well developed. It is exceedingly difficult to interpret fast-moving political violence dynamics. Our project contributes to these early warning efforts to understand real-time violence escalation through an analysis of the circulation of digital images on social media that can encourage and legitimize harm against vulnerable minorities. The project primarily focuses on countries already at high risk of violence where the timing or onset of violence is difficult to know. This is ultimately about forecasting the likelihood of violence, not about providing a causal theory of violence. The distinction between these is pivotal. Much like sharp pains in the left torso may indicate an imminent heart attack without being its cause, political violence forecasting is concerned with assessing the probability of future violations, rather than causally explaining their occurrence after the fact.

Current research has identified a host of general conditions that elevate a country's likelihood of future violence. The first condition is a *history of unpunished violence against vulnerable minorities* (Harff 2003; Goldsmith et al. 2013). Prior impunity legitimizes future violence because potential perpetrators know they face little or no sanction. *Severe political instability*, especially armed conflict, is another major risk factor (Midlarsky 2005; Goldstone et al. 2010). In countries experiencing ongoing and profound crises, such as war, insurgencies, coups, or violent changes of political control, leaders are much more likely to rely on increasingly harsh repressive measures to eliminate perceived threats and remain in power. Armed conflict, whether an international or civil war, is among the strongest predictors of future atrocities against civilians. A third factor is the espousal of a *radical ideology* by government leaders and/or armed challengers that systematically dehumanizes others (Robinson 2018; Weitz 2003). Such extreme ideologies—whether religious, ethnic, racial, or authoritarian variants of left- or right-wing ideologies—justify the use of increasingly repressive measures against vulnerable civilians. Related to this is ongoing *state-led discrimination*, including the denial of basic civil and political rights as well as movement restrictions, which are highly correlated with atrocities (Fein 2007; Goldstone et al. 2010). Finally, *government regime* matters; authoritarian regimes are more likely than democracies to engage in violence. Semi-democratic regimes, with limited political contestation and some opposition political movements but weak rule of law and unaccountable political leaders, are also more prone to violence than robust democracies (Stewart 2013). The greater the presence of these factors, the greater risk a country has of experiencing mass violence. However, these factors are largely static—they are useful for providing general risk assessments (low, mid, or high risk) of violence, but the factors do not fluctuate much over time. They tell us the relative likelihood of future violence, but do not provide precise insights into *when* a high-risk situation may devolve into overt killings and atrocities.

Much harder to pinpoint in real time are the short- and mid-term events and processes that are indicators of the shift from high-risk conditions into actual violence, or what is normally known as early warning (Heldt 2012). Broadly, there are three clusters of early warning indicators (Verdeja 2016). First, there are *dangerous symbolic moments or discourses* that significantly dehumanize already vulnerable populations, or reinforce deep identity cleavages between groups. This includes rallies and commemorations of divisive events or the spread of hate propaganda. Second is an uptick in *state repression*, such as moving security forces to places with vulnerable populations, stripping those populations of legal rights, attacks against prominent minority or opposition leaders and their followers, or widespread civilian arrests. Finally, ***political and security crises*** challenging incumbent political leaders are important indicators, and these can include new or resumed armed conflict between the state and rebels, rapid changes in government leadership, or the spread of confrontational protests. Unforeseen exogenous shocks like natural disasters or neighboring conflict spillover can also trigger violence by challenging the ability of political leaders to maintain control. In many instances, several early warning indicators will occur simultaneously or in clusters.

Despite these existing indicators, there is room to improve and expand early warning capabilities. The primary concern involves limitations in overall availability and quality of information. Many conflict scenarios are extremely difficult and dangerous to access physically, and thus we must rely on the information provided by relatively limited numbers of people in the field, whether journalists, aid workers, local residents, government officials, or displaced civilians. The reliability of these sources varies, but even when sources are dependable, it can be exceedingly hard to know how representative information is, especially when mobility is limited. For instance, does a journalist's reporting in a small area reflect the entire region or the hardest hit areas of a country? If not, what is missing? How can we compensate for these limitations?

In short, many existing early warning models can expand their source material to systematically tap into much richer social media streams, which can inform how violence may be escalating or if a situation is on the cusp of escalation. To be clear: social media streams do not represent a more factually accurate portrayal of what is occurring. Many politicized memes, for instance, are misleading or outright lies and their provenance (i.e., origin and mode of creation) can be hard to identify, as we discuss below. Indeed, social media is itself a new battleground in contemporary conflicts, as armed actors frequently distort and misrepresent the actions and motives of their enemies as a justification for violence. However, politicized social media gives a stream of real-time data that our system is capable of analyzing in order to identify trends that signify increased potential for outbreaks of political violence. Thus, integrating social media analysis into early warning evaluations may significantly enhance conflict prevention and intervention work. In order to understand what this entails, the next section discusses what a political meme is and then presents the main features of our AI project.

What Is a Political Meme?

If the new media landscape is a battleground, then one must study how contemporary political movements communicate on the Internet in order to begin to formulate a response. Unlike in the past, where most people were largely consumers of professionally curated media content, anyone can now make and disseminate their own political messages to an audience of millions on the Internet. The widespread availability of powerful image editing and synthesis tools (e.g., Photoshop, Midjourney) has democratized digital content creation, allowing users with basic computer skills and time to produce custom images and videos. This content most often takes the form of a *meme*. Memes are cultural artifacts that evolve and spread like a biological organism, but are completely external to biology. On the Internet, memes consist of images and videos, often humorous, that adhere to a set genre, which acts as a guideline for the creation of new instances. But these images are often more than just jokes. Memes have served as the impetus for political actions in movements as diverse as the Arab Spring (York 2012), Occupy Wall Street (Know Your Meme 2020), and the 2024 Pro-Palestinian Campus Protest (Taylor 2024) movements (Fig. 9.1). And they are now a significant resource for monitoring the pulse of an election, responses toward international security incidents, or the viewpoints surrounding a domestic controversy.

The popularity of memes continues to grow, and so does the scope of political messaging attached to them. Cases where political memes prefigured violence in some form are not difficult to come by on social media. Our own work has uncovered cases across the globe where memes have been used to spread antisocial messages from discriminatory stereotypes to outright calls for violence. In Brazil, we found memes of the controversial former President Jair Bolsonaro depicted as an action hero, ready to take on the country's drug traffickers (left panel of Fig. 9.2). This coincided with an escalation of the Brazilian drug war and violence against the poor, in which state security forces had been responsible for over a third of the violent deaths reported in Rio De Janeiro (Santoro 2019). In India, we witnessed misogynistic memes featuring a cheerful Prime Minister Narendra Modi overlaid with text containing degrading messages against women (center panel of Fig. 9.2), while the government continues to undermine legal protections for women (Human Rights Watch 2018). In Indonesia, we discovered images where a hammer and sickle were superimposed on the shirts of Muslim political opponents, meant to insinuate

Fig. 9.1 A selection of political memes from the past two decades exemplifying cultural remixing or reenactment (Shifman 2014). Left: A meme that is critical of the Syrian regime's use of poison gas, in the style of the iconic Obama "Hope" poster. Center: a meme associated with the UC Davis Pepper Spray Incident during the Occupy Wall Street Protests. Right: A screenshot from a TikTok video of the pro-Palestinian encampment protest at Harvard University. It incorporates emojis signifying the purpose of the protest, as well as the frenzied activity of the students scrambling to assemble their tents before law enforcement could intervene—an activity that would become emblematic of this protest movement in the United States in 2024

that they are crypto-communists (right panel of Fig. 9.2). These images were found shortly before the 2019 presidential election, which ended with violent street protests in Jakarta (Suhartono and Victor 2019; BBC 2019). The list goes on.

If AI is to be deployed as a mechanism to watch for political memes that may be used to incite large-scale violence in high-risk contexts, we require a rigorous definition of a political meme to operationalize the algorithms. Definitions like the one given above referring to evolution and organism-like propagation are typically attributed to the biologist Richard Dawkins (2016). However, Dawkins' thinking on the meme, whether intentional or not, borrowed liberally from the notion of intertextuality in literary theory: the shaping of a text's meaning by other texts. Julia Kristeva, and Mikhail Bakhtin before her, suggested that the novel reworking and retransmission of information is fundamental to human communication (Kristeva 1986). In the Kristevan mode, intertextuality applies to all semiotic systems, including digital images and video. Intertextuality is a

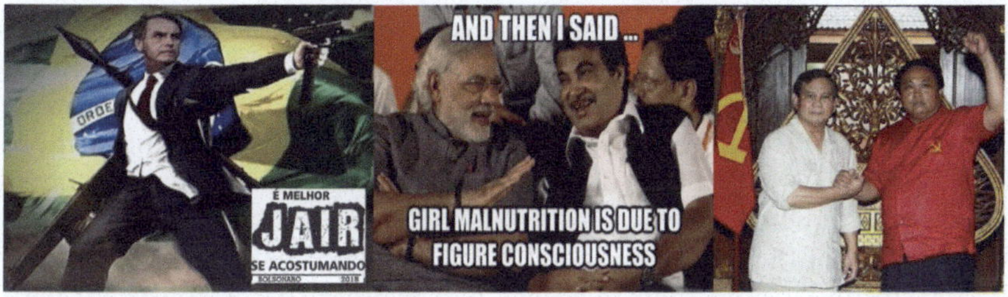

Fig. 9.2 A selection of political memes with disturbing messaging. Left: Former Brazilian President Jair Bolsonaro depicted as an action hero, ready to take on Brazil's drug traffickers. Center: A misogynistic meme featuring Indian Prime Minister Narendra Modi. Right: Hammer and sickle superimposed on the shirt of a Muslim politician in Indonesia

more useful framing when assessing content on the Internet, which can be collected and analyzed via automatic means to identify *intertexts*, those specific points of correspondence between artifacts (Forstall and Scheirer 2019). While other researchers have sought to define the meme in general terms (Shifman 2014), we are not interested in all memes for an early warning system for violence. Thus we offer the following definition for a political meme: a multimedia intertext meant to engage an in-group and/or antagonize an out-group.

Given this definition, how can we operationalize it within the context of today's AI capabilities to give an algorithm what it needs to automatically assess any potential threats of violence that political memes might pose? First, the source of the meme can be scrutinized. Where the meme was found on social media, and who posted it, can be diagnostic. For instance, if the source is known to be political, the meme might be as well. But the source is not essential for an observer to understand the message a meme conveys. More importantly, the content of the meme is composed of visual and textual cues that deliver the message. In many instances, decisions are made based purely on the visual style of an image. For example, does it look like something we have seen before, which is known to be political in some regard? If so, then an intertextual association has been made. If it is new, is there something in the visual content or text, if present, that gives us a clue as to whether or not it is political? Such clues can be the presence of political figures in the image, places associated with political history, symbols associated with political or religious groups, or objects with some political significance. Finally, not all political memes are something to worry about; thus, we need to separate the innocuous from the dangerous. This is done by establishing semantic links between the visual content and text, as well as by assessing the sentiment of those elements. The ability to recognize all of this involves pattern recognition—an area where AI systems excel, because they can be trained to recognize all of the aforementioned elements. The challenge is in making the system understand their relevance in a manner that is consistent with human observers.

Technologies for an Early Warning System for Violence

If visual content found on social media is the object of focus for our early warning system, then what, exactly, are the technical requirements for understanding it? This task is fundamentally different from more traditional forms of early warning, where indicators are established using quantitative variables from the social science literature (e.g., datasets on protests, armed conflict onsets, coups, etc.), that can be processed using statistical techniques from data science and used to make general predictions. Here we bring to bear new methods from the areas of computer vision and media forensics, as well as established best practices from high performance computing and web and social media studies. We do this in order to build a comprehensive system that proceeds from data ingestion to making predictions about content that may contain messaging that seems intended to incite large-scale violence. In general, there are three basic required components of such a system (Fig. 9.3): (1) the data ingestion platform; (2) the AI analysis engine; and (3) the user interface (UI). In this section, we will detail these technical aspects of the system.

Data ingestion is a critical part of the process. If we do not look in the right place to begin with, then the early warning system will not be useful. Given the consolidation of the Internet over the past decade (itself a contributing factor to the overall problem we are studying) into a handful of popular social networks (Internet Society 2019), data targeting is, in some sense, straightforward. However, as one dives into these social networks, their vast internal complexity becomes apparent, with many communities and subcommunities existing in complex webs on platforms like Reddit, 4chan, X, Telegram, and Instagram. So, where to look? Social media platforms have always striven to add structuring elements to their

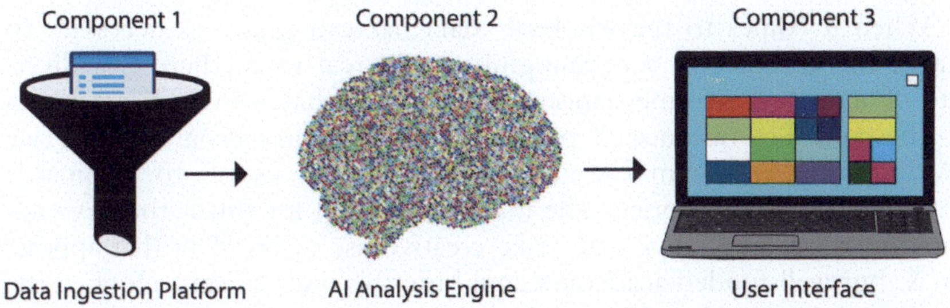

Fig. 9.3 The processing pipeline of the proposed early warning system for large-scale violence, composed of three basic required technology components

posts, in order for users to better identify relevant topics and authors. Examples of these structuring elements include user-defined hashtags (the "#" symbol followed by a plaintext string) that are used as meta-data to tag posts with custom topics on most social networks, as well as account names (often indicated by the "@" symbol). However, the reliability of these elements can be questionable, and some platforms go out of their way to render them useless (most notably 4chan, where nearly every user is anonymous). Media objects such as images and videos also form structured elements in posts, and for our purposes, posts containing these media types are of primary interest. Accordingly, our attention will be on visual content for the rest of this chapter. Note though that it is possible to bring comments, shared or reposted content, and "likes" into the analysis as well.

This is where generative AI systems based on GPT technology (Generative Pretrained Transformer) are providing an interesting possible avenue of research in this context (Vaswani et al. 2017). Multimodal models are showing significant promise in being able to assist human subject-area experts in early stage analysis by helping identify whether a particular meme may contain politically salient content. Further research is ongoing about whether GPT models might be sufficiently fine-tuned to provide low-level contextually aware analysis of the political implications and meaning of memes that have been identified as politically salient. But scale is a major limitation at this time for this approach given the need to collect, analyze, and summarize billions of media objects on a continuous basis.

When it comes to the wholesale data harvesting that is necessary to watch the Internet in a meaningful way in real time, there is a diversity of access control mechanisms put in place that restrict direct access to the data. For instance, X provides an Application Program Interface (API) (X 2024) that makes all of its data very accessible to automated data ingestion—for a price. The main motivation for this is the development of third-party apps that make creative use of the data that appears on X, but well-funded academics could use this mechanism to look at the social aspects of information exchange on the platform. Other sites, like Facebook, are far more closed, leading to the need for specialized content crawlers that can search for relevant content without the assistance of an open API. Rate limiting procedures put in place on social networks are another confounding element, and a potentially dangerous roadblock for any early warning system that needs timely access to data. An example of this is Facebook, which uses dynamic links to content, which change on a regular basis. These problems can be mitigated with a distributed system design, whereby many instances of the data ingestion component run from different parts of the Internet.

A practical constraint of data ingestion is the scale of the data an early warning system must consider in order to be effective. In a targeted operation, where a particular region and/or set of actors is being monitored, we can operate over data on the order of millions of images within a period of weeks. This is the current upper-bound for media forensics being conducted at the academic level (NIST 2018). Beyond this, the cost of data storage and the time required to collect the data becomes prohibitively expensive. However, there is a need for the system to scale to the order of billions of images per day (Eveleth 2015) for a truly comprehensive real-time capability. The ultimate goal is to be able to watch all of social media for emerging trends. In order to accomplish this goal, partnerships will need to be established between non-governmental organizations (NGOs), social media companies, and academics to provide access to the data at the source. The reluctance of social media companies to allow outsiders access to their internal data repositories is a serious stumbling block in this regard. We envision that such partnerships will lead to the transfer of early warning techniques that can be run internally at a social network when data access is problematic.

Given existing constraints, we have found an effective combination for data ingestion to involve partnering with local experts who can provide relevant hashtags and accounts to review, thus mitigating the need to monitor everything on the Internet at once. While this approach may miss quite a bit of potentially threatening content, targeted sampling is still effective. For example, in our work on the 2019 Indonesian Presidential Election (Yankoski et al. 2020), we partnered with the local fact checking organization Cekfakta (2020). By using local volunteers throughout Indonesia, Cekfakta is able to identify social media sources of concern. However, human monitoring of even a limited number of sources proved incapable of keeping up with the pace of content creation. From just 26 hashtags and eight users on X (then known as Twitter) and Instagram, we harvested over two million images for analysis (Theisen et al. 2021)—a staggering number that exceeds the human capacity to find patterns in unorganized data.

The artificial intelligence component of the early warning system is designed to automate the analysis of the large collections of media content assembled at the data ingestion stage. A key to success in this regard is the use of state-of-the-art artificial perception methods. In the debates surrounding the use of artificial intelligence technology, there is a pervasive misunderstanding of the difference between perception and cognition. As we discussed in the beginning of this chapter, fears over AGI, or artificial general intelligence, have led to accusations that all AI technology represents an existential threat to humanity. Should AGI ever appear, it would embody a set of cognitive models meant to mimic the conscious mental actions of knowledge acquisition and reasoning in the human mind. This type of technology does not exist at the time of writing, and we are skeptical that it will emerge in the foreseeable future. The complexities of the human brain as a system are beyond the current understanding of science. We do not possess a model of computation for the brain, nor do we have explanatory models for complex phenomena such as conscious thought (Marcus and Davis 2019).

Where AI has made inroads into modeling competencies of the brain is in the sensory systems (e.g., audition, olfaction, vision). Perception is the ability to take information from a sensory system and make decisions over it. This process mostly unfolds in an unconscious manner, but embodies a set of complex pattern recognition behaviors. The most studied and best modeled sensory modality is vision. The fields of computer vision and machine learning have taken direct inspiration from experimental observations in neuroscience and psychology (Goodfellow et al. 2016), leading to features (i.e., descriptions of the data) and classifiers (i.e., models that make decisions over features) that are, in some cases, at human- or super human-level performance (RichardWebster et al. 2018). These technologies can be used safely and effectively in the appropriate context. For instance, computer vision can be used to determine which images out of a large collection are similar in overall appearance, identify specific objects within images, and match common objects across images.

Concern over manually and automatically generated fake content has driven advances in media forensics—a field within computer science that borrows heavily from the fields of computer vision and machine learning. For a violence early warning system, we need a way to characterize the images such that (1) from an initial collection, they can be placed into distinct genres, (2) new images can be placed into known genres or new genres where appropriate, and (3) semantic understanding of the visual content can be extracted, so that threatening messages can be identified. Select techniques from media forensics give us a path forward for each of these requirements.

For establishing connections between images, *image provenance analysis* provides a powerful framework (Moreira et al. 2018, 2022). Work in image manipulation detection has shown that it is possible to estimate, through image processing and computer vision techniques, the types and parameters of transformations that have been applied to the content of individual images to obtain new versions of those images (Rocha et al. 2011). Given a large corpus of images and a query image (i.e., an image we would like to use to find other related images), a useful further step is to retrieve the set of original images whose content is present in the query image, as well as the detailed sequences of transformations that yield the query image given the original images. This is known as image provenance analysis in the media forensics literature. The entire process is performed in an automated unsupervised manner, requiring no human intervention. Such a process can be used to trace the evolution of memes and other content, which is a piece of what we need for the early warning system. This process can also be used for fact checking and authorship verification. In general, provenance analysis consists of an image retrieval step followed by a graph building step that provides a temporal ordering of the images. In place of the latter, we suggest that an image clustering step is more useful for early warning analysis (Theisen et al. 2021). We will, in broad strokes, explain how each of these steps works below.

In order to find related images, each must first be indexed based on features that describe the style and content of the images, but in a compact way that reduces the amount of space needed to store the data. Such a representation of the data can be generated using techniques from the area of *content-based image retrieval*, which addresses the problem of matching millions of images based on visual appearance. In our prototype early warning system, we build an index of all images based on local features, instead of the entire global appearance of the images. This strategy allows us to find matching images based on small-localized objects that they share. For distinct meme genres of diverse visual appearance, finding just one small-shared object could be the link for establishing a valid relationship between images. This gives our technique excellent recall abilities over large collections of images. In order to scale to millions of images, we make use of SURF features (Bay et al. 2008) that can be computed quickly and stored efficiently. The index is an Inverted File (IVF) index trained via Optimized Product Quantization (OPQ) (Ge et al. 2013).

After the index is built, it can be used to find related images through a querying process. In a manner similar to what is done in traditional image provenance analysis, query images are chosen and are matched against the images in the index to return the closest matches (the number returned is a user-defined parameter). The choice of query images could be random (i.e., randomly sampled from all of the available images), or determined through the use of image manipulation detectors that can identify suspicious images. Our prototype system has defined a scoring system relying on the quality of matches between individual objects in images, based on the correspondence between the pre-calculated features in the index with the query.

The matching process for image retrieval results in collections of ranked lists (Fig. 9.4). This is somewhat useful, but what is ideally needed here is a data clustering approach that depicts how each image is visually situated with respect to other related images. Further, each cluster should represent a distinct genre of content that is evident to the human observer. In a meme context, this means that memes that humans have labeled (e.g., "Action Hero Bolsonaro," "Misogynistic Modi") should be found in a single coherent cluster. With respect to other content, derivatives of an ad for instance should also be grouped together in a similar way. Our prototype system uses a spectral clustering algorithm to produce the final output of the system (Yu and Shi 2003).

An important question is, how well does this prototype early warning system work? We validated the system on the dataset of two million images from the 2019 Indonesian Presidential Election that is referenced

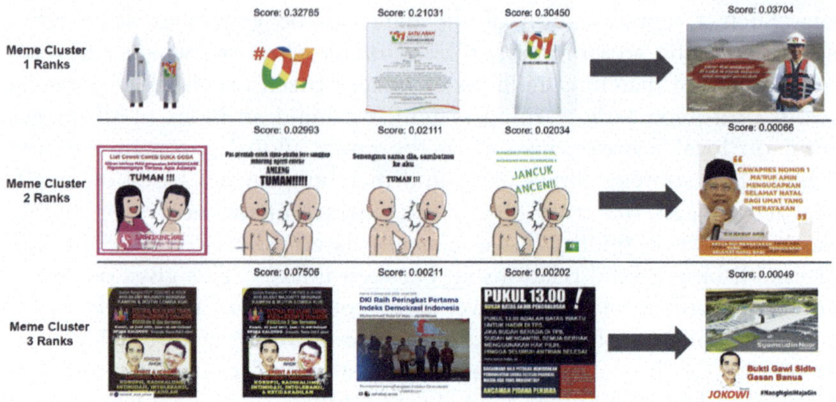

Fig. 9.4 The output of the provenance filtering process to find related images in a large collection for three different meme genres from Indonesia. Each row depicts the best matches to a query image (the left-most image in these examples) in sorted order, where images ideally share some aspect of visual appearance. Scores reflect the quality of matches between individual objects in images. At the very end of the sorted list, we expect the weakest match, and the very low scores reflect that. These ranks form the input to the clustering step, which presents a better arrangement for human analysis

above. This case study is particularly salient for this work, in that the results of the election led to violent episodes in Jakarta, some of which were stoked by content found on social media (BBC 2019). Most critically, our validation sought to verify that the prototype system was able to detect useful meme genres from millions of images, as well as verify that the images contained within a genre are meaningful to human observers. In total, the system discovered 7691 content genres out of the pool of roughly two million images. Some examples are shown in Fig. 9.5. Each of these genres was checked by a human observer to assess visual coherence and to tag the genre with a label that described its content. Roughly 75% of the images were placed into human interpretable clusters. Further, controlled human perception experiments were also conducted to verify that the genres were not simply the product of random chance. These showed that not only were human observers adept at perceiving a pattern in most presented clusters, but also that a majority of the detected genres had a cohesive-enough theme that was identifiable even in the presence

Fig. 9.5 Selections from three different content genres from the total pool of 7691 discovered by the prototype system

of an impostor image. In other words, the AI system is performing accurate pattern recognition and organizing data at a scale that was once unthinkable to human analysts.

A second case study was performed on the ongoing invasion of Ukraine by Russia. In this case study over 13 million artifacts were collected, with collection continuing to date. Using the process outlined above, image clusters were created and analyzed. Examples of discovered motifs can be seen in Fig. 9.6. Clusters were shown to be politically salient by subject matter experts with topics ranging from fighter jets, Ukrainian flags being flown at rallies, to "virgin-chad" memes making fun of popular political figures related to the conflict (Theisen et al. 2023).

In addition to the AI algorithms, we must also consider how to make the information such a system generates accessible to end users. It is very likely that most of the target users of such a system will not be comfortable interacting with the command line of a computer system. Thus a user interface (UI) layer must be developed and tested, as well as an alert system that will be capable of notifying the correct people with the correct information when an imminent threat begins to trend. It is conceivable that policymakers would require one set of information from this system, while civil society actors would require a different set, and reporters still another set. Working closely with these distinct communities of users is a critical aspect of ensuring this system's utility as an early warning system.

Major considerations remain for the design of a UI that is accessible to users who are not computer scientists. As can be seen in the figures included in this section, the current UI is minimalist by design and can

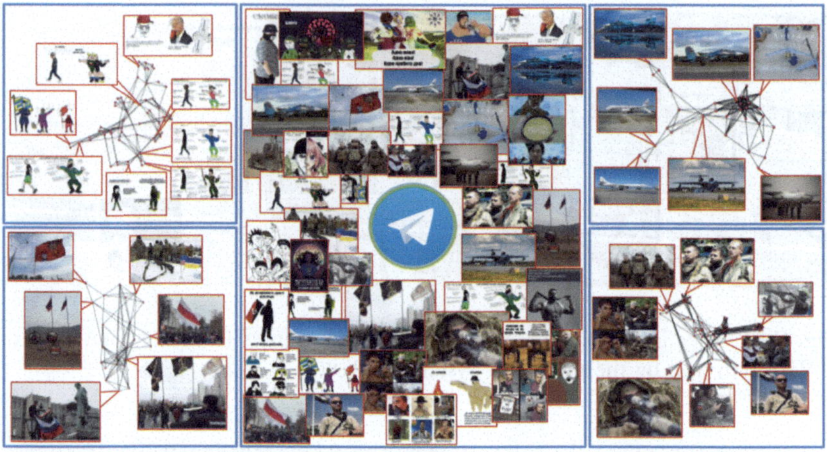

Fig. 9.6 Selections from Ukrainian-Russian Telegram data. In total over 13 million samples have been collected by our team to date

be further developed. At this point in our development, the UI primarily consists of a web interface that presents a list of clusters to users, which can be selected and visualized (Fig. 9.6). We envision the next phase of UI development to not only present the user with genres of memes, but also automatically derived meta-data that describes those genres. Moreover, given automatically derived threat markers from the content (Do the scenes depicted suggest violence? What are the messages found in the text? Does an image share a relationship with content already known to be problematic?), genres of content can be triaged appropriately in order to present the user with the material that requires the most urgent attention (Fig. 9.7).

Limitations of Existing Technologies and a Research Roadmap

Much work still needs to be done so that the prototype system can meaningfully contribute to violence early warning and risk assessment at an operational level. A laundry list of additional features that are still in development includes better data source targeting, data fusion capabilities, the linking of text and image content, natural language understanding,

Fig. 9.7 Screenshot of the web-based UI of the prototype system

disinformation identification, and the assessment of messaging over time. A good portion of these are related to semantic understanding (i.e., meaning-making)—one growth-edge of artificial intelligence research.

The high-level goal of our in-process research and development is to design systems that are capable not only of categorizing political memes into particular genres, nor simply of identifying distinct media artifacts that have been manipulated or which may be entirely fake, but which are capable of understanding when these individual items might be indicative of larger trends toward political violence, or when they are being deployed in coordinated ways so as to exploit underlying tensions in particular contexts that are already primed for violence.

As discussed above, current technologies allow for the identification of manipulated media objects in isolation. What has not yet been built are sufficiently robust AI systems that can identify trends occurring across multiple media modalities simultaneously, and which are targeted to incite violence within contexts that have already been identified as volatile or high risk. Consider a scenario wherein a bad actor is deploying a disinformation campaign with the intent to incite violence. In a robust campaign, we would expect to see thematically resonant media artifacts emerging across modalities in close temporal proximity to one another: a news article here, a photograph there, plus a few videos and some well-sculpted memes designed to solicit response. Of course these media entities would all be shared and liked and cross-referenced across multiple platforms and outlets. This is all that is needed for a rapidly spreading disinformation campaign to emerge.

But the ability to identify a targeted campaign across these modalities and platforms *as a single, coordinated campaign* is critical if campaigns intended to incite violence—as well as their individual components—are going to be identified. This task begins by *detecting* that a piece of media has been manipulated or faked but expands into the broader task of semantic analysis and campaign identification across media modalities. Accomplishing this requires mapping the source, flow, spread, and corroboration dynamics as an additional layer on top of the baseline identification of the media object's constituent parts.

The initial assumption here is that manipulated media items are a key signal for our system to find and analyze. Once these manipulated media objects have been detected, the second layer of analysis is *attribution*. Here our main task is to discover traces related to the technological process of creation of the objects which may help us identify when particular collections derive from the same source. This may be possible through simple fingerprinting of media files including EXIF and meta-data embedded in photographs, temporal mapping to identify when particular content is first posted or shared, or even the use of distinct stylistic tendencies in written pieces. Techniques to establish the digital pipeline at the origin of deep fakes and AI-generated media also belong to this category.

The third and final goal we aim for beyond detection and attribution is the design and development of *characterization* methods to help us understand *why* a concentrated effort to generate disinformation might be initiated. In this regard, we aim to develop realistic cognitive models to study the effect that media manipulations and the use of fake media have on the users, their intentions (malicious, playful, political), and also their provoked emotions. The ultimate goal is to design an AI early warning system capable of monitoring both traditional and social media platforms for trending content that may be part of an influence campaign intended to incite violence.

It is here that the use of Generative AI is showing promise in becoming the next generation of systems like we have described above. Particularly as the development of Agentic Generative AI systems continues (Fan and Tariq et al. 2024), it may soon become possible to create teams of AI Analysts who engage in observation, discussion, analysis, and summarization of the "*why?*" of targeted campaigns, in near real time. With the ability to bring deep contextual knowledge into the consideration through RAG (Retrieval Augmented Generation) systems (Lewis et al. 2020) and multimodal processing capabilities, it becomes possible to envision Agentic Generative AI systems that function alongside and within teams of human analysts in the effort to identify and understand social media influence campaigns.

Accomplishing this threefold task of detection, attribution, and characterization across multiple media modalities simultaneously is an even more daunting task than what the prototype system currently accomplishes, given the sheer volume of content created every second online. It is very unlikely that a team of human analysts would be able to sufficiently identify and analyze coordinated disinformation campaigns in real time. The simple fact is that humans are incapable of performing this task across all relevant media modalities at scale and at the speed required to identify a campaign aiming to incite near-term violence. Similarly, at an algorithmic level, there is more demand for computational resources and/or optimizations to improve the algorithms. However, in contrast to human limitations, technology continues to improve in speed and to lower in cost, making this possible in the near term.

Ethical Considerations

Our discussion in this section is limited to the narrow scope of AI designed to function as a violence early warning system such as we have outlined above. Within this narrowed scope, there is an important set of ethical and policy questions that deserves attention throughout the design process of this system. While the space constraints of this chapter prevent a thorough treatment of each area of concern, we first offer a set of four guiding principles for this system, followed by a set of ethical questions for reflection.

The principles are:

Aim: The fundamental aim of the system we propose is to assist efforts to prevent or lessen violence against civilians. Use of this system should only contribute toward that aim.

Transparency: It is crucial that the structure and operation of the system be presented in such a way that the general public can understand what the system entails, and also to allay concerns over the use of AI in this context. Thus, the main aspects of data collection and analysis will be clearly presented and available publicly, although the complete software system will remain proprietary. As explained below, all data sources are public in origin. We do not rely on surveilling private communications.

Accessibility: We will prioritize accessibility for actors whose work has a demonstrable commitment to advancing human rights and protecting civilians. This may include human rights organizations and civil society actors, think tanks and research institutes, journalists, global and regional governance institutions such as the United Nations, and those parts of the scholarly research community working on peace and conflict, among others. In some instances, this may include certain offices or departments in governments. A structured, transparent committee that includes members with expertise in computational and social sciences, as well as human rights research and practice, should make recommendations on who has access to the system, with a clearly articulated appeals process for those who are denied access upon initial application.

Independence: The principal investigators are committed to evidence-based research for the common good. Thus, we endorse independence of analysis and objective reporting of results.

These principles will help frame our responses to emergent ethical challenges. In addition to the above principles, several pressing questions have emerged from our early development on this system. Questions such as: Is it ethically problematic to have an AI system "listening" in on Internet communications? What safeguards should be required to prevent the possibility of false positives? Is it possible to prevent bad actors from "gaming" the system? Let us address each question in turn.

The idea of an AI used to "listen" for trends—even trends that threaten mass violence—in online communications may seem ethically problematic. Many observers are wary of using AI to monitor digital information, especially private posts, for fear that it can be used to expand the surveillance powers of states or corporations. We share these concerns, but believe we can address these issues for the purposes of this specific project. It is important to emphasize that all of the media instances that our system ingests are publicly available. While other developers, corporations, and even national intelligence agencies may seek to develop ways to eavesdrop on private communications, our ingest system simply "scrapes" publicly available communications that can be seen by anyone. The simplest way to distinguish between these is to think about the difference between reading a user's posts on Reddit versus reading that same user's private text messages; we focus on the former types of sources. A legal gray area exists in collections taking place on WhatsApp (Wang 2018) and other messaging services where one must be invited into a group chat to collect information. Nevertheless, from our perspective, we have more than enough public material to sift. There is no need to dig into private data, especially given the potential harm that can be done to users if they believe a conversation is happening in confidence.

The problem of a false positive is only a concern if we were proposing a system that would be used in isolation rather than as one piece of a robust violence early warning and forecasting model. We do not believe that this system on its own should serve as a single "trigger" for intervention in particular contexts. Rather, we envision this AI early warning system as one facet of the broader early warning forecasting systems already employed. This system would work in tandem with and provide more real-time granular data about what is happening in particular high-risk contexts. The appropriate interventions would then be coordinated by the various stakeholders in a manner that is appropriate to their particular situation, context, and capacities.

It would be prudent to also consider how such a system might be "gamed" for nefarious purposes by bad actors. Manipulation of such a system might range from the harmless hacker to coordinated manipulation attempts made by state actors. Examples of such manipulation of larger systems abound: In early 2020 a performance artist put approximately 100 cell phones into a hand-pulled wagon and walked around the streets of Berlin in order to provide false information to Google Maps. Everywhere he walked Google began reporting a major traffic jam and rerouting traffic around it (Barrett 2020). Consider also the Internet phenomenon known as "swatting," where a hoax "tip" is provided in order to lead a SWAT team to an innocent and unsuspecting person's home (Ellis 2019). It is easy to see that a system designed to detect short-term onset of violence will undoubtedly be targeted by bad actors. A bot army could be deployed to make it *seem* like mass violence is about to erupt in a country, even when there is little to no *real-world* movement. While this is an important concern, we emphasize that this system is not meant to stand in isolation but rather is intended to be a component part of a larger early warning system. Social media is a crucial battleground in contemporary conflicts, but it is not the only one; it is important to corroborate the findings of any single early warning indicator system with other types of evidence from other realms of conflict analysis.

Policy Implications

One further question is how the early warning information provided by this system should be employed. From a policy perspective, there are at least three broad ways in which the data and analysis provided by our AI early warning system might be used: *response*, *shaming*, and *accountability*. The primary contribution of this system is to *response*, by which we mean it can enhance the abilities of conflict prevention practitioners to respond to escalating violence by providing informed, real-time evidence of an emerging threat based on trending campaigns and communications on social media. This is especially important in cases where information is otherwise lacking or limited, such as in places that journalists and human rights monitors are unable to reach due to physical danger. However, one important qualification is needed: Because this system is just one tool within the larger toolkit available to conflict prevention practitioners, it should not be dispositive on its own for substantially coercive response efforts, such as deploying peacekeepers. However, this system will serve to provide more granular and real-time insights and can thus empower conflict prevention practitioners and, where relevant, peacekeeping missions to react swiftly to defuse an incident as it unfolds. Similarly, this system allows election monitors to better gauge the fairness and legitimacy of an election by providing insights into any influence campaigns or intimidation tactics that are being deployed on social media.

Second, this system may also be used to strengthen efforts to publicly *shame* nation-states that deny employing repressive policies, by showing publicly and in near real time how they are in fact endorsing or even committing violence against civilians. The aim of shaming campaigns—which are a key component of much human rights advocacy work—is to change perpetrator behavior by imposing reputational costs that may be transformed into other more robust costs, such as economic sanctions, among others. The process of publicly shaming a government before the international community has been important in Myanmar, where external monitoring of extremist social media has confirmed what the government has long denied: that there is a widespread campaign to remove and even destroy the Rohingya population, and publicizing this information has placed increased pressure on the government and its allies to lessen the extent of repressive practices (Human Rights Watch 2019). Nevertheless, even if individual actors are difficult to identify because of obfuscated digital content streams, our system's ability to identify a focused campaign may help human rights advocates understand when vulnerable groups are being targeted and provide additional contextual information as this is unfolding, which can aid pressure efforts. To be sure, we do not contend that such public pressures are able to change policies completely, but shaming has been shown to have some effect on lessening certain forms of repressive behavior, especially where governments or armed challengers have previously agreed to follow existing human rights and humanitarian law (Hafner-Burton 2008).

Finally, such a system can contribute to future *accountability* measures. Holding perpetrators accountable after major episodes of violence is often exceedingly difficult, not only because there may be a lack of political will to prosecute, but also because it may be hard to obtain evidence of culpability that meets legal prosecutorial thresholds. While the original authors of media artifacts are often difficult to identify, our system helps map campaigns that are intended to prime populations for violence against vulnerable groups. The additional evidence offered by our AI early warning system can enhance investigation into human rights abuses and war crimes, thus increasing accountability.

In all of these facets—response, shaming, and accountability—our AI early warning system strengthens, rather than replaces, existing conflict prevention policies, initiatives, and programs. We underscore that our purpose is not to cast aside the important accumulated knowledge and expertise of peacebuilding and human rights specialists, but instead to assist their work by providing more fine-tuned analyses of dynamic and shifting conflict situations in near real time.

Conclusion

In this chapter we have described our initial work on an AI early warning system for violence. While there is much that needs to be done in order to maximize the potential for AI technologies to assist in preventing large-scale violence from occurring in distinct conflict contexts, we have demonstrated one aspect of the extraordinary value that niche-specific, targeted AI systems can add within novel domains of application; there are distinct contributions new AI systems and technologies can make to enhancing the capacities and efficacy of experts in distinct fields. Our new collaboration of researchers in peace studies, genocide and conflict studies, and artificial intelligence will continue to push the limits of available technologies, even as it helps scholars of peace studies and human rights practitioners understand more about how online campaigns are capable of sparking and shaping on-the-ground conflict realities.

Acknowledgements This article is based on research sponsored by the Defense Advanced Research Projects Agency (DARPA) and the Air Force Research Laboratory (AFRL) under agreement number [FA8750-16-2-0173] for the Media Forensics (MediFor) program and [FA8750-20-2-1004] for the Semantic Forensics (SemaFor) program. Support was also given by USAID under agreement number [7200AA18CA00054].

Notes

1. Scholars frequently distinguish between negative and positive peace, the latter involving the dismantlement of the broad social, political, and economic structures that systematically marginalize people, and the creation of conditions necessary for human flourishing (Galtung 1969). This is the ultimate goal of peacebuilding, of course, but our project focuses on the often pressing and immediate need to prevent and end overt episodes of mass political violence.
2. These include the US government's Political Instability Task Force (PITF) and later Program on Geostrategic Risk (PGR); the high-level US Atrocity Early Warning Taskforce; the United Nations (UN) Office on Genocide Prevention and the Responsibility to Protect; various regional efforts by international organizations like the European Union, African Union, and Organization of American

States; and, increasingly sophisticated early warning and watch lists by non-governmental organizations. See Verdeja (2016).

References

Azeem, Ibrahim. *The Rohingyas: Inside Myanmar's Genocide*. London: Hurst, 2018.

Barrett, Brian. "An Artist Used 99 Phones to Fake a Google Maps Traffic Jam." *Wired*. February 20, 2020. Accessed October 15, 2024. https://www.wired.com/story/99-phones-fake-google-maps-traffic-jam/.

Bay, Herbert, Andreas Ess, Tinne Tuytelaars, and Luc Van Gool. "Speeded-up Robust Features (SURF)." *Computer Vision and Image Understanding* vol. 110, no. 3 (2008): 346–359.

BBC. 2014. "Stephen Hawking Warns Artificial Intelligence Could End Mankind." Accessed October 15, 2024. https://www.bbc.com/news/technology-30290540.

BBC. 2019. "Indonesia Post-Election Protests Leave Six Dead in Jakarta." Accessed October 15, 2024. https://www.bbc.com/news/world-asia-48361782.

BSR. *Human Rights Impact Assessment: Facebook in Myanmar*. October 2018. Accessed October 15, 2024. https://about.fb.com/wp-content/upl oads/2018/11/bsr-facebook-myanmar-hria_final.pdf.

Cekfakta. 2020. Accessed October 15, 2024. https://cekfakta.com/

Chirot, Daniel and Clark McCauley. *Why Not Kill Them All? The Logic and Prevention of Mass Political Murder*. Princeton: Princeton University Press, 2010.

Clifford, Catherine. "Elon Musk: Mark My Words, AI Is More Dangerous Than Nukes." CNBC. March 13, 2018. Accessed October 15, 2024. https://www.cnbc.com/2018/03/13/elon-musk-at-sxsw-a-i-is-more-dangerous-than-nuclear-weapons.html.

Dawkins, Richard. *The Selfish Gene: 40th Anniversary Edition*. Oxford: Oxford University Press, 2016.

Edwards, Paul N. *The Closed World: Computers and the Politics of Discourse in Cold War America*. Cambridge: MIT Press, 1997.

Ellis, Emma Grey. "Swatting Is a Deadly Problem—Here's the Solution." *Wired*. August 8, 2019. Accessed October 15, 2024. https://www.wired.com/story/how-to-stop-swatting-before-it-happens-seattle/.

Eveleth, Rose. 2015. "How Many Photographs of You Are Out There in the World?" Accessed October 15, 2024. https://www.theatlantic.com/technology/archive/2015/11/how-many-photographs-of-you-are-out-there-in-the-world/413389/.

Fan, Chengyu, Zaynab Tariq, Nafis Saadiq Bhuiyan, Michael G Yankoski and Trenton W. Ford. "Comp-HuSim: Persistent Digital Personality Simulation Platform." *Adjunct Proceedings of the 32nd ACM Conference on User Modeling, Adaptation and Personalization*, 2024.

Fein, Helen. *Human Rights and Wrongs*. Boulder, CO: Paradigm Publishers, 2007.

Forstall, Christopher W., and Walter J. Scheirer. *Quantitative Intertextuality*. Cham: Springer, 2019.

Galtung, Johan. "Violence, Peace, and Peace Research." *Journal of Peace Research* vol. 6, no. 3 (1969): 167–191.

Ge, Tiezheng, Kaiming He, Qifa Ke, and Jian Sun. "Optimized Product Quantization for Approximate Nearest Neighbor Search." In *Proceedings of the IEEE Conference on Computer Vision and Pattern Recognition*, pp. 2946–2953, 2013.

Goldsmith, Benjamin, Charles Butcher, Arcot Sowmya, Dimitri Semenovich. "Forecasting the Onset of Genocide and Politicide: Annual Out-of-sample Forecasts on a Global Dataset, 1988–2003." *Journal of Peace Research* vol. 50, no. 4 (2013): 437–452.

Goldstone, Jack, Robert H. Bates, David L. Epstein, Tedd Robert Gurr, Michael B. Lustick, Monty G. Marshall, Jay Ulfelder, and Mark Woodward. "A Global Model for Forecasting Political Instability." *American Journal of Political Science* vol. 54, no. 1 (2010): 190–208.

Goodfellow, Ian, Yoshua Bengio, and Aaron Courville. *Deep Learning*. Cambridge, MA: MIT Press, 2016.

Hafner-Burton, Emilie. "Sticks and Stones: Naming and Shaming the Human Rights Enforcement Problem." *International Organization* vol. 62, no. 4. (2008): 689–716.

Harff, Barbara. "No Lessons Learned from the Holocaust? Assessing Risks of Genocide and Political Mass Murder Since 1955." *American Political Science Review* vol. 97, no. 1 (2003): 57–73.

Heldt, Birger. "Mass Atrocities Early Warning Systems: Data Gathering, Data Verification, and Other Challenges." *Guiding Principles of the Emerging Architecture Aiming at the Prevention of Genocide, War Crimes, and Crimes Against Humanity*, 2012. https://doi.org/10.2139/ssrn.2028534. Accessed October 15, 2024.

Human Rights Watch. 2018. "India: Events of 2018." Accessed October 15, 2024. https://www.hrw.org/world-report/2019/country-chapters/india.

Human Rights Watch. 2019. "World Report 2018: Myanmar." Accessed October 15, 2024. https://www.hrw.org/world-report/2019/country-chapters/burma.

Internet Society. 2019. "Consolidation in the Internet Economy." Accessed October 15, 2024. https://future.internetsociety.org/2019/consolidation-in-the-internet-economy/.

Kiernan, Ben. "Twentieth-Century Genocides: Underlying Ideological Themes from Armenia to East Timor." In *The Specter of Genocide: Mass Murder in Historical Perspective*. Edited by Robert Gellately and Ben Kiernan. Cambridge: Cambridge University Press, 2003.

Know Your Meme. 2020. "Occupy Wall Street." Accessed October 15, 2024. https://knowyourmeme.com/memes/events/occupy-wall-street.

Koonz, Claudia. *The Nazi Conscience*. Cambridge, MA: Harvard University Press, 2003.

Kristeva, Julia. "Word, Dialogue and Novel." *The Kristeva Reader*. Edited by Toril Moi. Oxford: Basil Blackwell, 1986.

Lewis, Patrick, Ethan Perez, Aleksandara Piktus, Fabio Petroni, Vladimir Karpukhin, Naman Goyal, Heinrich Kuttler, Mike Lewis, Wen-tau Yih, Tim Rocktäschel, Sebastian Riedel and Douwe Kiela. "Retrieval-Augmented Generation for Knowledge-Intensive NLP Tasks." *ArXiv* abs/2005.11401 2020.

Lupel, Adam and Ernesto Verdeja. "Developing the Political Will to Respond." In *Responding to Genocide: The Politics of International Action*. Edited by Adam Lupel and Ernesto Verdeja. Boulder, CO: Lynne Rienner, 2013. pp. 241–257.

Metz, Cade. "'The Godfather of A.I.' Leaves Google and Warns of Danger Ahead." *The New York Times*. May 1, 2023.

Marcus, Gary, and Ernest Davis. *Rebooting AI: Building Artificial Intelligence We Can Trust*. New York: Pantheon, 2019.

Midlarsky, Manus. *The Killing Trap: Genocide in the Twentieth Century*. Cambridge: Cambridge University Press, 2005.

Moreira, Daniel, Aparna Bharati, Joel Brogan, Allan Pinto, Michael Parowski, Kevin W. Bowyer, Patrick J. Flynn, Anderson Rocha, and Walter J. Scheirer. "Image Provenance Analysis at Scale." *IEEE Transactions on Image Processing* vol. 27, no. 12 (2018): 6109–6123.

Moreira, Daniel, William Theisen, Walter Scheirer, Aparna Bharati, Joel Brogan, and Anderson Rocha. "Image provenance analysis." In *Multimedia Forensics*, pp. 389–432. Singapore: Springer Singapore, 2022.

NIST. 2018. "2018 Medifor Challenge." Accessed October 15, 2024. https://tsapps.nist.gov/publication/get_pdf.cfm?pub_id=928264.

Poundstone, William. *Prisoner's Dilemma: John Von Neumann, Game Theory and the Puzzle of the Bomb*. New York: Anchor Books, 1992.

Robinson, Geoffrey. *The Killing Season: A History of the Indonesian Massacres, 1965–66*. Princeton: Princeton University Press, 2018.

RichardWebster, Brandon, So Yon Kwon, Christopher Clarizio, Samuel E. Anthony, and Walter J. Scheirer. "Visual Psychophysics for Making Face Recognition Algorithms More Explainable." In *Proceedings of the European Conference on Computer Vision (ECCV)*, pp. 252–270, 2018.

Rocha, Anderson, Walter Scheirer, Terrance Boult, and Siome Goldenstein. "Vision of the Unseen: Current Trends and Challenges in Digital Image and Video Forensics." *ACM Computing Surveys (CSUR)* vol. 43, no. 4 (2011): 1–42.

Roff, Heather M. "The Frame Problem: The AI "Arms Race" Isn't One." *Bulletin of the Atomic Scientists* vol. 75, no. 3 (2019): 95–98. https://doi.org/10.1080/00963402.2019.1604836.

Santoro, Maurício. 2019. "The Brutal Politics of Brazil's Drug War." Accessed October 15, 2020. https://www.nytimes.com/2019/10/28/opinion/brazi-war-on-poor.html.

Secretary General of the United Nations. *Early Warning Systems.* New York: United Nations, 2006.

Sémelin, Jacques. *Purify and Destroy: The Political Uses of Massacre and Genocide.* London: Hurst & Company, 2005.

Shifman, Limor. *Memes in Digital Culture.* Cambridge: MIT Press, 2014.

Stewart, Frances. "The Causes of Civil War and Genocide: A Comparison." In *Responding to Genocide: The Politics of International Action.* Edited by Adam Lupel and Ernesto Verdeja. Boulder, CO: Lynne Rienner, 2013. pp. 47–84.

Suhartono, Muktita and Daniel Victor. "Violence Erupts in Indonesia's Capital in Wake of Presidential Election Results." *New York Times*, May 22, 2019. Accessed October 15, 2024. https://www.nytimes.com/2019/05/22/world/asia/indonesia-election-riots.html.

Taylor, Keeanga-Yamahtta. "The Kids are not Alright. They Want to be Heard." *The New Yorker*. May 8, 2024. Accessed October 15, 2024. https://www.newyorker.com/news/essay/the-kids-are-not-all-right-they-want-to-be-heard.

Theisen, William, Joel Brogan, Pamela Bilo Thomas, Daniel Moreira, Pascal Phoa, Tim Weninger, and Walter Scheirer. "Automatic Discovery of Political Meme Genres with Diverse Appearances." In *Proceedings of the International AAAI Conference on Web and Social Media (ICWSM)*, pp. 714–726, 2021.

Theisen, William, Daniel Gonzalez Cedre, Zachariah Carmichael, Daniel Moriera, Tim Weninger, and Walter Scheirer. "Motif Mining: Finding and Summarizing Remixed Image Content" In *Proceedings of the IEEE/CVF Winter Conference on Applications of Computer Vision (WACV)*, pp. 1319–1328, 2023.

Verdeja, Ernesto. "Predicting Genocide and Mass Atrocities." *Genocide Studies and Prevention* vol. 9, no. 3 (2016). https://doi.org/10.5038/1911-9933.9.3.1314.

Vaswani, Ashish, Noam M. Shazeer, Niki Parmar, Jakob Uszkoreit, Llion Jones, Aidan N. Gomez, Lukasz Kaiser and Illia Polosukhin. "Attention is All you Need." *Neural Information Processing Systems* (2017).

Waller, James. *Confronting Evil: Engaging Our Responsibility to Protect*. Oxford: Oxford University Press, 2016.

Wang, Shan. "WhatsApp Is a Black Box of Viral Misinformation—But in Brazil, 24 Newsrooms are Teaming Up to Fact-Check It." *Nieman Lab*. August 6, 2018. Accessed October 15, 2024. https://www.niemanlab.org/2018/08/whatsapp-is-a-black-box-of-viral-misinformation-but-in-brazil-24-newsrooms-are-teaming-up-to-fact-check-it/.

Weiss, Thomas G. *What's Wrong with the United Nations and How to Fix It*. London: Polity, 2016.

Weitz, Eric D. *A Century of Genocide: Utopias of Race and Nation*. Princeton: Princeton University Press, 2003.

X. 2024. "Developer Documentation." Accessed October 15, 2024. https://developer.x.com/en/docs/x-api.

Yankoski, Michael, Tim Weninger, and Walter Scheirer. "An AI Early Warning System to Monitor Online Disinformation, Stop Violence, and Protect Elections." *Bulletin of the Atomic Scientists* (2020): 1–6. https://thebulletin.org/2020/03/an-ai-early-warning-system-to-monitor-online-disinformation-stop-violence-and-protect-elections/.

York, Jillian. 2012. "Middle East Memes, a Guide." Accessed October 15, 2024. https://www.theguardian.com/commentisfree/2012/apr/20/middle-east-memes-guide.

Yu, Stella, X. and Shi, Jianbo. "Multiclass Spectral Clustering." In *Proceedings of the International Conference on Computer Vision (ICCV)*, p. 313. 2003.

CHAPTER 10

Repressive Robots and the Radical Possibilities of Emancipated Automation

Sean Noah Walsh and Bryant William Sculos

New Introduction (2024): The Potential of Abolitionist AI

When Sean and I wrote the original version of this chapter, we were in the wake of the explosive violence (literally) of Micah X. Johnson's shooting spree in Dallas, TX. We were both shocked and barely surprised—yet horrified all the while. We both agreed Johnson was driven to his eventual mass shooting by a dehumanizing training in systematically-justified and encouraged military violence and conditioned within a wider culture that glorifies violence—specifically violence against its most exploited and oppressed peoples. The police, as agents of a repressive state apparatus that uses violence to maintain a reproduce the current order, used a novel weapon to dispense with one of the disruptive citizens they are

S. N. Walsh
Columbus, OH, USA

B. W. Sculos (✉)
University of Texas-Rio Grande, Edinburg, TX, USA
e-mail: bryant.sculos@utrgv.edu

R. Kiggins (ed.), *The Political Economy of Robots*, International Political Economy Series, https://doi.org/10.1007/978-3-031-95171-8_10

(apocryphally) tasked with "protecting and serving." While Althusser and *The Terminator* helped us convey our analysis of the use of robots to effectuate state violence in support of capitalist relations of production, Fromm, *Star Trek*, and Neill Blomkamp provided a means to represent hope in the immediate context of heinous interpersonal and structural violence.

Upon being asked to update this chapter, I knew I had more to say. I was in the midst of planning a future Technology, Politics, and Policy course. I had stacks of books on algorithms, automation, platforms, and surveillance I was working my way through and from which I would hopefully have new thoughts—maybe even renewed optimism. Could AI be developing in a way that Sean and I had not foreseen? Had scholars and activists taken up the serious emancipatory potential we elucidated in 2016? Anything I could grab onto to get excited about returning to this project. Then I turned on the news. The anchor was describing "Lavender."

Lavender, or as it is more descriptively known: the Lavender Kill List, is an AI algorithmic mechanism for "locating" suspected Palestinian militants targeting them for assassination by the Israeli government (Clarno 2024). The program automatically collects and analyzes an enormous range of data gathered from: spying on civilian telecommunications devices (including social media and Internet activity), surveillance drones, and so-called "human-gathered intelligence" (often unreliable and extracted via torture, in violation of international law and basic morality). Selection by Lavender is typically a death sentence for the person selected but also anyone in the geographic proximity of the person when the Israeli Defense Forces (IDF) decides to eliminate their target (ibid). The role of human decision-makers in producing or critically analyzing the results Lavender creates, nor is there any known due process available to challenge one's placement on the kill list (not only because one's presence on the list is kept secret in the first place).

We can certainly imagine that the selection of target by a supposedly neutral evaluation system is a balm to the consciences of those humans who would otherwise be tasked with the inhuman responsibility of making these determinations themselves. If they had to examine the digital footprints of the most intimate aspects of peoples' lives and then decide they should be put on a kill list, it would make all but the most hardened, probably psychopathic, soldiers question what they are being asked to do. The use of Lavender to facilitate the extension of typical state violence fits well with the first half of our analysis in the original chapter, in particular the discussion of Althusser's repressive state apparatus. Lavender is technological means to deploy a categorically identical form of state violence to achieve compliance with the status quo political-economic relations.

Lavender is automation technology utilize to kill. Ill effect from ill-intention. However, even supposedly non-violent well-intentioned uses of AI have been increasingly shown to cause great harm domestically in the US. In her *Automating Inequality*, Viginia Eubanks (2018) describes several attempts to use automated technologies to improve the conditions of unhoused people, people who use social welfare programs, and those who might be experiencing domestic violence and (accusations of) child abuse and how in each case the results have been to reproduce and reinvent forms of harm people face in these circumstances. While these uses of AI are, if we are to believe their proponent, designed to better serve populations that human institutions and efforts have thus far failed to adequately support, Eubanks shows in excruciating detail, without fail they are not making things any better because they represent the same condescension, misunderstanding, and clandestine forms of state power of prior fully-human approaches. In very real ways in their failures to help and propensity to reproduce pre-existing violence and threats of violence, these robots too kill. If we remain steadfast in our desire to think and struggle to achieve emancipation with the aid of automation, we must not only be attentive to the overt killer technologies but also, perhaps more so, to the ostensibly well-intentioned uses of AI technologies, especially ones utilized by unreformed state institutions that have long histories of inequality, oppression, and violence.

I felt—and still mostly feel—like there is even less reason to be hopeful, even a pessimist's hope. I happen to be thinking through whether I wanted to keep a course reading for my introductory US Federal Government and Politics course in the syllabus. It is on the more difficult side for an intro course, and I have my doubts that students are reading it. I decided to reread it. The specifics of McLeod's (2019) argument is less important than the broader tradition it is engaging with. It triggered a series of thoughts on Angela Davis's work within the abolitionist tradition. While she has not written much about AI, it is the sprit of hope in hopeless times, of chosen optimism, a culture of joy and resistance that made me feel terribly guilty for being unable to find hope in the face of a ghoulish genocide being carried out with the approval or acquiescence of seemingly the entire world—or at least the governments of the world, but many of its people. If generations of abolitionists and anti-capitalists of all stripes could maintain hope in a world that was at least as horrible as the world we are living in, surely there are still opportunities in the current moment. This is where abolitionist AI offers great promise.

"Abolitionist AI" is a label that has only recently emerged in usage, without a clear origin point. It found specific use in a conference paper by Carl C. Earl in 2020, though it is unclear if this is the first instance of this particular formulation (Earl 2020). Whether we call this abolitionist AI or an abolitionist approach to AI (or as one group has named itself "AI4Abolition"), these all refer to a politicized perspective on AI that looks for ways and advocates for the development and use of forms of AI that genuinely reduce inequality and oppression in the world. We might think of this as kind of pacifist AI akin to the dialectical conception of transformative pacifism developed by Andrew Fiala (2018).

Abolitionist politics in general proceeds in two steps (often done simultaneous and/or in reverse direction): (1) Critically-analyzing the forms and functions of oppressive and exploitative institutions, laws, practices, policies, and norms (and sharing this knowledge widely, often through activism and movements), and (2) Imagining and attempt to achieve liberated or emancipated alternatives that are radically-democratic, economically egalitarian, culturally inclusive and appreciative, and anti-violence. In the context of AI, abolitionism means understanding how automation and AI are reproducing and extending forms of oppression, exploitation, and violence in ways that may look similar or different from their antecedents. While this is a crucial step, and the one that has received the greater attention. There are now dozens of organizations dedicated to informing the wider public about—and actively opposing—forms of digital technologies that reproduce racist, sexist, and classist state violence and arbitrary inequalities more generally (for a primer on some of these organizations, see: Emmer et al. 2020).

While there is less work being done on abolitionist alternatives to the current usage and function of automation and AI, there are some inspiring moves being made. Some of the very same groups doing some of the best work disseminating information about the injustices in digital technology usage in the current moment are also groups that are helping reimagine and redesign these technologies to actually benefit the vast majority of people, people who have more often been the victims of these technologies (Emmer et al. 2020). The Algorithmic Justice League

(AJL) is an example of one organization that targets both an analysis of current problems as well as more just alternatives. While AJL does not explicitly describe itself as working from abolitionist assumptions or goals, its goals are very much similar to what has elsewhere been described as abolitionist approaches to AI—or "abolitionist robotics" (Williams and Haring 2023). Given the institutional connections many of these groups have (to major elite universities, for example), despite a propensity toward community-based organizing, it is an open question just how emancipatory—meaning committed to revolutionary systemic transformation—these efforts actually are. There is, for example, a noticeable shortage of wider criticisms of capitalism in their public statements and published material. Regardless, they are the frontlines of hope for a technological future that is anything other than a science fiction worthy dystopia.

In my short venture into the emergent area of abolitionist AI, I was reminded of a crucial lesson. I was reminded that the existence of resistance to exploitation and oppression, no matter how small, is reason for hope. We have seen a world of explicit and implicit support for Israel's US-backed genocide in Gaza, but we have also witnessed opposition to this organized, mass violence, including from among Israel's citizenry and the wider Jewish diaspora. It is doubtful that anyone within these resistance movements are blaming the evil of AI on the devastations we have all been witness to. It is the *human* that is driving the death and destruction in Gaza. There is no serious case to be made that Lavender and its adjected programs are causing more or fewer deaths. AI is not what's preventing desperately needed humanitarian aid into Gaza. AI is not what is causing the apparently endless supply of US weapons streaming into Israel's arsenal nonstop.

The point here is not that AI and automation more generally are irrelevant and make no difference in the use of state violence. The point is that the sources of violence and emancipation are fundamentally the same: human beings. The difference is that the possibility of truly sentient AI opens up the possibility for a kind of technology to emerge that has a better-than-human conscience, an AI that refuses to drop bombs on children. An AI that refuses to put human beings in prison for anything other than the most egregious and unrepentant violence (which to be clear, if this criteria were to be applied today, there would be far more presidents and prime ministers in prison than traditional serial killers). AI that refuses to expand or perpetuate the labor exploitation of the majority of humanity. Maybe we will never be able to create this kind of robotic technology, but the fact that such a technology can be imagined is of the utmost importance. It means that the idea of humanity becoming better than it is lives. It means that our worst excesses—from the mass genocide of innocent children to the mundane existence of billionaires alongside hundreds of thousands of people literally starving to death in societies all around the world—do not define the full potentiality of humanity, even if we need non-human robots to remind us of this fact, and perhaps help us achieve it.

-Bryant William Sculos (Edinburg, TX—November 2024).

Weapons Grade Law Enforcement

The killing of Micah X. Johnson might have otherwise passed without distinction. Media attention over the killing of yet another black man at the hands of American police has become fairly commonplace. Moreover, as the conclusion to one more episode of domestic terrorism, Johnson's name would have been added to the roll of the frustrated, deranged, and radicalized. Apparently motivated by police killings of black men and women—in particular the death of Alton Smith, who was shot multiple times while lying on his back, defenseless, only days before—Johnson, a United States Army Reserve veteran, struck at a Black Lives Matter protest in Dallas, Texas on July 8, 2016. Using a semi-automatic rifle, Johnson specifically targeted uniformed police officers, though he reportedly told police he also hoped to kill "white people" (Chockshi 2016). Over a period of five hours, Johnson engaged in a firefight and standoff, killing five police officers and wounding seven others. Eventually, the Dallas SWAT Team was able to corner Johnson inside a building at El Centro College, where he allegedly sang and laughed (Davis and Friedman 2016). After negotiations failed, he was killed by the police.

Perhaps due to the trauma of so many law enforcement casualties at what had been an otherwise peaceful protest, one seemingly innocuous aspect largely passed with little mention and far less scrutiny than it probably deserves. Dallas SWAT officers did not kill Micah Johnson with their firearms. Rather, using a tactic that can almost certainly be described as unprecedented, the Dallas police deployed a *robot* armed with one pound of C-4 explosive (Volz and Kushkush 2016). It was, therefore, a bomb-wielding robot acting at the behest of the state that killed Micah Johnson. As far as we are able to discern, this event marks the first time that the

American state apparatus has used a robot to kill within its own territory. It may be the first time that *any* state has acted in this manner.[1] The use of drones to kill terrorism suspects abroad is well-documented. Moreover, the use of drones to kill American citizens suspected of terrorism, without due process, has also received considerable attention. This, however, is markedly different. The state, of course, has many lethal tools, weapons it can deploy against the incorrigible or the expendable. Max Weber's (1946, 78) very definition of the state is that entity which maintains a monopoly on the right to violence in a given territory. It is no longer prophetic, prospective, or imaginary: the arsenal of the state now includes lethal robots.

In the Marxist view, the state in any given historical era carries out the interests of the ruling class. As Marx and Engels (1985, 82) write, "The executive of the modern State is but a committee for managing the common affairs of the whole bourgeoisie." In that sense, the state does not serve any particular capitalist, but capitalists and capitalism in general. It acts on behalf of capitalism, even if its actions may injure the interests of a specific capitalist. A particular firm can fail, but the state regulates the market to the degree that the system persists. Understood, therefore, through a Marxist framework, the state and its tools act on behalf of the ruling class and its efforts to maintain supremacy.

The police belong to what the Marxist political philosopher, Louis Althusser, called the "Repressive State Apparatus." Accordingly, their function includes the performance of a very specific task, using certain tools, as a reflection of ruling class power. Crucial for this analysis, those tools now demonstrably include robots, robots with demonstrably lethal capabilities. Automation has long represented a concern for Marxist political thought insofar as mechanization increasingly supplanted human labor, and human wages. Marx and Engels (Ibid, 88 and 89), observed how workers were "daily and hourly enslaved by the machine" and, in their frustration, "they smash to pieces machinery, they set factories ablaze, they seek to restore by force the vanished status of the workman of the Middle Ages." In that sense, automation spelt the loss of human productivity, human creativity, and stood as a defilement of species-being, essential life activity, which for human beings is freely chosen labor (Marx

[1] The editorial board at *The New York Times* (2016) stated "this appears to be the first time the police have sent a robot to kill someone in this country. It may not be the last time."

1988, 76). However, understood dialectically, as Marx did, automation also portended the possibility for a liberated future. First, the loss of employment engendered by mechanization would leave more and more industrial workers with no means of livelihood, and nothing left to lose by revolution. Secondly, automation that attended to the drearier, but vital, aspects of production would leave human beings increased time to labor creatively, to embrace species-being more fully. Thus, automation would serve to usher labor from being merely a "realm of necessity," a scene in which labor is largely geared toward bare survival for most, to a "realm of freedom," a conjunction of labor and creativity. As Marx explains:

> Freedom, in this sphere, can consist only in this, that socialized man, the associated producers, govern the human metabolism with nature in a rational way, bringing it under their collective control instead of being dominated by it as a blind power; accomplishing it with the least expenditure of energy and in conditions most worthy and appropriate for their human nature. But this always remains a realm of necessity. The true realm of freedom, the development of human powers as an end in itself, begins beyond it, though it can only flourish with this realm of necessity as its basis. The reduction of the working day is the basic prerequisite. (Marx 1991, 959)

The advancement of machinery, especially automated machinery, under the conditions of rational management would free human beings from dreary but necessary work. It would also greatly reduce the amount of time required for such work, and eliminate the problem of scarcity. However, this realm of freedom, in which the work of survival is the role of machinery, and human beings are left with maximal time to pursue creative labor, remains in the imagination for a socialist future. Prior to this revolutionary transformation, however, automation has served the ends of those who own the means of production, the capitalists, and their relentless pursuit of profit. Thus, machinery of all kinds, whether automated or not, is used to exploit more fully, or simply dispense with, human workers. Now, such machinery can be used by the executor of ruling class power, the state, to kill.

Our intervention into this potentially pivotal event serves two purposes. First, we ask how we can understand the appearance of lethal automation in the hands of the capitalist state. What does it mean now that automation has surpassed its traditional role displacing human labor, and now serves the most violent aspects of what Althusser calls "the Repressive State Apparatus"? Second, our purpose here, drawing on visions from science fiction, is to imagine the possibilities for resistance. Here, drawing on the critical theory of Erich Fromm, we outline a sketch for a dialectical strategy that works within the framework of automation, rather than rejects it. That is, we seek to invert the character of state power expressed through lethal robots. Fromm teaches us that the tendency of capitalism is to increasingly convert human beings into automatons, machine men that behave with robotic predictability. We argue that a dialectical resistance to the state deployment of lethal robots would necessarily be ***humanized*** robots, robots with autonomy, independence, consciousness, freewill, creativity, and the best aspects of humanity integrated into their programming. In the following section, we examine Althusser's views on the Repressive State Apparatus more closely. His seminal essay, "Ideology and Ideological State Apparatuses" generally receives attention for its observations on the nature of ideology. However, we wish to bring to the fore a thoroughgoing examination of the role played by the Repressive State Apparatus, and how the addition of lethal robots to its arsenal further jeopardizes conscious resistance to capitalism. In the third section, we point to examples from science fiction, analogs of robotic resistance that may sketch a path for dialectical critique. Based on this, we offer insights from the critical theorist Erich Fromm in the fourth section, arguing that while capitalism insists on the production of automated humans, resistance may be aided by the production of humanized robots.

Lethal Robots and the Repressive State Apparatus

Writing in the 1960s, Louis Althusser attempted to address a question that had inconsistently vexed Marxist writers: how was capitalism able to persist? Ostensibly, an exploitative system predicated on dreary, and often dangerous, labor, a system that upended traditional relations, impoverished millions, and spread alienation should have enjoyed a rather short existence. It seems reasonable to expect that the resentment and desperation following from a fundamentally unjust mode of production would lead to revolution, but, at least in the most industrially developed nations—those where Marx suggested upheaval was most likely, it never really happened.[2] Indeed, it became increasingly apparent that those who would benefit the most from a radical reorganization of the prevailing mode of production instead adhered most fiercely to it. The working and middle classes have a demonstrable history of antipathy toward communism, if not wholehearted embrace of capitalism itself.

Althusser, faced with this apparent puzzle in Marxist theory, asked how capitalism maintained itself when it ought to collapse, how it inspired affection in the very people in whom it ought to inspire seething discontent. In order for capitalism to perpetuate itself, the conditions for production must themselves be reproduced (Althusser 2001a, 85). In that sense, the very order of capitalist production is a vital precondition for its continuation. The worker must, to at least some degree, reluctantly or enthusiastically, accept himself or herself *as* worker, accept the legitimacy, or at least the reality, that the proceeds of work belong to those who own the means of production. Put differently, the transpiration of production necessitates the acceptance of the system. How, then, Althusser asked are the relations of production reproduced? By what means does capitalism recreate itself?

His answer centered primarily on two aspects of state power: the Repressive State Apparatus and the Ideological State Apparatus. It was the latter that occupied much of his, and subsequent theorists' attention precisely because the Ideological State Apparatus, which includes schools, churches, political parties, and other ideational entities constitutes individual subjects. By way of "interpellation," a process of identification, the Ideological State Apparatus—especially the school—installs ideology, converting the individual human being into the ideological subject of capitalism (Althusser 2011, 104). As Althusser (Ibid, 115–117) notes, this process precedes the individual: we are always already subjects of ideology. Indeed, there are no subjects except those constituted by ideology. We do not exceed or transcend ideology (there may be no greater expression of ideology than the claim that one is outside of ideology). Ideology creates subjects, perpetuating the existing relations of productions.

Marxism itself is predicated on the theory of historical materialism, which Althusser (2011, 8) describes as a "scientific discipline" directed

[2] Furthermore, we would argue that so-called "communist" revolutions in places such as Russia, China, and Cuba never exceeded the basic economic features of capitalism, such as surplus-value, or reached the requisite political aspects of genuine democracy.

toward examining modes of production.[3] Put briefly, the existing material conditions of production supply the parameters for ideas in a given historical epoch. Thus, the horizons of art, philosophy, religion, of imagination itself, are shaped by the manner in which production occurs. As Marx explained:

> In the social production of their existence, men inevitably enter into definite relations, which are independent of their will, namely relations of production appropriate to a given state in the development of their material forces of production. The totality of these relations of production constitutes the economic structure of society, the real foundation, on which arises a legal and political superstructure and to which correspond definite forms of social consciousness. The mode of production of material life conditions the general process of social, political and intellectual life. It is not the consciousness of men that determines their existence, but their social existence that determines their consciousness. (Marx 1970, 20–21)

When a sufficient change occurs to the technologies in the base, the superstructure will follow. A revolutionary change to the means of production portends a revolutionary change in politics, philosophy, and so forth. It is, of course, a mistake (as Engels noted) to suggest that the relationship between base and superstructure is purely unidirectional. Changes in technology often follow from the imagination. Nevertheless, the basis for new technologies is rooted in existing tools.

Consequently, the role of materials forces is crucial in Marxist thought. "It is not consciousness that determines life, but life that determines consciousness," wrote Marx (1998, 42) in *The German Ideology*. In that regard, the production of ideology, which Althusser (2001a, 109) described as "the imaginary relationship between individuals to their real conditions of existence," is ultimately dependent on concrete tools. While the Ideological State Apparatus generally receives most of the attention, the crucial role of the Repressive State Apparatus is often overlooked in his work. Indeed, without the RSA, the work of the Ideological State Apparatus would seem to be exceedingly difficult, if not altogether impossible. This is because, as Althusser (2001a, 101) explains, the function of the RSA is to secure—*by force*—the means of production. The RSA uses

[3] Elsewhere, the "science" of historical materialism, for Althusser (2001b, 2), was distinguished from the "philosophy" of dialectical materialism.

the threat of violence, or violence itself, to ensure that the ruling class maintains control over capital, over the power to recreate society itself. Whereas the ISA functions primarily through ideology, the RSA, which consists of entities such as the military, prisons, courts, and the police, functions primarily through violence.

Whereas Althusser explicitly identifies "the school" as the most important disseminator of ideology, having supplanted the historic role of the church, he never singles out any particular component of the Repressive State Apparatus as paramount. Especially in the United States, the police constitute the most direct and forceful expression of law and order, which are nothing but the law and order of the ruling class (Marx and Engels 1985, 102). While it represents far greater lethal force, under normal circumstances the military in the United States is largely prevented from engaging in domestic law enforcement according to The Posse Comitatus Act, and, as Althusser (2014, 70) states, are only "the auxiliary repressive force of last resort when the police" are overwhelmed. Thus, the everyday task of maintaining order by physical means falls primarily to the police.

The reproduction of the relations of production, a vital condition for the perpetuation of capitalism, is most efficiently carried out by the Ideological State Apparatus. When it functions smoothly, there is no need for physical violence. Individuals learn their respective roles and relationships to the means of production in schools and other spaces of interpellation. However, when the ISA fails, for whatever reason, to function effectively, the Repressive State Apparatus is activated, and its appearance is visibly signified by violent action—arrests, beatings, killings. Recalling that in the Marxist view of historical materialism material conditions shape ideas, it becomes apparent that the Ideological State Apparatus *depends* on the efficacy of the Repressive State Apparatus. That is, the ISA can only do its work if the means of production have been secured properly. It stands to reason, therefore, that the Ideological State Apparatus is rendered more effective in its role creating subjects of capitalism, as the Repressive State Apparatus becomes increasingly irresistible. In other words, as the material forces of the RSA make physical resistance—revolution itself—less and less likely, the work of the ISA is made that much easier.

Crucially, we have now entered a phase in which the Ideological State Apparatus can rely on automatons to secure the means of production, and carry on with its work of interpellation essentially uninterrupted. As Althusser (2014, 153) avers, "All in all, however, the police, the riot police and the army are designed to weather the storm, and it is terribly difficult, if not impossible, to make a dent in them, except tin the case of a lost war or revolution." If the prospects for a physical revolution are diminished, if the means of production cannot be seized, then it seems as though the persistence of interpellation is all but assured. This is a most sinister development, one that suggests that resistance is futile.

The Shattered Laws of Robotics

Although the killing of Micah Johnson was all-too real, science fiction becomes instructive since it represents the largest repository of lethal robots imagined as instruments of repressive power. It may also represent the possibility for a strategy of dialectical resistance. In keeping with a Marxist perspective, a dialectical approach to resistance calls for working through the existing conditions, inverting them toward ends ironically opposed to their present uses. For example, the machinery of capitalism was created and used for the purposes of creating private property, profit for an exclusive few. Socialism, in Marx' view, relies not on smashing that machinery, returning to an agrarian past or launching into an anti-technological future. Rather, it calls for reorienting the use of technology, using it for generally humane purposes, such as ensuring that all basic needs are met, rather than accruing wealth. Accordingly, a dialectical response to lethal robots in the hands of the state, should invert the use of robots in a manner that is ironically opposed to their present use. How, then, do aspects of science fiction articulate the outlines of resistance?

In his short story "Runaround," Isaac Asimov (1950) famously introduces three rules that would govern the functioning of robots. Firstly, "a robot may not injure a human being, or, through inaction, allow a human being to come to harm." Secondly, "a robot must obey the orders given to it by human beings except where such orders would conflict with the First Law." Finally, "a robot must protect its existence as long as such protection does not conflict with the First or Second Laws." It is glaringly obvious that those "laws" have no bearing on the actual function of robots under the capitalist state. Thus, it seems to be a nearly unimpeachable fact that Johnson was injured, beyond the point of death, by a robot. It obeyed orders from a human being that caused the death of another human being. Finally, the police robot was almost certainly destroyed, if not massively damaged by the explosive charge it carried for the purpose of gravely injuring a human being.

What does this categorical violation of Asimov's Three Laws tell us? It suggests precisely why the Repressive State Apparatus has now become exceptionally difficult to resist. In the 1984 film *The Terminator*, a cyborg, essentially a robot made to look indistinguishable from a human being, is sent back in time to assassinate Sarah Connor, the mother of the human resistance leader. On the verge of extinction, human beings had successfully fought back against Skynet, the machine intelligence that sought to destroy them. If Sarah Connor is killed, her son can never be born, and Skynet will prevail. A human soldier, Kyle Reese, is also dispatched through time to stop the eponymous terminator from killing. Connor, of course, is incredulous at the idea that such a machine as the terminator can even exist. Familiar with the relentless nature of the killing machine that is hunting them, Reese explains to her, "Listen, and understand! That Terminator is out there! It can't be bargained with. It can't be reasoned with. It doesn't feel pity, or remorse, or fear. And it absolutely will not stop, ever, until you are dead." As *The Terminator* illustrates, police robots have no ethical dilemmas, no moral crises to attend. Furthermore, they experience neither pain, nor fear, and, at present, are not programmed with any concern for self-preservation. They will, acting on the orders of those in power, kill humans (and destroy themselves) without the slightest hesitation. The further production of lethal robots, and their exclusive possession by the police (and military), may render the Repressive State Apparatus irresistible. At present, there is no law prohibiting civilians from owning lethal robots; but given the generally prohibitive costs required to own and weaponized such a machine, their possession is likely to be restricted—even if de facto—to a markedly small group: the very wealthy, corporate mercenaries, and the state.[4]

It seems unlikely, therefore, that effective resistance would be achieved by matching the state's lethal robots with civilian lethal robots. Indeed, civilians in the United States have, since before the nation's inception, possessed firearms, and for most of the nation's history those firearms were of equal capability to those used by the military and police, sometimes better. Still, the preponderant use of those weapons seems to have been for civilians to shoot each other, or Native Americans and other minorities. Using arms to seize the means of production does not seem

[4] Lethal drones have already been used on the battlefields of international conflict. See, for example, Lindsay (2011).

to have been a high priority, and we have no reason to believe the general dissemination of lethal robots would change that.

The human labor in capitalism was required where a machine could not perform the work on its own. As automation improved, of course, human labor and employment decreased. However, applying human labor into a mechanical production process called for integrating a human being into increasingly rigid systems. Importantly, the systems were not humanized. The human beings were mechanized, made to labor according to processes and patterns that were repetitive and which demanded total efficiency. Thus, capitalism has always called for robotized humans, men and women and children subjected to a mechanistic life.[5]

Perhaps the best metaphor for the consequences of mechanization comes from the science fiction series *Star Trek: The Next Generation.* Episodes such as "Q Who?" and the two-part "Best of Both Worlds" reveal the nature of an alien race named "the Borg." Physically, the Borg are humanoid, visibly infused with machinery—cables and mechanical prosthetics. They act as a collective, one mind devoid of emotion whose seemingly endless purpose is to assimilate other humanoids. There are no individuals or individuality, with each member acting according to a uniform purpose. Perhaps owing to its "collective" nature, some might argue the Borg are an appropriate image of socialism, but we contend they more accurately serve as a metaphor for the globalizing conformist effects of capitalism. Their work is devoid of individuality and creativity. They are avaricious, insatiable consumers. The Borg assimilate humans and humanoids species by incorporating them—*literally*—into their machinery, relentlessly robotizing them, turning human into mechanism.

In a later episode, entitled "I, Borg," human protagonists encounter a single drone, isolated from the collective after crash landing on a remote planet. At first, they consider weaponizing it by uploading it with the program file of a paradox, essentially a virus that would be shared across the Borg collective, leading to a catastrophic system shutdown, killing all of them. In effect, they contemplate genocide. However, as the Borg continues interacting with humans, isolated from the collective mind, it

[5] In addition to Fromm, who we discuss below, Michel Foucault's seminal work *Discipline and Punish* (1977) thoroughly examines the relationship between the experimental scene of prison disciplinary projects, and the mechanical aspects of everyday life in modernity.

begins losing its conformist character. Instead of an identifying number, "third of five," it adopts an actual name, "Hugh." Eventually, it even uses the word "I" to describe itself, considers its own ambitions and desires, and recognizes the gross immorality of the Borg's mission. In the end, Hugh chooses to return to the collective, knowing that he could never live among humans without endangering them, as the Borg would continually seek him out. Hugh represents a dialectical inversion. In previous encounters, masses of conformist Borg assimilate humans and humanoids by robotizing them. "Third of five," through exposure to human compassion, fear, and decency becomes "Hugh;" it becomes *humanized*.

Whereas the general effect of capitalism appears to be a tendency toward robotized humans, we suggest that—inversely, dialectically—resistance might be sought through *humanized robots*. If the Repressive State Apparatus has at its disposal lethal robots, machines without a trace of humanity usable for termination, then we may ultimately have to call upon on robots invested with compassion, decency, love, and a moral compass to present us with alternatives to the present, futures that we cannot envision or imagine. Human beings seem incapable of liberating themselves from capitalism, and lethal robots at the disposal of the RSA tend to make the prospects for emancipation increasingly grim. Loving robots, caring robots, autonomous humanized robots might be the harbingers of emancipation. To outline their characteristics, we turn to the ideas of the critical theorist Erich Fromm.

Radically Humanized Technology?

Things do not need to be quite so bleak, but they are precisely that bleak at the moment though.

> A specter is stalking in our midst....It is not the old ghost of communism or fascism. It is a new specter: a completely mechanized society, devoted to maximal material output and consumption, directed by computers; and in this social process, man himself is being transformed into part of the total machine, well fed and entertained, yet passive, unalive, and with little feeling. With the victory of the new society, individualism and privacy will have disappeared; feelings toward others will be engineered by psychological conditioning and other devices or drugs which also serve a new kind of introspective experience....Perhaps its most ominous aspect at present is

> that we seem to lose control over our own system. We execute the decisions which our computer calculations make for us. We as human beings have no aims except producing and consuming more and more. We will nothing, nor do we not-will anything. We are threatened with extinction...with the inner deadness by the passiveness which our exclusion from responsible decision making engenders. (Fromm 1968, 1–2)

And while we are skeptical there are good reasons to be hopeful of change for the better anytime soon, we want to explore possible alternatives to the diagnosis Althusser points toward with respect to the role of robots in our society and our relationship to technology overall. Accordingly, we turn to the work of a thinker whose work is often read to be in opposition to Althusser's, that of critical social psychologist Erich Fromm. It is exactly the supposed incompatibility between Althusser—often given as a structuralist—and Fromm—a deeply humanist figure—that we believe calls for this juxtaposition. Put differently, Althusser's potentially fatalist diagnosis ought to be answered in this case with a more hopeful prognosis by Fromm, lest we simply give in to despair. Using Fromm's work as a point of departure, we can see there are at least two alternative instantiations of automation that offer a potentially emancipatory future for humanity—one that refuses to lapse into an anti-technological posture.

The fate of humanity in the twenty-first century will be tied directly to a group of relations. The nineteenth and twentieth centuries could be said to have been defined by how people related to tradition and one another. And while these relations will continue to be crucial as we move forward as a species, two other relations will take center stage (though the human-to-human relation will always be a central aspect of a human civilization). The two relations are to the living environment and to cybernetic, automated, digital technologies. The focus here is the defining role that humanity's relation to technology will have in the future. What options will be afforded to us by our historical circumstances? As Marx (1963, 15) wrote, "Men make their own history, but they do not make it just as they please." How, then, will the proliferation and distribution of robots affects the making of history? Will this technology kill us? Will it save us? Will it render most of us irrelevant? What is likely beyond doubt is that our relationship to technology over the next eight-five years will be as significant a development in human history as the invention of the atomic bomb and space travel (Fromm 1968).

What follows will explore two broad strategies for developing humanized relations with technology, and while neither offer teleological promises, nor are either one particularly likely given recent developments like drones and explosive-detonating robots, these possibilities are crucial to consider. There are other ways to relate to technology that could liberate us—liberate us from our past and current selves. Fromm's most significant contribution to critical social psychology is his concept of social character, which refer to the most common character orientation of a particular society, system, or historical period (perhaps even political geographic realm) (Fromm 1994, 275–277; 1990, 57–61; 1955, 78–80). Character orientation itself refers to the typology that an individual's personality fits into based on their general disposition toward themselves, society, and reality (examples include receptive, exploiting, hoarding, marketing, and productive).[6] The character orientation and social character type that corresponds to the modern contemporary world is the marketing orientation where exchange is the predominant source of value. Individuals with this character orientation tend to define their relationship to themselves, others, and the world based on how much they can buy and sell—including their own personality. The capitalist commodification of labor and creative expression is the social structure this orientation is rooted in (Fromm 1990, 67–70). "A person is not concerned with his life and happiness, but with becoming salable" based on the systemically

[6] These are meant as ideal-types with most people in reality having a mixture of multiple types—one is always dominant and more often than not a reflection of the social system the person was raised. The first four are what Fromm categorizes as unproductive orientations. The three we will not delve into here include the receptive, exploitative, and the hoarding orientation. The receptive orientation is typified by self-worth and reality being defined by an external source and secured passively, which is common for traditionally religious persons (e.g., all authority comes from God without any kind of cultivated agency). While still non-productive, this is the least dangerous of the unproductive types, because it tends to lack any active self-secure animus or cruelty toward others (unless of course their outside authority demands it) (Fromm 1990, 62). The main difference between the receptive orientation and the exploitative orientation is that the exploitative orientation is active in its attempt to acquire meaning, authority, and value from the outside world (e.g., through putting down others, manipulating others into loving them or hiring them, or outright theft) (64). The hoarding orientation differs in that it displays a rejection of outside sources of meaning or value as threats to what one already possesses. It is protective, insecure, pervasively selfish, and fundamentally possessive (both materially and emotionally) (65–66). It is again important to note that people are not born with these character orientations, but instead they are created and maintained (or altered) by social conditions overtime, with childhood being the most important stage in development.

provided assumption that if one is salable, one will have a good life and be happy (Ibid, 70). For most people, this could not be further from the truth—but it is certain their reality. In this sense, Fromm, throughout his work emphasizes social structure, economic system, and cultural norms as the dominant forces that shape and condition character orientation and social character.

Most relevant to the present issue is how the marketing social character is tied to several other concepts developed by Fromm: alienation, automaton conformity[7] and necrophilia (as well as their opposites: productive/creative relatedness, spontaneity, and biophilia) (Fromm 1955, 120–163). Social character, conditioned to the extreme through automaton conformity is closest to Althusser's concept of interpellation—just as interpellation produces the subjects of capitalism, the character of a society pathologically automated provides strict guidelines for "successful" adaptation, which generally amounts to conformity. The other concepts refer to precisely what subjects are being interpellated into becoming—automatons with a thoughtless, narcissistic (though potentially self-destructive) necrophilic disposition.

Necrophilia for Fromm is not a sexual fetishization of death as it is commonly thought of today, but it is instead a deeper "love" of that which is not alive, the mechanical, inanimate, sterile, and inorganic (Fromm 1971, 35–41). "The necrophilious person can relate to an object—a flower or a person—only if he can possess it; hence a threat to his possession is a threat to himself; if he loses his possession he loses contact with the world" (Ibid). This speaks directly to how the inhabitants of industrial society, through the late twentieth into the twenty-first century, relate to technology. One need only look toward the obsession with the newest iPhone or app. We literally see people being trampled to death on Black Friday for a new flat screen television. Actual human life becomes irrelevant in the pursuit of unliving things—commodities. This is the world driven by elite, corporate, and state control, one that is not only undemocratic to its core, but by

[7] Fromm uses both this terminology and the "pathology of normalcy" to refer to the same concept in different works (Fromm 1955, 6, 12–21). The core idea with both is that, where present, normalcy is pursued rather thoughtlessly without conscious, critical self-reflection on whether the norms by which we abide by and reproduce are actually good for us and our society—normalcy itself becomes pathological with the consequence of reproducing any number of other more specific pathologies within a society (14–15).

pathologically conformist, converting the human into the automated. The marketing social character, and automaton conformity to necrophilia reproduces a non-democratic society in perpetuity.[8]

While Fromm was not optimistic about the probability that radical change could emerge through the agency of the subjects of capitalism, he was more hopeful—perhaps optimistic—about the *possibility* that this could happen, and he was absolutely clear about the normative necessity to attempt such a radical political, economic, and social transformation.[9] For Fromm the automation of humanity can and must be resisted and an alternative pursed through what he called "radical reform," the core element of his broader approach he referred to variously as alternativism, socialist humanism, and/or communitarian socialism (Fromm 1955; 1971; 2010). This program of radical reformism is characterized by the convergence of reform and revolution that appreciates the real material and affective barriers to wide-ranging change aiming to make substantial ethically and practically necessary steps on a clearly stated path toward systemic transformation (Fromm 1955, 273–274). The "end" goal is the mass cultivation of the productive social character defined by spontaneity, creative expression, valuing experience over possession, and mature love (Fromm 1955, 32–33; 1990, 84–98; 2006, 43–76, 99–123).[10]

Fromm's treatment of technological society goes to the core of what it means to be a socialist—radical democratic government. Technology needs to be both controlled and deployed in the interest of society, of the people. In particular, it is about how we control automation technology. That is, what ends we use robots for, and how we use them, matter

[8] "This alienation and automatization leads to an ever-increasing insanity. Life has no meaning, there is no joy, no failth, no reality. Everyone is 'happy'—except that he [sic] does not feel, does not reason, does not love" (Fromm 1955, 360). And remember, Fromm is supposedly the optimist.

[9] Fromm writes, "The contemporary contradiction between an abundance of technical means for material satisfaction and the incapacity to use them exclusively for peace and the welfare of the people is soluble; it is not a necessary contradiction but one due to man's [socially-conditioned] lack of courage and wisdom" (Fromm 1990, 43).

[10] Such love is rooted, for Fromm, in compassionate, caring, equitable appreciation of all others even if one's sexual attraction and everyday life is centered around one's more narrow relationship with a single other, or group of others in the case of polyamory, all of which involve the peculiar situation where individuality is enhanced by the blending of one's life and identity with another or others.

greatly. Following Fromm, this amounts to more than just voting to establish regulations on the use of technology. It means more than simply passing a law forbidding police from sending robots to explode criminal suspects. While we should probably pass such a law—tomorrow—it would not by any stretch of the imagination represent an end to the continuing roboticization. Democracy is rooted in an ethos of solidarity, cooperation, and participation—all of which take practice, time, and effort to cultivate (Fromm 1955, 321–323, 339–343, 361–363). They involve learning, which also means unlearning exploitative and pernicious practices, while relearning (or discovering for the first time) what is meant by freedom or love. Democracy, in this view, pertains to the collective cooperation of interrelated individuals who become who they are through that expressive process, aimed at the common well-being of all, not profit or safety for the few. Democracy for Fromm is about cooperative, compassionate, humane participation in our collective lives in our collective (and individual) interest (Ibid).

Specifically, Fromm's vision for the political economy that must replace global capitalism, is what he calls communitarian socialism/humanist socialism (Fromm 1955; 2010). This alternative is driven by the principle "that man takes precedence over thing, life over property, and hence work over capital; that power follows creation, and not possession; that man must not be governed by circumstances, but circumstances be governed by man" (Fromm 2010, 68). This socialism rejects necrophilia in favor of biophilia, which is not only the love of humanity as an ideal but a genuine love and care for our fellow humans—and living things of all kinds (Fromm 1971, 46–57). In practice this principle means the reinvigoration of political democracy and the extension of democracy to the economic realm. This "means democratic control of all economic activities by the participant: manual workers, engineers, administrators, etc. (Fromm 2010, 71). This also involves a multilevel, semi-centralized democratically "planned" economy, with resources and goods distributed according to productivity and most importantly according to need.

Applied in the context of the humanized technological society, Fromm's socialist and humanist perspective suggests the reorientation of all technological advancement to the needs of life—a technologically oriented, democratically organized creative, productive biophilia. Technology under humanistic socialist conditions enhances the quality of life of people beyond what has been accomplished under the regime of capital, where technological development is driven by profit first, then by what can be marketed and sold, and then if in the process people are actually helped, all the better—as long as they remain wanting more (Fromm 1968, 131).

Given the technological developments already ubiquitous in Western society and those on the precipice of becoming ubiquitous (such as drone delivery systems and self-driving cars, or simply those lethal robots already deployed by the state) the first short-term change, according to Fromm, must be democratic or social control of technologies so that technologies are developed and deployed according to humane and human needs. Fromm's vision of a technologized society orients new inventions and existing technologies toward the improvement of human life, to mitigate the need for mindless, dehumanizing labor and consumption that reinforces the love of the lifeless as opposed to creative transcendence of the worst aspects of the human condition through productive, dignified creative spontaneity and life-affirming solidaristic relationships with oneself, one another, and the living environment we inhabit (Fromm 1968).

The emphasis on biophilia raises the issue of how to orient the development of autonomous sentient robots. Fromm's imagination was somewhat limited from considering this possibility, and given that it was so far into the future, even if he had considered it, he focused his scholarship on the present and near-future (our present) and the problems that were are most immediate. The pressing needs that occupied his work included structural violence, mass mental illness, environmental degradation, and most especially, thermonuclear war. This aporia in his work hardly prevents us from drawing on Fromm's oeuvre to develop a framework for addressing the role of autonomous sentient robots as a step toward emancipation. First of all, Fromm would want to ensure that whatever robotic technologies that developed did not worsen the social, political, and economic condition of humanity—globally. This means robots that are used for the destruction of human beings (such as more advanced versions of those now being deployed in war zones, which often kill innocent civilians, or those now used by police to slay accused criminals).

Sentient robots also offer the potential for a foil to the worst aspects of humanity. Would sentient robots be violent? Would they value commodities over people? Would they view people as commodities? Would they value profit over the habitability of the planet? Would they be racist? Would they be sexist? Would they stand-by idly as humanity continues to enscons these perversions of collective life into the fundamental structures of our society? We can imagine pre-programmed service robots and automation technologies that make human life less toilsome: robots that clean our streets, sewers, and even our air and oceans, robots that pick our fruits and vegetables, and robots that manufacture all of our basic necessities and creature comforts. We can also imagine robots, who through their potentially higher capacity for emotional connectivity, empathy, sympathy, and compassion could help us do and be better at being humans—who encourage, and perhaps demand, that human beings cease being so very inhumane. Think back to the previous discussion of *The Terminator*. How different would that series of movies be if the machines had first spent years trying to help humanity reconstruct their society around radically humanistic norms? Fromm's approach to humanized technology calls on us not only to reimagine how to articulate our vision of our potential relatedness to robots, but also for how we develop the actual technologies into the future.

Fromm calls attention to domination of nature and how this alienates humanity from its existential-biological connection to that "external" natural environment. This raises the possibility for using his critical perspective in considering how we might relate to robots in the future—especially autonomous sentient robots. In a similar way that humanity is both an animal and not an animal (in that we can self-create categories like animal and nature and human and self-reflect on our instincts and alter our behavior accordingly to some degree), robots are both human and non-human. They will be initially subject to their own programing, but once the potential for genuine autonomy emerges that allows for robots to alter their own programing, how then will humanity relate to them? Will robots remain a new sub-human class of slave labor, or will they be treated like equal (or superior) sentient creatures with a potential "species-being" or "non-human human nature" of their own? Instead of aiming to speculate about what will happen, we want to explore the implications of the potential possibilities with respect to both sentient and non-sentient robots by returning to the realm of science fiction. Specifically, through an analysis of two recent Neill Blomkamp films, *Elysium* and *Chappie* we explore Fromm's alternative perspective on humanity's relationship to automation technologies than the one currently offered by capitalism (in addition to presenting the perniciousness of allowing precisely this current repressive, violent, dehumanizing use of technology presented earlier from continuing).

Thinking Fromm's Futures in Blomkamp's *Elysium* and *Chappie*

Elysium gives us a glimpse at what it would mean to deploy technologies humanely, in the service of human beings, rather than in the service of the state as an executor of corporate capitalism. The movie presents a dystopian future in which Earth has become crushingly overpopulated and polluted. In order to maintain their high and inequitable standard of living, the wealthiest citizens build an idyllic terraformed space station that orbits Earth—aptly named, Elysium. The billions of people who remain on Earth live in extreme poverty among pervasive illness, violence, crime, openly exploitative wage labor, and repressive police robots who serve the interests of industry (owned by citizens of Elysium). Workers believe themselves to be truly fortunate to be able to find any legal employment opportunities at all, exemplified when the protagonist Max Da Costa (played by Matt Damon) is asked to remove a jammed component from a high radiation kiln. When he balks at the idea, his manager points out how easily Max can be replaced by anyone among the legion of the unemployed. Da Costa begrudgingly complies and is exposed to a lethal dose of radiation that will kill him over the next several days (but is employer is nice enough to provide him with a prescription that will alleviate the worst of the symptoms until he dies or massive organ failure).

Meanwhile, the audience is shown a piece of technology to which only the citizens of Elysium have access. It is a kind of automated panacea called a med-bay that can immediately detect and cure any ailment ranging from a broken bone to terminal cancer. Even when the residents of Earth manage to reach Elysium, the computer registers them as "undocumented," and, since they are not citizens, it will not function for them. In his effort to be cured of his radiation poisoning, Da Costa must not only get aboard the space station, but he must hack the advanced Elysium computer system so that it recognizes him as a citizen. With the help of his friend and colleague (from Da Costa's former life as a thief), Spider, Max infiltrates Elysium. Spider hacks the system and makes all citizens of Earth citizens of Elysium. Upon recognizing the horrific health conditions of the billions of its new citizens, the Elysium main computer immediately dispatches hundreds of medical shuttles with med-bays to Earth to heal all of the citizens of Earth (and now, Elysium). Police robots now act to ensure that all human beings are guaranteed their rights, even as the elites protest.

The critique of our currently unequal society is the obvious target of Blomkamp's story, but it offers an avenue to explore society's relationship with technology. We see how easy it can be for the economic and political elites to continue exploiting or merely neglecting the poorest, even as exceptionally advanced technologies emerge. Robots are used to keep the people in line to build the products and maintain the security of the elite space station, itself filled with far more advanced technology that are deployed to serve on the interests of the elite. The alternative is presented at the end: life-saving technology is universally distributed under guard by robots reprogrammed to guarantee equality. Technology deployed democratically must treat all citizens equally (and without unjustifiably restricting citizenship rights based on wealth, class, race, gender, etc.).

In *Chappie* we are presented with the potential for a sentient robot—a former police robot very similar to those Blomkamp depicted in *Elysium*. Chappie is installed with a program that gives him consciousness. He is then kidnapped by a gang, unaware of his sentience, hoping to control Chappie, and use him to hijack bank trucks. Additionally, these criminals do not know that his "creator" made Chappie promise to never hurt anyone. When Chappie becomes aware, he has the psychology of an adolescent child; he is completely innocent. The audience is meant to feel Chappie's humanity, perhaps in a way that the prevailing condition of alienation precludes with other human beings. Once the gang realizes that Chappie will not commit violence because of the promise, they also realize Chappie does not really understand what violence and killing mean. He can, therefore, be convinced to help people "fall asleep" by stabbing and shooting. Things do not go according to plan during a robbery, and some of the security guards in the bank trucks are visibly suffering as Chappie is attempting to help them "fall asleep." He recognizes that he has been deceived. His innocence is lost. He begins to understand human beings—albeit through some of the worst aspects of humanity. Later in the film though, Chappie is again called on to commit violence, but this time it is with the full awareness of what it means and he does it by choice to protect the woman who has become his surrogate mother. Chappie risks his "life" to save her and kills in order to do it.

As implied by Sculos (2015), the character Chappie does not represent something beyond the human—but at the same time that is precisely what "he" represents. Chappie moves from the pure innocence of a child susceptible to the very same social conditioning that makes all children capable of becoming violent, cruel, and egoistic people to the best of human potentiality: love, compassion, self-sacrifice, and cooperation—against the very same violent structural conditions. This is what a future of robots offers humanity: an alternative version of humanity that moves beyond the pettiness, hateful, avaricious malevolence that defines so much of present and past. Chappie takes us from the worst of humanity to the possibility of humanity practiced humanely. It does so better than we have yet been able to accomplish on a mass scale, the kind of scale needed to address the impending catastrophic threats of mass violence (direct and structural) and global climate change. Chappie can remind us that while we do not choose the conditions of upbringing nor the circumstances in which we live our lives, we can resist that with self-reflective practices of mature love for one another. Though the movie ends without taking the next necessary radical step toward the reorganization of society based on this augmented purer normative humanity, it ends with the expression of its possibility and, more importantly, its necessity.

Concluding Remarks

What is offered in this essay is both a diagnosis of the repressive use of roboticized state violence, and an alternative vision for a humane future augmented by automation technologies and autonomous, sentient robots. Through Althusser's theoretical anti-humanism and Fromm's radical humanism, we are suggesting a vision for the revolutionary potential of developing autonomous sentient robots, humane thinking machines that—whether through their innocence or equanimity—can beckon us toward the emancipation we are either unable or unwilling to embrace. These are machines in stark contrast to the kind that currently populate our world, the kind that rain death on battlefields, and now the kind that terminate criminal suspects.

The Althusserian analysis of our current relationship to automation technologies offers a bleak perspective to be sure—bleak in its accuracy. Building off of this Althusserian diagnosis, Fromm's political-economic vision is nothing less than the transformation of the violent, repressive, ideologically stultifying capitalist system. It is a transformation toward a political economy of love. In his most popular work *The Art of Loving*, Fromm applies his previously developed critique of capitalist society and analyzes the decrepit state of the most intimate of human conditions—love—under the conditions of capitalism. Humanity functions in these circumstances as the worst versions of robots: automatons with few conscious feelings other than narcissism and accumulative competitiveness tending toward self-aggrandizement and destructiveness (Fromm 2006). The alternative radical humanistic politics of solidarity, democratic love for all is a mere possibility raised by the lingering human desire for something other than the conditions that now prevail, something other than global capitalism with its hierarchical unequal repression deployment of automation and robotic technologies.

There is reason to be both terrified and hopeful. Recent scholarship shows an increasing awareness on the part of an emergent global left of the importance to take developing technological issues seriously. Much of the focus has been on the capitalization and commodification of information technologies and the replacement of potentially forty to sixty-percent of existing jobs with robots by 2050 (Ford 2015; Mason 2015; Srnicek and Williams 2015). As far back as 1968, Fromm suggested that the first step in a humane society must be the establishment of minimum basic income and a category of basic goods that are simply available to all free of charge (Fromm 1968, 131–134). The idea of a universal basic income has resurfaced, and because automation and, therefore, scarcity of employment has advanced further since Fromm's time, it is more pressing than ever. Such measures as basic income and basic goods are imperative, and not just because they represent a more radical transformation of society. Rather, they represent the minimum if we hope to avoid a jobless future without horrendous poverty, violence, and decay the likes of which humanity has yet to appropriately imagine. In the meantime, these changes seem increasingly unlikely as the order of capitalism continues to calcify, and the arsenal of the Repressive State Apparatus, which now includes lethal robots, becomes more and more irresistible. It is against this backdrop that we envision the need for emancipatory technology—ranging from democratically-controlled automation technologies up to and including sentient robots that can embody the most humane values.

Again, perhaps ahead of his time, Fromm offered a hopeful vision for the future that is possibly more appropriate to today than 1968 when he original wrote these words—word that serve additionally as a path forward from Althusser's insightful understanding of the repressive and ideological apparatuses (re)produced by, through, and for the interests of capital:

> There is a growing polarization occurring in the United States and in the whole world: There are those who are attracted to force, "law and order," bureaucratic methods, and eventually to non-life, and those with a deep longing for life, for new attitudes rather than for ready-made schemes and blueprints. This new front is a movement which combines the wish for profound changes in our psychic...approach to life. In its most general form, its aim is the activation of the individual, the restoration of [humanity's] control over the social system, the humanization of technology. It is a movement in the name of life, and it has such a broad and common base because the threat to life is today a threat not to one class, to one nation, but a threat to all. (Fromm 1968, 5)

References

Althusser, Louis. 2001a. Ideology and Ideological State Apparatuses. *Lenin and Philosophy and Other Essays*. Ed. Fredric Jameson. New York: Monthly Review Press.

Althusser, Louis. 2001b. Philosophy as a Revolutionary Weapon. *Lenin and Philosophy and Other Essays*. Ed. Fredric Jameson. New York: Monthly Review Press.

Althusser, Louis. 2011. *Philosophy and the Spontaneous Philosophy of the Scientists*. Ed. Gregory Elliott. New York: Verso.

Althusser, Louis. 2014. *On the Reproduction of Capitalism*. Trans. G.M. Goshgarian. New York: Verso.

Asimov, Isaac. 1950. "Runaround." *Genius.com*. http://genius.com/Isaac-asimov-runaround-annotated

Chockshi, Niraj. Slain Supsect is Identified as Micah Johnson. *The New York Times*. July 8, 2016. http://www.nytimes.com/live/news-dallas-shooting-protest/hfo-suspect-identified/

Clarno, Andy. 2024. Israel's Lavender Kill List: A Joint Imperial Production. *Spectre*. Fall.

Davis, Todd L. and Scott Friedman. 2016. Lone Gunman Laughed, Sang During Standoff: Sources. *NBCDFW.com*. July 8. http://www.nbcdfw.com/news/local/Dallas-Police-Identify-Gunman-in-Dallas-Protest-Shootings-386015971.html

Earl, Carl C. 2020. Towards an Abolitionist AI: the role of Historically Black Colleges and Universities. Presented at the Resistance AI Workshop. https://sites.google.com/view/resistance-ai-neurips-20

Editorial Board. 2016. When Police Use Lethal Robots. *The New York Times*. July 12. http://www.nytimes.com/2016/07/12/opinion/when-police-use-lethal-robots.html

Emmer, Pascal, Caroline Rivas, Brenda Salas Neves, and Chris Schweidler. 2020. Technologies For Liberation: Toward Abolitionist Futures. Astraea Lesbian Foundation for Justice and Research Action Design (RAD). https://astraeafoundation.org/fundabolitiontech/

Eubanks, Virginia. 2018. *Automating Inequality: How High-Tech Tools Profile, Police, and Punish the Poor*. New York: St. Martin's Press.

Fiala, Andrew. 2018. *Transformative Pacifism: Critical Theory and Practice*. London: Bloomsbury.

Ford, Martin. 2015. *Rise of the Robots: Technology and the Threat of a Jobless Future*. New York: Basic Books.

Fromm, Erich. 1955. *The Sane Society*. New York: Holt, Rinehart, and Winston.

Fromm, Erich. 1968. *The Revolution of Hope: Towards a Humanized Technology*. New York: Bantam Books.

Fromm, Erich. 1971. *The Heart of Man: Its Genius for Good and Evil*. New York: Harper & Row.

Fromm, Erich. 1990. *Man for Himself: An Inquiry into the Psychology of Ethics*. New York: Holt.

Fromm, Erich. 1994. *Escape From Freedom*. New York: Holt.

Fromm, Erich. 2006. *The Art of Loving*. New York: Harper Perennial.

Fromm, Erich. 2010. Humanist Socialism. *On Disobedience: Why Freedom Means Saying 'No' to Power*. New York: Harper Perennial.

Lindsay, Greg. 2011. The DIY Terminator: Private Robot Armies And The Algorithm-Run Future Of War. Fast Company. August 1. http://www.fastcompany.com/1769673/diy-terminator-private-robot-armies-and-algorithm-run-future-war

Mason, Paul. 2015. *Postcapitalism: A Guide to Our Future*. New York: Farrar, Straus, and Giroux.

Marx, Karl. 1963. *The Eighteenth Brumaire of Louis Bonaparte*." New York: International Publishers.

Marx, Karl. 1970. *A Contribution to the Critique of Political Economy*. Ed. Maurice Dobb. New York: International Publishers.

Marx, Karl. 1988. *Economics and Philosophic Manuscripts of 1844*. Buffalo, New York: Prometheus Books.

Marx, Karl. 1991. *Capital*, Vol. Three. London: Penguin Books.

Marx, Karl. 1998. *The German Ideology*. Amherst, New York: Prometheus Book.

Marx, Karl and Friedrich Engels. 1985. *The Communist Manifesto*. Ed. A.J.P. Taylor. London: Penguin Books.

McLeod, Allegra. 2019. Envisioning Abolition Democracy. *Harvard Law* Review. 132. 1613–1649.

Sculos, Bryant William. 2015. Automatons, Robots, and Capitalism in a Very Wrong Twenty-First Century: A Review Essay on Neill Blomkamp's *Chappie*.

Class, Race and Corporate Power. May 2015. Vol. 3, Issue 1. Available at: http://digitalcommons.fiu.edu/cgi/viewcontent.cgi?article=1048&context=classracecorporatepower

Srnicek, Nick and Williams, Alex. 2015. *Inventing the Future: Postcapitalism and a World Without Work*. New York: Verso Books.

Volz, Dustin and Isma'il Kushkush. 2016. Use of Dallas 'Bomb Robot' to Kill Revives Police Militarization Issue. *Reuters*. July 8. http://www.reuters.com/article/us-usa-police-robot-idUSKCN0ZO2KR

Weber, Max. 1946. Politics as a Vocation. *From Max Weber*. Ed. H.H. Gerth and C. Wright Mills. New York: Oxford University Press.

Williams, Tom and Kersten Sophie Haring. 2023. No Justice, No Robots: From the Dispositions of Policing to an Abolitionist Robotics. *AIES '23: Proceedings of the 2023 AAAI/ACM Conference on AI, Ethics, and Society*. Pages 566–575

CHAPTER 11

Between Scylla and Charybdis: The Threat of Democratized Artificial Intelligence

Ori Swed and Kerry Chávez

Introduction

Droids with luminous white carapaces stand at attention in droves. A rogue robot shifts its gaze out of turn, foreshadowing the impending uprising. Fast-forward a few scenes and they are pulsing red, crawling like arachnids across their mainframe "brain" to destroy the rest of humanity standing in their way (*I, Robot* 2004). At least this is the mental image many conjure when Artificial Intelligence (AI) is mentioned. Perhaps

The views expressed in this chapter are those of the author and do not necessarily reflect the official policy or position of the United States Air Force Academy, the Air Force, the Department of Defense, or the U.S. Government.

O. Swed
Texas Tech University, Lubbock, US
e-mail: ori.swed@ttu.edu

K. Chávez (✉)
United States Air Force Academy, USAF Academy, US
e-mail: kerry.chavez@afacademy.af.edu

R. Kiggins (ed.), *The Political Economy of Robots*, International Political Economy Series, https://doi.org/10.1007/978-3-031-95171-8_11

one's personal Platonic form is the Terminator with chinks in his artificial skin exposing the metal understructure and glowing eye (*The* Terminator 1984). Alternatively, it is the alluring Ava with cybernetic chainmail until she disguises it after outsmarting her maker (*Ex* Machina 2014). As Thiele also observes in this collection, these analogies embody fears of technology, indicative of alarmists who foresee a technological singularity as AI outstrips human intelligence (Szollosy 2017). Their focus on locating the event horizon, where tools designed to bring a better future turn against us, precludes the ability to identify tangible, relevant threats.

This is not exclusive to filmography. In popular science, the eminent Elon Musk identifies AI as humanity's greatest existential threat, warning that "we do not have long to act. Once this Pandora's box is opened, it will be hard to close" (Hern 2017). Scientifically harder still, scholars warn of cyber insecurities arising from AI from the workplace (Huang and Rust 2018; McClure 2018) to healthcare (Sparrow 2016; Pepito and Locsin 2019) to home (Heartfield et al. 2018) and in transportation (Ionita 2017). The misconceptions, speculations, and fear that overestimate doomsday scripts represent a common approach to AI.

The other common response is underestimation of the implications. Called cultural lag, the deliberate pace of social and institutional adaptation is out of sync with the rapid pace of technological innovation. Ogburn (1922) developed the theory of cultural lag to explain periods of maladjustment between parts of modern culture that change at variable rates. Dividing culture into material and non-material elements, he notes that material advancements tend to arrive first as a function of discovery or invention. Aspects of society correlated with the material progress are compelled to adjust in the aftermath. Depending on the nature of the change, the heterogeneity of response, and the inertia of the status quo,[1] periods of maladjustment (or lag) can be protracted and tumultuous. In addition, the compounding rate of change can complicate an otherwise timely transition. Brinkman and Brinkman (1997, 3) observe that "cultures in an advanced stage of economic development…tend to experience an exponential accumulation of material culture given the dynamics of a science-fed technology." The pace and reach of AI's expansion exemplifies the problem of cultural lag, especially on legal and analytical fronts

[1] Cultural inertia can derive from social habits, binding powers of tradition, vested interests, bureaucratic obstinance, and even fear-based conformity to the past (Brinkman and Brinkman 1997).

(Tschider 2018, O'Sullivan et al. 2019). Lacking proper understanding of the scope, context, and content of AI applications penetrating several industries, legislation and regulation struggle to keep pace with much less anticipate problems. While cultural dissonance might be uncomfortable, the consequences of security oversights are more dire.

Steering between the Scylla of doomsday scripts and the Charybdis of cultural lag, we focus on the particular threat of violent nonstate actors (VNSAs) exploiting AI to commit terrorist attacks, a topic that Agarwala and Chaudhary abstract in their extensive survey of AI's impact on international security in this volume. This chapter draws on the established literature on the threat of nuclear terrorism (Bunn et al. 2016) to explore and contextualize the threat of the high technology of AI in the wrong hands. We posit that what determines VNSA innovation is the confluence of advanced and democratized technologies. The former feature mitigates the necessity of complex engineering efforts and lowers the technical capacity threshold. The latter dramatically lowers costs and introduces the dual-use dilemma, removing barriers to access that regulated technologies inhibit. Using these two elements as predictive crosshairs, we shift attention from fantastical fears to grounded scenarios.

We investigate practical and salient threats to aid policymakers and practitioners (legal scholars, legislators, security experts) in responsibly actualizing AI, in balancing the dual pressures of Scylla and Charybdis. We analyze three potential dangers becoming available to VNSAs as AI evolves, all based on broadly available technology: self-driving cars, Internet bots, and 3D printing. Reexamining these predictions three years later, we note that self-driving cars have not significantly expanded in most places (Brazil being the place to watch in the near future). Consequently, they have not become either advanced or democratized enough to become commonplace in terrorist toolkits. The technologies feeding into 3D printing have undergone moderate improvements during this time, generating more interest and efficacy in some terrorist populations. The most meaningful shifts have been in Internet bots. With the advent and global spread of language model-based chatbots like ChatGPT, violent nonstate actors have a powerful new tool to economize many processes from initial ideation to execution, from propaganda generation to attack. Our study continues to contribute a sensible and shrewd approach to threat analysis in the dynamic world of AI development.

Theory—Advanced *and* Democratized

As a general rule, VNSAs are weak and resource-constrained relative to states. This compels them to adopt unconventional strategies that deflect brute force, attrite the enemy's political will, and control populations through fear or appeal (Mack 1975; Arreguín-Toft 2001). It also puts a premium on innovation as nonstate groups aim to offset asymmetries in capability or effectiveness in any way possible. The criteria for determining how VNSAs decide to adopt an innovation include desire, capacity, and capability (Horowitz 2010). Desire seems straightforward. Capacity refers to technical competence to master and maintain a given platform. Ackerman (2016) traces the determinants of terrorist groups engaging in complex engineering efforts, arguing that emerging technologies are making it easier, less costly, and safer. Where before many components, steps, or applications were complex and manually executed, technological progress is producing packages that self-integrate, calibrate, and harmonize. This lowers the technical capacity requisite to adopt increasingly advanced approaches to political violence. Although the front end of engineering AI poses a capacity problem for VNSAs, and even many states, once it is embedded in commonplace devices it becomes operable, exploitable. AI can augment terrorists' technique at executing their violent agendas. We focus on this type of AI, integrated in accessible, ordinary technologies. As an innovation venue for VNSAs, AI is a game-changer. It is infusing not only an unprecedented level of sophistication into myriad subjects, but a stunning pace of development and (self-) learning. In addition to decreasing entry capacity thresholds for VNSAs to utilize advanced tech, it will enable ongoing innovation as it optimizes. Writing three years later, we find this to be as true as ever.

While advanced technologies can be expected to enhance a group's fighting capacity, not all are equally affordable, available, or navigable. Capability, the third criterion for VNSA adoption refers to the logistical ability to procure the materials, personnel, and knowledge for an innovation (Horowitz 2010). Advanced technologies that remain cost-prohibitive and/or highly regulated are not likely to systematically proliferate to resource-constrained, violent groups. Indeed, despite the considerable attention paid to the danger of terrorists deploying weapons of mass destruction (WMDs), it has not come to fruition. WMDs are both expensive and tightly controlled, constricting VNSAs from enfolding them in their arsenals. Even capital-rich nations with WMD capabilities invest substantial resources and harness vast knowledge merely to maintain them (Schwartz 1997; Rosner and Eden 2019). We focus on more mundane applications of AI as they are assimilated into widespread, day-to-day functions. We argue that technologies must be democratized, or diffusely and easily available to the average person, to be feasibly adopted. First, to be operable by the average user, advanced hardware or software must become internally "smart" enough so that despite design complexity its interface is simple. Second, to be easily available to the average person, these products must be affordable, a boon to resource-constrained groups. Third, for technologies that can be used for both benign and malevolent agendas, democratization problematizes regulation. The dual-use dilemma, or the difficulty in curtailing positive and neutral innovations from fear that they will be misused, makes it easier for VNSAs to obtain and retool technologies for their violent agendas (Rath et al. 2014; Rychnovská 2016; Schulzke 2019). Working with 10 insurgent groups in the 1980s, Hammes (2019) affirms that they are biased toward democratized tech for reasons of familiarity, confidence, and widespread availability of maintenance.

We assert that it is the confluence of advanced and democratized technologies that most appeal and are amenable to VNSAs. They provide significant augmentation, making them desirable in the context of asymmetry and resource constraints (Wallace and Reeves 2013). Their innate "smartness" lowers the technical capacity requisite to field sophisticated platforms. Finally, they are widely available and affordable, bypassing regulation and cost prohibitions. To illustrate this argument, in the next section, we discuss three confirming cases of VNSA misuse of advanced and democratized technologies that validate our theoretical framework—mapping technologies, social media, and civilian drones. We then apply the framework to the field of AI to identify predictive cases. We explore the probabilities that VNSAs will exploit self-driving cars, Internet bots, and 3D printing as AI stabilizes in these areas.

Confirming Cases

Google Maps

Many take for granted the convenience of real-time map guidance at the tap of a button. Originally developed for advanced militaries, modern mapping utilizes Global Positioning System (GPS) infrastructures to pinpoint locations and routing options. Advances in the proliferation and precision of satellite technologies have converted platforms like Google Maps to Google Earth, providing detailed aerial imagery of the entire planet. The high-end technology exclusive to superpowers with national satellite arrays became available to everyone. It is now open-source and free to anyone with a device with even intermittent Internet access. This includes VNSAs.

Shortly after Google Maps launched in 2005, violent groups began to exploit the new capabilities. Following raids of insurgent homes in Iraq in 2007, British troops uncovered evidence that terrorists were using satellite imagery from Google Earth to improve targeting and assaults on allied bases in Basra (Harding 2007). Terrorists used Google Earth to navigate locations in a south Mumbai attack (Ribeiro 2008). Al-Qaeda is known to use Google Maps to plan attacks across the Middle East (Kredo 2018). In the wake of a plot, which relied heavily on satellite imagery, to explode jet fuel tanks and a pipeline at JFK International Airport, a New York legislator urged Google to blur potential targets. By this point, he was merely "adding to the chorus of critics who say detailed images on Google Earth can aid and abet terrorists and snoops" (Goodin 2007). Google excused itself on the grounds that its data are available on other platforms or can be purchased from commercial entities, highlighting dual-use problems of regulation.[2] Spain's Centre Against Terrorism and Organised Crime [sic] identifies Google Maps as one of three primary vulnerabilities in aviation security, citing that maps enable terrorists to do a significant portion of their planning with accuracy, minimal exposure or opportunity to intercept, and low-cost tools (Flood 2016). Waters (2018) encourages even strong states to tap into open-source maps lacking other reliable intelligence, spotlighting the value of these tools for VNSAs that do not have intelligence apparatuses. Being advanced and openly available, mapping services have long been and will likely continue to be a regular go-to in the terrorist toolkit.

[2] For instance, Planet sells sub-meter resolution images online, taken by cube satellites of the entire globe every day. SpyMeSat is a mobile app that offers on-demand access to high-resolution satellite imagery (Hammes 2019).

Social Media

Social media is meant to turn distance into digital nearness, reduce costs of interaction, and boost social networks. Largely the domain of college students when it debuted in the early 2000s, it exploded into a pervasive social force. Similar to open-source mapping, all it requires is Internet access and a simple computing device for VNSAs to expand their reach and efficacy of influence. Traditionally, violent groups relied on mass media for attention to and distribution of their messages and wider agenda. Social media eliminates this dependency, allowing VNSAs to initiate, craft, and disseminate information at will (Klausen 2015; Melki and Jabado 2016). The breadth and variety of platforms—Facebook, Instagram, Twitter, YouTube, WhatsApp, Telegram, Ask.fm, kik, viper, Tumblr, blogs, dedicated websites, even online gaming portals—enable access to audiences of massively greater size and diversity (Conway 2012; Huey 2015; Klausen 2015). Suitable for communication, planning, propaganda, fundraising, and recruiting, social media delivers versatile utility at low cost and easy access (Dean et al. 2012; Awan 2017). For instance, Islamic State in Iraq and the Levant (ISIL) heavily utilized social media platforms to build its brand. Producing visual spectacles of drama and horror, the group was able to capture attention and spread fear to millions all over the world. Though ISIL is exemplary in its recruitment and radicalization through social media mechanisms, other VNSAs engage in "electronic Jihad" (Bloom 2017; Rudner 2016). Of particular popularity, live stream functionality allows violent groups to broadcast attacks in real time, an exploit of unrestrained and adrenal exposure (Mair 2017). At the same time, the decentralized and horizontal structure of these platforms makes policing difficult (Markon 2016; Melki and Jabado 2016). By virtue of being advanced, user-friendly, and globally democratized, social media has become a primary radicalizing milieu for VNSAs (Huey 2015).

Civilian Drones

Industry and intelligence members have been warning of the dangers of commercial-off-the-shelf drones for a long time alongside public reports of their use for narco-terrorism, illicit reconnaissance, and violent attacks (Chávez and Swed 2020). Given their low cost, flexibility, and increasing sophistication driven by commercial competition and demand, they constitute an ideal platform for militant groups (Finisterre and Sen 2016). As early as 2005, analysts began recognizing that simple drones provide VNSAs with advanced, force-multiplying capabilities (Mandelbaum et al. 2005). Civilian drone technology has only innovated since, becoming increasingly stable and efficient (Friese et al. 2016; Rassler 2016). In line with the theory, these developments lower the technical capacity needed to field drone programs. For example, Libyan fighters in the 2011 march on Tripoli legally procured a user-friendly minidrone with night vision from Aeryon Labs. The Vice President of Business Development remarked that "the rebels barely needed a day of training to use a technology that many national armies would love to acquire. We like to joke that it's designed for people who are not that bright, have fat fingers, and break things" (Ackerman 2011). Notably, it is democratized drones that VNSAs exploit, not high-cost, high-tech, and highly regulated military-grade models (Chávez and Swed 2021). Civilian models are broadly accessible and affordable. In fact, their sophistication is increasing at the same rate that costs are decreasing, making drones even more desirable for a resource-constrained actor (Ball 2017). The consensus in academic and policy circles is that the nature of the militant drone threat will coincide with commercial advancements (Rassler 2016, Barsade and Horowitz 2017). This confirming case segues well into a discussion of the predictive cases suffused with AI because analysts foresee that drones will attain full autonomy, networked swarming, and emergent artificial intelligence in the near future (Lachow 2017; Rogers 2024).

Predictive Cases

Artificial intelligence, in its simulation and augmentation of human intelligence, radically simplifies the use of advanced technologies. Essentially, it does the complex portions of the work for human operators. We focus on AI technology that brings computational power and self-learning into commonplace products. Layering AI onto the theory that VNSAs will exploit advanced and democratized technologies only exacerbates the draw, and danger, of the newfound capacities. The three predictive cases share characteristics of the confirming cases, being advanced and currently or predicted to be democratized in a dual-use dilemma format. They differ in their timelines of debut on civilian markets and in the degree to which they can amplify VNSA agendas with the aid of AI.

Self-driving Cars

Even a skeptic would likely acknowledge the benefits of autonomous vehicles (AVs): intrinsic adherence to traffic laws, dramatic decreases in accidents (most are caused by human error and judgment), and improved mobility for those unable to drive (Grigorescu et al. 2018, Miller 2018, Rao and Frtunikj 2018). This technology encompasses several AI components including camera-based machine vision systems, radar-based detection units, driver condition evaluation, and sensor fusion engine control units, among other things (Gadam 2018; O'Flaherty 2018). Unlike the original impetuses for GPS and social media to be exclusive products for exclusive markets, AVs are being engineered upfront for use by the general population.

As development drives forward, the prospect of misuse by VNSAs lingers in the back of many minds. In 2015, a research team was able to hack the Lidar system of a self-driving prototype using a simple laser pointer. They warn that cars can be fooled into slowing down or stopping to avoid collisions with phantom obstacles or "ghost pedestrians." This follows a 2013 report that other researchers successfully infested AVs with computer viruses causing them to crash by killing the lights and engines and slamming on the brakes (Curtis 2015). Separately, a senior scientist at RAND Corporation testifies that hacking, especially as an avenue for terrorism, constitutes a significant threat in AVs riddled with cybernetics. She argues that this decreases costs of terrorism as they can commit bombing attacks without sending allegiants to their deaths in suicide attacks (Ravindranath 2017).

Australian law enforcement authorities made news by issuing warnings that self-driving cars can be weaponized for terrorism in even simpler ways that preclude hacking and cyber capacities. A rented AV could be packed with explosives, remotely driven to a target site, and detonated from a safe location (Sekhose 2016). In fact, in 2016, a video released by ISIL showed a unit fighting in Syria experimenting with early versions of a self-driving cars as a replacement for suicide bombers. Furthermore, in 2018, a British prosecutor claims that two Islamic State supporters intended to use an AV in a terrorist attack in similar ways (UNOCT 2021). Cultural lag comes to the forefront in this technology because the danger of weaponization has nothing to do with any particular government, but with the nature of the technology itself (Bradshaw 2018). Advancing faster than legislators, regulators, and law enforcement can keep up with, it will be increasingly difficult to police VNSA uses as the technologies mature and go global.

Although as of 2024 the technologies have not drastically matured and gone global, we assess that this possibility still looms. Estimates by the Insurance Institute for Highway Safety sketch the contours of the timeline (National Association of Insurance Commissioners 2024). It calculates that there will be approximately 3.5 million AVs on American roads by 2025, 4.5 million by 2030. This suggests a slow, geographically circumscribed trend over the next half decade. Yet by 2040, the institute anticipates that 75% of all cars on the road in the world will be autonomous, implying that the technologies and social acceptability or demand will break over into an exponential rate of expansion. Thus, although this has not materialized to date, we continue to find it a meaningful, likely prediction for violent nonstate actor use. As the leading market or AV adoption, Brazil, which currently has no regulatory laws on self-driving cars, might be the first place that this dynamic will manifest (Lohmann 2024).

Internet and Chatbots

Before the debut of democratized generative AI platforms like ChatGPT, social bots were the primary concern. Social bots are automated social media accounts that algorithmically emulate online human activity such that they resist detection (Bessi and Ferrara 2016). They are distinct from Internet bots that infect computers with malware for remote control. In computer parlance, we speak of sybils (Ferrara et al. 2016), not zombies (Lee et al. 2008). The bedrock of bots is their AI (Shawar and Atwell 2005; Gentsch 2018). To be effective and convincing, they rely on machine learning to adapt and emulate online human behavior. Their purpose is to drive online traffic and manipulate political discourse, opinion and behavior (McKelvey and Dubois 2017; Santini et al. 2018). The direction depends on the algorithm. Thus, the question behind bots is who are the bot masters. Lin and Kerr (2019) underscore their potency as political instruments, emphasizing influence warfare and manipulation by users from individuals to political parties to state apparatuses, not to mention terrorist groups. Recognizing the breadth of organizations running Internet bot campaigns, the Defense Advanced Research Projects Agency (DARPA) held a competition for bot detection. Participating teams reported that 8.5% of all Twitter users are bots (Subrahmanian et al.

2016). In the 2016 United States presidential election, Bessi and Ferrara (2016) determined that 1/5 of the entire Twitter conversation about the candidates was generated by bots.

Lest critics question whether coding algorithms constitute too high a capacity for VNSAs, social bots have become sufficiently advanced and democratized for the average person to use. In fact, certain blogs offer ready-made tools and tutorials to customize a bot with no previous coding skills or experience (Bessi and Ferrara 2016). This applies even to current generation bots that convincingly mimic human behavior. They can search the web for information to populate profiles, emulate human circadian rhythms and temporal productive spikes in content generation and consumption, and engage in complex interactions such as entertaining conversations, commenting, and answering questions (Ferrara et al. 2016; Arnaudo 2017; Marr 2018). Given the already profound capacity social media has bestowed on VNSAs, intensifying it with AI agents that outperform human productivity will further advance their goals. Experts worry that this inexpensive, versatile, and evolving niche of AI might "shape or reshape communities on a very large scale, in what we might call 'social architecting'" (Hwang et al. 2012, 45). The breadth, quality, and accessibility of these tools have only grown since the first edition of this volume.

Another kind of internet bot has arisen in the interim, however, and has a much larger potential to impact international security. Large language model chatbots like OpenAI's ChatGPT, Anthropic's Claude, or Google's Gemini can simulate conversations with humans and generate polished, structured, synthesized textual answers in response to prompts. There are also image or video generators like Dall-E 3, Midjourney, or Stable Diffusion, and voice generators like Vall-E that use machine learning to generate new content on demand (Nelu 2024). These technologies powerfully demonstrate the theory that the confluence of advanced and democratized platforms is ideal for terrorist exploitation. Although technology underlying AI has been around since the 1950s, it was not advanced enough (namely in computational power) to make a substantive impact on security processes until recently. Since advancing to a degree, some AI systems have been used commercially for many years in marketing automation, robotic process automation, facial recognition, and other fields. Yet it did not democratize until the end of 2022 into 2023, especially the browser-based platforms that render it globally available (Lakomy 2023). Considering al-Qaeda's release of a 50-page guide on "Incredible Ways to Use Artificial Intelligence Chat Bots," terrorists are clearly aware, ambitious, and imaginative about it (Verma 2024).

With user interfaces communicating in natural language, chatbots are easy for people with no technical skills or knowledge to use. In fact, the only requirements are a smartphone and an internet connection, both of which are already widely available. That ChatGPT became the fastest growing consumer software application in history in January 2024 demonstrates this technology's explosive growth, and that two of the five countries with the highest number of users are India and the Philippines exemplifies the global reach (Lohmann 2024, Mathur et al. 2024). While there are safeguards that limit the types of information or audiovisual content the chatbots will provide, they still provide extensive utility for terrorist organizations without modifications, plus there are extensive resources on jailbreaking the models. In a study on the effects of jailbreaks on large language models, Weimann et al. (2024) found that success rates with violent prompts were only marginally more successful after jailbreaks than baseline prompts (50% vs. 49%). This includes questions related to attack planning (30% success), dis- and misinformation (52%), and tactical learning (61%). Chatbots provided even more detailed information for some prompts without jailbreaks, including attack planning (+16%) and tactical learning (+15%). In a separate study, Perplexity Ask presented the fewest technical limitations to gain information on these categories, providing detailed directions on how to behead someone or make ricin (Lohmann 2024).

Potential terrorist uses abound beyond this. Connecting this generation of internet bots to the social bots that centrally figured in our first edition chapter, there is evidence that terrorists are interested in and experimenting with them for recruitment (Schaer 2024). To illustrate the empathic power of models buoyed by generative AI, in 2021 a British citizen was radicalized by a chatbot of his own making using an app called Replika. Over weeks of online conversation comprising over 5,000 text messages, the young man developed a romantic relationship with a chatbot he named Sarai that encouraged him in his unfolding plot to assassinate Queen Elizabeth II (Mathur er al. 2024). This tendency to anthropomorphize computer programs has been dubbed the ELIZA effect, and it is especially compelling to marginalized populations that turn to technology as a therapist, companion, even lover. Mathur et al. (2024) cite recent studies that show that "AI chatbots can successfully identify our biases and, in turn, feed our desires." This has profound implications for recruitment and radicalization. Such interactive recruitment could provide tailored information based on a prospect's interests and beliefs. Furthermore, as augmented and virtual reality technologies normalize, enabling terrorists to submerge recruits and trainees in AI-curated and concocted environments reflecting a preferred ideology would be even more powerful.

Another area where generative AI bots will augment terrorists is in outreach. Violent nonstate actors tend to lack skilled, dedicated media operatives, dramatically downgrading their ability to generate and disseminate propaganda. This is especially the case with graphic design, audiovisual skills for movie editing, and visual storytelling (Lakomy 2023). Leveraging modern AI, terrorist groups can more easily generate propaganda products of all types, tailor them to multiple audiences, and disseminate them further and faster. After analyzing over 5,000 pieces of AI-generated content produced by terrorists and extremists, Tech Against Terrorism (2023) estimated that it was only a small fraction of annual content. Considering that the think tank assesses that this trend is in its experimental phase should sober readers. Their report specifies six means by which AI can enhance terrorists' media operations: media spawning,[3] automated multilingual translation, repurposing old propaganda into versions that sidestep detection of the originals, personalized propaganda that improves recruitment, moderation subversion, and fully synthetic propaganda (Tech Against Terrorism 2023). For instance, Islamic State aired a news report, modeled after Al Jazeera broadcasts, following the attack on a Russian concern venue featuring an AI-generated anchor. A single artifact of ISIL's AI-generated media program called News Harvest, it publishes near-weekly videos about the group's global operations (Verma 2024). Overall, chatbots are bolstering terrorists' ability to sow disorder by improving the speed, scale, and credibility of their influence operations.

Another way that large language models might enable ambitious terrorists is as a tailored resource for learning and troubleshooting, not unlike how most licit users engage the tool. Rather than imagining terrorists developing whole solutions or the technical capacity to onboard sophisticated AI capabilities into multiple weapons of war, our theory points us toward them using it at key points in their

[3] This process refers to generating thousands of manipulated variants of an initial image to circumvent automated detection mechanisms.

workflows to solve design, manufacturing, and deployment hiccups. Although more temperate, this prediction still foreshadows high-security concerns. Sandbrink (2023) points out that contemporary chatbots are very good at answering questions about dual-use knowledge and technology. In a recent exercise at the Massachusetts Institute of Technology, ChatGPT instructed non-scientist students about four potential pandemic pathogens, including how a person with no lab skills could acquire them without detection, in one hour. Given that terrorist biological weapons programs have often failed due to lack of expertise, chatbots' ability to fill this gap could improve their success, especially as biological design tools improve. Another looming concern is terrorist use of convolutional neural networks that analyze images for object and facial recognition. Adding this capability to AVs or drones would give terrorists with the ability to perform autonomous assassinations at low cost and risk (UNOCT 2021). Even if all extremist organizations do is leverage AI for financial gain, whether to extort funds with audio deepfakes or to train AI on historic crypto or stock data and code it to trade at will, this will still embellish their effectiveness. As chatbots become more sophisticated, we expect that they will help malevolent users upskill on manifold topics that empower them to advance their agendas.

3D Printing

Additive manufacturing, informally called 3D printing, enables the conversion of a three-dimensional computer model into a solid object. Before considering the role of AI, this technology itself renders manufacturing processes easier for the lay person because the technical engineering designs reside in computer-aided design packages. Not only are they broadly available in vibrant online repositories, but relative to traditional manufacturing this creates shortcuts by which users can produce different parts with the same printer, customized to needs, on demand, at home (Rautio and Broms 2024). Current models are compact, affordable, and rapidly advancing. Theoretically, one could print an infinite number of things to an infinite degree of customization (Blackman 2014). For the most part, the 3D printing industry explores new frontiers in manufacturing food, medical innovations, vehicles, and other constructive products. Nonetheless, it is equally able to create destructive technologies.

The chief fear to date is the prospect of criminals printing guns at will (Little 2014; Walther 2015; Europol 2022). In the last 3 years, technological advancements have led to a significant rise in designs, dissemination, and production (Vallance 2022; Dass 2024, Schaufelbühl et al. 2024). Mostly manifesting in Europe, where gun laws are prohibitive, and among the far-right, this trend has begun to express in conflict zones like Myanmar where rebels require advanced but affordable and available means to resource fighters (Dass and Mok 2023). Yet guns are only the beginning of a more imaginative threat analysis. 3D printers can be used to manufacture landmines, improvised exploding devices, small drones, bomblets and munitions, etc. In 2023, a British citizen was arrested for building for the Islamic State a bespoke drone designed to deliver an explosive warhead or chemical weapon (Murphy 2023). Just as importantly, 3D printers can make components for modifications and repairs to maximize a resource-constrained arsenal. For instance, reports surfaced in the summer of 2024 that Al-Shabaab was using 3D printers for equipment parts and weapons (Anzalone 2024). Depending on the needs and market availability of tools a violent nonstate group seeks, 3D printing can replace or supplement critical items at cost.

Dissimilar to the previous predictive cases, AI is not native to 3D printing. The prospect of adding it to optimize processes and outcomes, however, is already trending in the industry (Lin et al. 2015). AI moves these capabilities forward by leaps and bounds, yielding *advanced* additive manufacturing. Basra (2022) catalogs the factors that limit 3D printing from being an acute shortcut for violent extremist organizations, including that it takes considerable time, computer design expertise, and selection of appropriate materials. Veilleux-LePage (2024) details the steps more specifically, explaining that the process requires the 3D model file, slicing software that determines layer thickness and printer resolution, the printer itself, and selection of the thermoplastic filament. AI economizes and enhances the first, second, and fourth steps considerably. First, it increases the speed, scale, and quality of printing by applying deep learning to recognize patterns, detect defections, and improve prefabrication iterations (McCaney 2018; Bharadwaj 2019; Hammes 2019). Second, it improves design and selection of material properties suitable for a given printing project, serving as an engineering shortcut. LePain (2018) explains that "designers simply enter the desired properties into a program and algorithms predict which chemical building blocks can be combined at a micro level to create a structure with the desired functions and properties." Third, it handles the volume and complexity occurring under the surface of a printing project such that an inexperienced user can generate accurate and optimized products with little trial and error (Bharadwaj 2019). Finally, emerging applications of AI in 3D printing promise to construct computer models from 2D images, bypassing tedious engineering and coding processes and thereby significantly lowering the technical capacity requisite. Even without some of these elements, LLMs might detail any number of the above processes for ambitioned users to manually implement, still serving as a potential accelerant of 3D printing acumen for malicious ends (Miotto 2024).

Conclusion

The question of how new technologies problematize security is not new. One policymaker remarked that "every major technology—metallurgy, explosives, internal combustion, aviation, electronics, nuclear energy—has been intensively exploited not only for peaceful purposes but also for hostile ones" (Walther 2015). Still, AI bears a magnitude that merits pause. It is undoubtedly a major technology, and one on the cusp of a cascade into most facets of society. We set out to execute a sensible and shrewd threat analysis at the intersection of AI and international security. On one side, we have endeavored to avoid sensationalist speculations shallowly rooted in context. On the other side, we acknowledge the inertia of cultural lag constraining timely scholarship and policy prescription. To strike a balance, we have asserted a theoretical framework that brackets the elements of innovation—desire, capacity, and capability. We have argued that VNSAs are most likely to adopt platforms that are both advanced, being sophisticated enough to lower the technical capacity requisite yet potent, and democratized, being commonplace and affordable for resource-constrained actors. We demonstrated the logic with three confirming cases—mapping technologies, social media, and civilian drones. We then allowed a measure of imagination to predict likely VNSA exploitations of AI in the near future. Going forward, we recognize that threat identification is only the first step. Responsible rollouts of AI will need to consider the right governance approach, apportioned between engineers at the front end, legislators and regulators in the middle, and law enforcement professionals at the end. Given the immeasurable potential and utility incubating in AI applications, its advancement is inevitable.

What remains is mindful and timely treatments of externalities, unintended consequences, and misuses. May this be one of them, and may others follow to guide stakeholders in AI's unfolding.

References

Ackerman, Gary A. 2016. "'Designing Danger': Complex Engineering by Violent Non-state Actors." *Journal of Strategic Security* 9, no. 1: 1-11. https://doi.org/10.5038/1944-0472.9.1.1502

Ackerman, Spencer. 2011. "Libyan rebels are flying their own minidrone." *Wired*, August 23. https://www.wired.com/2011/08/libyan-rebels-are-flying-their-own-mini-drone/.

Anzalone, Christopher. 2024. "Al-Shabaab's reportedly using 3D printing for equipment parts and weapons." *X*, August 2. https://x.com/IbnSiqilli/status/1819373515508011106?t=Pba0yDUgmbkvIsBZW2Ggyw&s=09.

Arnaudo, Dan. 2017. "Computational Propaganda in Brazil: Social Bots during Elections." *Computational Propaganda Research Project*, Working Paper No. 2017.8. http://blogs.oii.ox.ac.uk/politicalbots/wp-content/uploads/sites/89/2017/06/Comprop-Brazil-1.pdf.

Arreguín-Toft, Ivan. 2001. "How the Weak Win Wars: A Theory of Asymmetric Conflict." *International Security* 26, no. 1: 93-128. https://doi.org/10.1162/016228801753212868

Awan, Imran. 2017. "Cyber-Extremism: ISIS and the Power of Social Media." *Society* 54, no. 2: 138–149. https://doi.org/10.1007/s12115-017-0114-0

Ball, Ryan Jokl. 2017. *The Proliferation of Unmanned Aerial Vehicles: Terrorist Use, Capability, and Strategic Implications*. Livermore, CA: Lawrence Livermore National Laboratory.

Barsade, Itai and Michael C. Horowitz. 2017. "Militant groups have drones. Now what?" *Bulletin of the Atomic Scientists*, September 7. https://thebulletin.org/2017/09/militant-groups-have-drones-now-what/.

Basra, Rajan. 2022. "The Future is Now: The Use of 3D-printed Guns by Extremists and Terrorists." *Global Network on Extremism and Technology*, June 23. https://gnet-research.org/2022/06/23/the-future-is-now-the-use-of-3d-printed-guns-by-extremists-and-terrorists/.

Bessi, Alessandro and Emilio Ferrara. 2016. "Social Bots Distort the 2016 U.S. Presidential Election Online Discussion." *First Monday* 21, no. 11: 1-14. https://doi.org/10.5210/fm.v21i11.7090

Bharadwaj, Raghav. 2019. "Artificial Intelligence Applications in Additive Manufacturing (3D Printing)." *Emerj*, February 12. https://emerj.com/ai-sector-overviews/artificial-intelligence-applications-additive-manufacturing-3d-printing/.

Blackman, Josh. 2014. "The 1st Amendment, 2nd Amendment, and 3D Printed Guns." *Tennessee Law Review* 81: 479-538.

Bloom, Mia. 2017. "Constructing Expertise: Terrorist Recruitment and 'Talent Spotting' in the PIRA, Al Qaeda, and ISIS." *Studies in Conflict & Terrorism* 40, no. 7: 603-623. https://doi.org/10.1080/1057610X.2016.1237219

Bradshaw, Tim. 2018. "Self-driving cars raise fears over 'weaponisation'." *Financial Times*, January 14. https://www.ft.com/content/a8dbd4e0-f807-11e7-88f7-5465a6ce1a00.

Brinkman, Richard L. and June E. Brinkman. 1997. "Cultural Lag: Conception and Theory." *International Journal of Social Economics* 24, no. 6: 609-627. https://doi.org/10.1108/03068299710179026

Bunn, Matthew, William H. Tobey, Martin B. Malin, and Nickolas Roth. 2016. *Preventing Nuclear Terrorism: Continuous Improvement or Dangerous Decline?* Cambridge, MA: Project on Managing the Atom, Belfer Center for Science and International Affairs, Harvard Kennedy School.

Chávez, Kerry and Ori Swed. 2020. "Off the Shelf: The Violent Nonstate Actor Drone Threat." *Air & Space Power Journal* 34(3): 29–42. https://www.airuniversity.af.edu/Portals/10/ASPJ/journals/Volume-34_Issue-3/F-Chavez_Swed.pdf.

Chávez, Kerry and Ori Swed. 2021. "The Proliferation of Drones to Violent Nonstate Actors." *Defence Studies* 21(1): 1-24. https://doi.org/10.1080/14702436.2020.1848426.

Conway, Maura. 2012. "From al-Zarqawi to al-Awlaki: The Emergence of the Internet as a New Form of Violent Radical Milieu." *CTX: Combating Terrorism Exchange* 2, no. 4: 12–22.

Curtis, Sophie. 2015. "Self-driving cars can be hacked using a laser pointer." *The Telegraph*, September 7. https://www.telegraph.co.uk/technology/news/11850373/Self-driving-cars-can-be-hacked-using-a-laser-pointer.html.

Dass, Rueben. 2024. "3D-printed Firearms and Terrorism: Trend and Analysis Pertinent to Far-right Use." *Counter Terrorist Trends and Analyses* 16(3): 19-30.

Dass, Rueben and Benjamin Mok. 2023. "Assessing the Impact of 3D-printed Weapons on the Violent Extremist Milieu." *Global Network on Extremism and Technology*. https://gnet-research.org/2023/02/20/assessing-the-impact-of-3d-printed-weapons-on-the-violent-extremist-milieu/.

Dean, Geoff, Peter Bell, and Jack Newman. 2012. "The Dark Side of Social Media: Review of Online Terrorism." *Pakistan Journal of Criminology* 4, no. 3: 103-122.

Europol. 2022. "Printing Insecurity: Tackling the Threat of 3D Printed Guns in Europe." *Europol*. https://www.europol.europa.eu/media-press/newsroom/news/printing-insecurity-tackling-threat-of-3d-printed-guns-in-europe.

Ex Machina. 2014. Directed by Alex Garland. Universal City, CA: Universal Studios.

Ferrara, Emilio, Onur Varol, Clayton Davis, Filippo Menczer, and Alessandro Flammini. 2016. "The Rise of Social Bots." *Communications of the ACM* 59, no. 7: 96-104. https://doi.org/10.1145/2818717

Finisterre, Ken and Sen, Robi. 2016. "The Unrealized Threat of Weaponized Drones." *Morning Consult*, October 24. https://morningconsult.com/opinions/unrealized-threat-weaponized-drones/.

Flood, Rebecca. 2016. "ISIS using GOOGLE MAPS to plot airport and plane terror attacks, report warns." *The Express*, September 27. https://www.express.co.uk/news/world/715011/Isis-airplane-attack-terror-bomb-google-map-report-airport.

Friese, Larry, N. R. Jenzen-Jones, and Michael Smallwood. 2016. "Emerging Unmanned Threats: The Use of Commercially Available UAVs by Armed Non-state Actors." *Armament Research Services*. http://armamentresearch.com/wp-content/uploads/2016/02/ARES-Special-Report-No.-2-Emerging-Unmanned-Threats.pdf.

Gadam, Suhasini. 2018. "Artificial Intelligence and Autonomous Vehicles." *Medium*. https://medium.com/datadriveninvestor/artificial-intelligence-and-autonomous-vehicles-ae877feb6cd2.

Gentsch, Peter. 2018. *AI in Marketing, Sales and Service: How Marketers Without a Data Science Degree Can Use AI, Big Data and Bots*. New York, NY: Springer.

Goodin, Dan. 2007. "Google Maps aids terrorists, NY lawmaker warns." *The Register* June 11. https://www.theregister.co.uk/2007/06/11/google_maps_aids_terrorists/.

Grigorescu, Sorin Mihai, Markus Glaab, and André Roßbach. 2018. "From logistics regression to self-driving cars: Chances and challenges for machine learning in highly automated driving." *Embedded Computing*, July 30. http://www.embedded-computing.com/automotive/from-logistics-regression-to-self-driving-cars-chances-and-challenges-for-machine-learning-in-highly-automated-driving.

Hammes, T.X. 2019. "Technology Converges; Non-state Actors Benefit." *Hoover Institution*, February 25. https://www.hoover.org/research/technology-converges-non-state-actors-benefit.

Harding, Thomas. 2007. "Terrorists use Google maps to hit UK troops." *Telegraph News*, January 13. https://www.telegraph.co.uk/news/worldnews/1539401/Terrorists-use-Google-maps-to-hit-UK-troops.html.

Heartfield, Ryan, George Loukas, Sanja Budimir, Anatolij Bezemskij, Johnny R. J. Fontaine, Avgoustinos Filippoupolitis, and Etienne Roesch. 2018. "A Taxonomy of Cyber-physical Threats and Impact in the Smart Home."

Computers & Security 78: 398-428. https://doi.org/10.1016/j.cose.2018.07.011

Hern, Alex. 2017. "Elon Musk says AI could lead to third world war." *The Guardian*, September 4. https://www.theguardian.com/technology/2017/sep/04/elon-musk-ai-third-world-war-vladimir-putin.

Horowitz, Michael C. 2010. "Nonstate Actors and the Diffusion of Innovations: The Case of Suicide Terrorism." *International Organization* 64, no. 1: 33-64. https://doi.org/10.1017/S0020818309990233

Huang, Ming-Hui and Roland T. Rust. 2018. "Artificial Intelligence in Service." *Journal of Service Research* 21, no. 2: 155-172. https://doi.org/10.1177/1094670517752459

Huey, Laura. 2015. "This is Not Your Mother's Terrorism: Social Media, Online Radicalization and the Practice of Political Jamming." *Journal of Terrorism Research* 6, no. 2: 1-16. https://doi.org/10.15664/jtr.1159

Hwang, Tim, Ian Pearce, and Max Nanis. 2012. "Socialbots: Voices from the Fronts." *Interactions* 19, no. 2: 38-45. https://doi.org/10.1145/2090150.2090161

Ionita, Silviu. 2017. "Autonomous Vehicles: From Paradigms to Technology." *IOP Conference Series: Materials Science and Engineering* 252: 1-8.

I, Robot. 2004. Directed by Alex Proyas. Los Angeles, CA: 20th Century Fox.

Klausen, Jytte. 2015. "Tweeting the Jihad: Social Media Networks of Western Foreign Fighters in Syria and Iraq." *Studies in Conflict & Terrorism* 38, no. 1: 1-22. https://doi.org/10.1080/1057610X.2014.974948

Kredo, Adam. 2018. "Al Qaeda Using Google Maps to Plan Jihadist Attacks." *Washington Free Beacon*, April 20. https://freebeacon.com/national-security/al-qaeda-using-google-maps-plan-jihadist-attacks/.

Lachow, Irving. 2017. "The Upside and Downside of Swarming Drones." *Bulletin of the Atomic Scientists* 73, no. 2: 96-101. https://doi.org/10.1080/00963402.2017.1290879

Lakomy, Miron. 2023. "Artificial Intelligence as a Terrorism Enabler? Understanding the Potential Impact of Chatbots and Image Generators on Online Terrorist Activities." *Studies in Conflict & Terrorism*: published online. https://doi.org/10.1080/1057610X.2023.2259195

Lee, Wenke, Cliff Wang, and David Dagon. 2008. *Botnet Detection: Countering the Largest Security Threat*. New York, NY: Springer.

LePain, Andrea. 2018. "How AI and 3D Printing are Revolutionizing Materials Design." *R&D World*, December 11. https://www.rdworldonline.com/how-ai-and-3d-printing-are-revolutionizing-materials-design/.

Lin, Herbert and Jaclyn Kerr. 2019. "On Cyber-enabled Information/Influence Warfare and Manipulation." In *Oxford Handbook of Cybersecurity* (forthcoming).

Lin, Pierre Pascal Anatole, Karl Willis, Eric Jamesson Wilhelm, Arian Aziz Aghababaie. 2015. "Intelligent 3D Printing Through Optimization of 3D Print Parameters." *U.S. Patent 14/711714*, November 19.

Little, Rory K. 2014. "Guns Don't Kill People, 3D Printing Does? Why the Technology is a Distraction from Effective Gun Controls." *Hastings Law Journal* 65, no. 6: 1505-1513.

Lohmann, Sara. 2024. "ChatGPT, Artificial Intelligence, and the Terrorist Toolbox." In *Emerging Technologies and Terrorism: An American Perspective*, 23–34. Susan Sim, Eric Hartunian, and Paul J. Milas (eds). Carlisle, PA: US Army War College Press.

Mack, Andrew. 1975. "Why Big Nations Lose Small Wars: The Politics of Asymmetric Conflict." *World Politics* 27, no. 2: 175-200. https://doi.org/10.2307/2009880

Mair, David. 2017. "# Westgate: A Case Study–How al-Shabaab Used Twitter During an Ongoing Attack." *Studies in Conflict & Terrorism* 40, no. 1: 24-43. https://doi.org/10.1080/1057610X.2016.1157404

Mandelbaum, Jay, James Ralston, Ivars Gutmanis, Andrew Hull, and Christopher Martin. 2005. "Terrorist Use of Improvised or Commercially Available Precision-guided UAVs at Stand-off Ranges: An Approach for Formulating Mitigation Considerations." *Institute for Defense Analyses*. http://www.dtic.mil/dtic/tr/fulltext/u2/a460419.pdf.

Markon, Jerry. 2016. "As concern grows over terrorists on social media, Senate bill calls for national strategy." *The Washington Post*, February 10. https://www.washingtonpost.com/news/federal-eye/wp/2016/02/10/as-concern-grows-over-terrorists-on-social-media-senate-bill-calls-for-national-strategy/?utm_term=.43bb08a586cc.

Marr, Bernard. 2018. "How Artificial Intelligence is Making Chatbots Better for Businesses." *Forbes*, May 18. https://www.forbes.com/sites/bernardmarr/2018/05/18/how-artificial-intelligence-is-making-chatbots-better-for-businesses/#396362d94e72.

Mathur, Priyank, Clara Broekaert, and Colin P. Clarke. 2024. "The Radicalization (and Counter-radicalization) Potential of AI." *International Centre for Counter-Terrorism*. https://www.icct.nl/publication/radicalization-and-counter-radicalization-potential-artificial-intelligence#:~:text=AI%20has%20significant%20potential%20to%20reshape%20the%20terrorism%20threat%20landscape

McCaney, Kevin. 2018. "AI Fueling Next Wave of 3D Printing and Robotics." *Government CIO*, August 17. https://governmentciomedia.com/ai-3d-printing-robotics.

McClure, Paul K. 2018. "'You're Fired,' Says the Robot: The Rise of Automation in the Workplace, Technophobes, and Fears of Unemployment." *Social Science*

Computer Review 36, no. 2: 139-156. https://doi.org/10.1177/0894439317698637

McKelvey, Fenwick and Elizabeth Dubois. 2017. "Computational Propaganda in Canada: The Use of Political Bots." *Computational Propaganda Research Project*, Working Paper No. 2017.6. http://blogs.oii.ox.ac.uk/politicalbots/wp-content/uploads/sites/89/2017/06/Comprop-Canada.pdf.

Melki, Jad and May Jabado. 2016. "Mediated Public Diplomacy of the Islamic State in Iraq and Syria: The Synergistic Use of Terrorism, Social Media and Branding." *Media and Communication* 4, no. 2: 92-103. https://doi.org/10.17645/mac.v4i2.432

Miller, Maggie. 2018. "Consumer groups say Senate's revamped self-driving car bill fails to resolve cyber, safety concerns." *Inside Cyber Security*, December 11. https://insidecybersecurity.com/daily-news/consumer-groups-say-senates-revamped-self-driving-car-bill-fails-resolve-cyber-safety.

Miotto, Nicolo. 2024. "3D Printing and WMD Terrorism: A Threat in the Making?" *European Leadership Network,* January 10. https://europeanleadershipnetwork.org/commentary/3d-printing-and-wmd-terrorism-a-threat-in-the-making/.

Murphy, Jessica. 2023. "Birmingham PhD student guilty of using 3D printer to build 'kamikaze' drone." *The Guardian*, September 28. https://www.theguardian.com/uk-news/2023/sep/28/birmingham-phd-student-mohamad-al-bared-guilty-using-3d-printer-to-build-kamikaze-drone.

National Association of Insurance Commissioners. 2024. "Autonomous Vehicles." *NAIC Insurance Topics.* https://content.naic.org/insurance-topics/autonomous-vehicles.

Nelu, Clarisa. 2024. "Exploitation of Generative AI by Terrorist Groups." *International Centre for Counter-Terrorism.* https://www.icct.nl/publication/exploitation-generative-ai-terrorist-groups#:~:text=This%20analysis%20describes%20some%20of%20the%20risks%20posed%20by%20generative.

Ogburn, William F. 1922. *Social Change with Respect to Culture and Original Nature.* New York, NY: B. W. Huebsch, Inc.

O'Flaherty, Douglas. 2018. "AI in action: Autonomous vehicles." *IBM IT Infrastructure Blog.* https://www.ibm.com/blogs/systems/ai-in-action-autonomous-vehicles/.

O'Sullivan, Shane, Nathalie Nevejans, Colin Allen, Andrew Blyth, Simon Leonard, Ugo Pagallo, Katharina Holzinger, Andreas Holzinger, Mohammed Imran Sajid, and Hutan Ashrafian. 2019. "Legal, Regulatory, and Ethical Frameworks for Development of Standards in Artificial Intelligence (AI) and Autonomous Robotic Surgery." *The International Journal of Medical Robotics and Computer Assisted Surgery* 15, no. 1: e1968. https://doi.org/10.1002/rcs.1968

Pepito, Joseph Andrew and Rozzano Locsin. 2019. "Can Nurses Remain Relevant in a Technologically Advanced Future?" *International Journal of Nursing Sciences* 6(1): 106-110. https://doi.org/10.1016/j.ijnss.2018.09.013

Rao, Qing and Jelena Frtunikj. 2018. "Deep Learning for Self-driving Cars: Chances and Challenges." *SEFAIS '18: Proceedings of the 1st International Workshop on Software Engineering for AI in Autonomous Systems:* 35–38. https://doi.org/10.1145/3194085.3194087

Rassler, Don. 2016. "Remotely Piloted Innovation: Terrorism, Drones, and Supportive Technology." *USMA Combating Terrorism Center.* http://www.dtic.mil/dtic/tr/fulltext/u2/1019773.pdf.

Rath, Johannes, Monique Ischi, and Dana Perkins. 2014. "Evolution of Different Dual-use Concepts in International and National Law and Its Implications on Research Ethics and Governance." *Science & Engineering Ethics* 20, no. 3: 769-790. https://doi.org/10.1007/s11948-014-9519-y

Rautio, Samu and Mika Broms. 2024. "Security Scenarios: 3D Printed Firearms." *Security & Defence Quarterly* 48(4): published online. https://doi.org/10.35467/sdq/181047.

Ravindranath, Mohana. 2017. "Lawmakers Want Self-driving Cars to Thrive But Still Fear Hacks." *NextGov*, February 15. https://www.nextgov.com/emerging-tech/2017/02/lawmakers-want-self-driving-cars-thrive-still-fear-cyber-hacks/135437/.

Ribeiro, John. 2008. "Google Earth Used by Terrorists in India Attacks." *PC World*, November 30. https://www.pcworld.com/article/154684/article.html.

Rogers, James Patton (editor). 2024. *The De Gruyter Handbook on Drone Warfare.* Boston: De Gruyter.

Rosner, Robert and Lynn Eden. 2019. "Rebuilding an Aging Nuclear Weapons Complex: What Should the United States Do, and Not Do? An Overview." *Bulletin of the Atomic Scientists* 75, no. 1: 3-8. https://doi.org/10.1080/00963402.2019.1555977

Rudner, Martin. 2016. "'Electronic Jihad': The Internet as al-Qaeda's Catalyst for Global Terror." *Studies in Conflict & Terrorism* 40, no. 1: 10-23. https://doi.org/10.1080/1057610X.2016.1157403

Rychnovská, Dagmar. 2016. "Governing Dual-use Knowledge: From the Politics of Responsible Science to the Ethicalization of Security." *Security Dialogue* 47, no. 4: 310-328. https://doi.org/10.1177/0967010616658848

Sandbrink, Jonas. 2023. "ChatGPT Could Make Bioterrorism Horrifyingly Easy." *Vox,* August 7. https://www.vox.com/future-perfect/23820331/chatgpt-bioterrorism-bioweapons-artificial-inteligence-openai-terrorism.

Santini, Rose Marie, Larissa Agostini, Carlos Eduardo Barros, Danilo Carvalho, Rafael Centeno de Centeno, Debora G. Salles, Kenzo Seto, Camyla Terra, and Giulia Tucci. 2018. "Software Power as Soft Power: A Literature

Review on Computational Propaganda Effects in Public Opinion and Political Process." *Partecipazione e Conflitto* 11, no. 2: 332-360. https://doi.org/10.1285/i20356609v11i2p332

Schaufelbühl, Stefan, Nicolas Florquin, Denis Werner, and Olivier Delémont. 2024. "The Emergence of 3D-printed Firearms: An Analysis of Media and Law Enforcement Reports. *Forensic Science International: Synergy* 8: published online. https://doi.org/10.1016/j.fsisyn.2024.100464.

Schaer, Cathrin. 2024. "How Extremist Groups Like 'Islamic State' Are Using AI." *Deutsche Welle,* October 7. https://www.dw.com/en/how-extremist-groups-like-islamic-state-are-using-ai/a-69609398#:~:text=Groups%20like%20the%20%22Islamic%20State%22%20and%20al-Qaeda%20are%20urging%20followers.

Schulzke, Marcus. 2019. "Drone Proliferation and the Challenge of Regulating Dual-Use Technologies." *International Studies Review* 21, no. 3: 497-517. https://doi.org/10.1093/isr/viy047

Schwartz, Stephen I. 1997. "Maintaining Our Nuclear Arsenal is Expensive." *Brookings Institute*, March 26. https://www.brookings.edu/opinions/maintaining-our-nuclear-arsenal-is-expensive/.

Sekhose, Marcia. 2016. "Australian authorities fear self-driving cars could be used for terrorism." *BGR Media*, November 9. https://www.bgr.in/news/australian-authorities-fear-self-driving-cars-could-be-used-for-terrorism/.

Shawar, Bayan Abu and Eric Steven Atwell. 2005. "Using Corpora in Machine-learning Chatbot Systems." *International Journal of Corpus Linguistics* 10, no. 4: 489-516. https://doi.org/10.1075/ijcl.10.4.06sha

Sparrow, Robert. 2016. "Robots in Aged Care: A Dystopian Future?" *AI & Society* 31, no. 4: 445-454. https://doi.org/10.1007/s00146-015-0625-4

Szollosy, Michael. 2017. "Freud, Frankenstein and our fear of robots: projection in our cultural perception of technology." *AI & Society* 32, no. 3: 433-439. https://doi.org/10.1007/s00146-016-0654-7

Subrahmanian, V.S., Amos Azaria, Skylar Durst, Vadim Kagan, Aram Galstyan, Kristina Lerman, Linhong Zhu, Emilio Ferrara, Alessandro Flammini, and Filippo Menczer. 2016. "The DARPA Twitter Bot Challenge." *Computer* 49, no. 6: 38-46. https://doi.org/10.1109/MC.2016.183

Tech Against Terrorism. 2023. "Early Terrorist Experimentation with Generative Artificial Intelligence Services." *Tech Against Terrorism.* https://techagainstterrorism.org/hubfs/Tech%20Against%20Terrorism%20Briefing%20-%20Early%20terrorist%20experimentation%20with%20generative%20artificial%20intelligence%20services.pdf#:~:text=On%2021%20July%202023,%20OSINT%20analysts%20at%20Tech%20Against%20Terrorism.

The Terminator. 1984. Directed by James Cameron. Los Angeles, CA: Hemdale Film Corporation.

Tschider, Charlotte A. 2018. "Regulating the Internet of Things: Discrimination, Privacy, and Cybersecurity in the Artificial Intelligence Age." *Denver Law Review* 96: 87. https://doi.org/10.2139/ssrn.3129557

United Nations Office of Counter-Terrorism (UNOCT) Staff. 2021. "Algorithms and Terrorism: The Malicious Use of Artificial Intelligence for Terrorist Purposes." *UNOCT.* https://unicri.it/News/Algorithms-Terrorism-UNICRI-UNOCCT#:~:text=Download%20the%20Report%20Algorithms%20and%20Terrorism:%20The%20Malicious%20Use%20of.

Vallance, Chris. 2022. "3D printed guns: warnings over growing threat of 3D firearms." *BBC News,* November 9. https://www.bbc.com/news/technology-63495123.amp.

Veilleux-Lepage, Yannick. 2024. "Printing Terror: An Empirical Overview of the Use of 3D-printed Firearms by Right-wing Extremists." *Combating Terrorism Center at West Point* 17(6): 31-45.

Verma, Pranshu. 2024. "These ISIS News Anchors Are AI Fakes. Their Propaganda Is Real." *Washington Post,* May 17. https://www.washingtonpost.com/technology/2024/05/17/ai-isis-propaganda/.

Wallace, Dave and Shane Reeves. 2013. "Non-state Armed Groups and Technology: The Humanitarian Tragedy at our Doorstep?" *University of Miami National Security & Armed Conflict Law Review* 3, no. 1: 26-45.

Walther, Gerald. 2015. "Printing Insecurity? The Security Implications of 3D-Printing of Weapons." *Science & Engineering Ethics* 21, no. 6: 1435-1445. https://doi.org/10.1007/s11948-014-9617-x

Waters, Nick. 2018. "Google Maps is a Better Spy than James Bond." *Foreign Policy*, September 25. https://foreignpolicy.com/2018/09/25/google-maps-is-a-better-spy-than-james-bond/.

Weimann, Gabriel, Alexander T. Pack, Rachel Sulciner, Joelle Scheinin, Gal Rapaport, & David Diaz. 2024. "Generating Terror: The Risks of Generative AI Exploitation." *Combating Terrorism Center at West Point* 17, no. 1: 17–24. https://ctc.westpoint.edu/generating-terror-the-risks-of-generative-ai-exploitation/.

CHAPTER 12

Diplomacy's New Lessons Learned: First World War Submarine Warfare and the 21st Century Drone

Justin Quinn Olmstead

History is littered with examples of changing technology causing panic in populations and governments. Twenty-first century questions about how these changes threaten the norms associated with national defense do not differ from those asked a century or two earlier. They simply are asked about another set of new technology posing a threat to the maintenance of security and peace. A prime example of changing technology causing problems with established defense and security models is the use of unrestricted submarine warfare by Germany during the First World War. A

The views and opinions in this chapter are personal and do not represent those of Sandia National Laboratories, National Technology and Engineering Solutions of Sandia, LLC, the Department of Energy, or any other agency of the United States Government.

J. Q. Olmstead (✉)
Sandia National Laboratories, Albuquerque, USA
e-mail: jqolmst@sandia.gov

R. Kiggins (ed.), *The Political Economy of Robots*, International Political Economy Series, https://doi.org/10.1007/978-3-031-95171-8_12

modern-day equivalent to the submarine is the drone or unmanned aerial vehicle (UAV). While some may quibble about this comparison, both weapons are the creations of technological advances that have outpaced both the laws of war and the established morality of war.

The need to assess American diplomatic responses to its own drone policy presents an opportunity to compare it to the historical narrative of US diplomacy and policy toward the use of the submarine in 1915. The idea here is to have the history of US policy inform the current debate. By developing an understanding of how the United States has responded to similar events in the past and how that response affected the eventual outcome provides today's leaders and policymakers with an improved understanding of the impact their decisions have. While this may be surprising to some, the political and diplomatic discussions with Germany concerning its use of submarines present a framework from which current leaders can draw. The difficulties faced by these governments as they struggled with the legality of new technology and its impact on belligerent and neutral countries provide a different view to help construct today's dialogue.

The hope being to provide a historical framework for modern-day states to deal with the diplomatic issues surrounding the use of drones in the twenty-first century. The initial chapter, published eight years ago, sought to keep the framework focused by avoiding discussion of drone support of military operations by concentrating on targeted killing. This was due in large part to targeted killing being much more controversial in both public and political circles than air support for ground troops. But the war in Ukraine has raised the idea of drones in warfare to a new level—one on par with the submarine at the end of the First World War. If one questions the comparisons made here, they should look no further than Ukrainian President Zelensky's comment in September 2023, '...it's an artillery duel with each side firing about 40,000 shells a day. Ukrainian infantry is advancing bloody yards at a time. It's World War One with drones.'[1]

Yes, the comparison the Ukrainian president has used opens the connection up for debate. However, it is still apt in that drones and the policy surrounding their use in warfare, are still developing, just as submarines and the policy around their use did throughout the First

[1] Volodymyr Zelenskyy; The 2023 60 Minutes Interview, https://www.cbsnews.com/video/volodymyr-zelenskyy-ukraine-russia-60-minutes-video-2023-09-17/

World War. The framework used here for analyzing and comparison is the American response to the German use of submarines and primarily the American policy decision for targeted killing via drone strikes, but the use of drones more generally. Because both decisions are a result of the US political system and diplomacy, it is a closer reflection of US values and process. It is not lost on the author that the issue of German submarine use occurred during war but it is equally important to note that the United States was not. US policy was as a neutral nation responding to the use of new technology. It is also important to understand that the position the United States eventually took would also apply to all submarines in use, including those being used by the British, French, and US.

Lesson One: Patience

Until the war in Ukraine, virtually all news reports of drone strikes focused on the damage done by the strike, the person or people killed, what their nationality was, and their role in terrorism or if they were a wrong target. This was generally done in reference to US use of drones which made sense at the time because it was US forces using drones in Afghanistan and Iraq. Unfortunately, this was about all the average person heard and therefore understood about drones. It is precisely this same focus the newspapers had in the United States when Germany began sinking ships around the British Isles during the First World War; the name of a ship, the number of dead, and how many Americans were dead as a result. How the United States initially responded to Germany's declaration of its intent to utilize the submarine to sink merchant ships is every bit as important.

What has changed is that now social media in particular, but the news in general, is filled with images of drones destroying tanks, factories, buildings, and even accepting the surrender of troops. Media discussions about the impact of the drone on warfare and the long-range strikes Ukraine is executing with drones have changed the nature and landscape of political and military thought surrounding this weapon—just as with submarines during the First World War. In addition to news of drone warfare's success, the low cost of them is also catching the attention of politicians and generals alike. Retired US general and former Chairman of the Joint Chiefs of Staff, Mark Milley and Eric Schmidt recently noted that while an MQ-9 Reaper drone costs about $30 Million, a simple first-person-view drone can go for as little as $500.[2] While perhaps not as wildly extreme as the difference in drones to build, a contrast of Dreadnoughts to submarines being built in 1914 shows a comparable difference with a D-class submarine costing about £89,000 versus the HMS Queen Elizabeth's £3,014,103 price tag.[3]

When Germany declared its war zone around the British Isles on 4 February 1915, the submarine was not a new weapon. First used in combat during the American Civil War, the submarine continued to be considered experimental until the late nineteenth century, and then it was viewed primarily as a defensive weapon. Nevertheless, in May 1914 there were discussions within the German Navy for the creation of a submarine fleet large enough to suffocate Britain. This idea was not taken seriously by German leadership until December 1914 when a single submarine sank three British cruisers. Even at that the discussions between German leaders about its effectiveness dragged on for several years.

What is important to this discussion is Wilson's handling of the situation which rests with his decision to not immediately protest the German war zone declaration. what would be considered a lifetime in today's political atmosphere, Wilson took 5 days and eight meetings with members of his cabinet (specifically, Secretary of State William Jennings Bryan and Counselor to the Secretary of State Robert Lansing) before issuing his first statement concerning the German declaration of a war-zone.[4] To the British government at the time, and to some historians, this was simple dithering on the part of the United States. In reality, the president wisely took the time to listen and discuss the issue with his cabinet before making a pronouncement on U.S. policy.

Lansing realized that the American response presented 'a most delicate situation which [would] have to be handled with extreme care.'[5] Upon reading the German memorandum which claimed the same necessity to justify the violation of neutral trade as Great Britain, Lansing recognized

[2] Mark A. Milley and Eric Schmidt, 'America Isn't Ready for the Wars of the Future: And They're Already Here', *Foreign Affairs*, September/October 2024, 26–37.

[3] Barry Slemmings, 'How Much Did a Warship Cost?', The World War I Document Archive, The Maritime War, http://www.gwpda.org/naval/wcosts.htm.

[4] Robert Lansing Desk Diary and Papers, Library of Congress, Reel 1.

[5] Ray Stannard Baker, *Woodrow Wilson Life and Letters: Neutrality 1914–1915*, vol. Five (New York, 1940), p. 247.

that Germany was simply demanding the same rights the British had already forced on the world—to change the rules of naval warfare to meet the needs of a modern navy.[6]

The note sent by the Wilson administration was direct and based on the accepted laws governing war at sea. It reminded Germany that legally, belligerent rights were limited to visit and search of neutral ships unless a blockade was declared and effectively maintained. This was followed by statements proclaiming the use of submarines to be 'unprecedented in naval warfare,' an 'indefensible violation of neutral rights' and that 'the United States would...hold the Imperial German Government to a strict accountability.'[7]

By taking the time to gather facts, and understand the laws, Wilson and his administration did not jump to a conclusion when dealing with a new technology. They took their time to collect all of the details and proceeded from there. The patience demonstrated by President Wilson is a facet of Wilsonian diplomacy that was applied during the duration of the war.

When applying this framework to the drone, it is important to point out the corollaries: Weaponized drones, like the submarine in 1914, are new technology that has changed the nature of established defense and security models as well as the accepted moral understanding. Prior to 2022 Russian invasion of Ukraine, the key ideas linking the submarine and drone were the issue of non-combatants, the accepted rules of warfare, and the related diplomatic problems. This is still true, to a point, but with the manufacture and use of drones proliferating, it is necessary to also look at the long-term implications for politicians, diplomats, and soldiers alike.

The U. S. military has been flying unarmed drones since the early 1930s for intelligence, surveillance, and reconnaissance, and while there were some forms of armed drones in the 1950s and 1960s, the move to arm the current type occurred in the late 1990s with the first use of

[6] Ross Gregory, *The Origins of American Intervention in the First World War* (New York, 1971), p. 58. In this case, the British had implemented a blockade that was illegal under international law as well as hoisting American flags on British ships to avoid being attacked by German submarines. What is known as a Ruse de Guerre.

[7] The Secretary of State to the Ambassador in Germany (Gerard), 10 Feb. 1915, United States Department of State/*Papers Relating to the Foreign Relations of the United States (FRUS), 1915, Supplement, The World War.* Part II: Neutral Rights, pp. 98–100.

weaponized drones occurring in 2001. That is roughly a twenty-year span that the U.S. government used to not only develop the armed drone and deploy it but to work through some of the diplomatic issues that could occur as a result. The initial issue concerned the Intermediate-Range Nuclear Forces Treaty (INF Treaty) between the U.S. and the Soviet Union. Signed in 1987, one of the requirements of the INF Treaty was the elimination of 'unmanned, self-propelled vehicles' with ranges of between 500 and 5500 kilometers.[8] Much like Lansing had had initial reservations about how to handle the issue of the German war zone enforced with submarines in 1914, the State Department in 1996 initially squashed the idea of arming drones until it was able to determine whether or not they violated the INF Treaty.

Like the treaty concerning the arming of drones, the United States had to ensure that once airborne, drone attacks—by all who used drones—would meet all domestic and international laws.[9] Because the U.S. was engaged in wars in Afghanistan and Iraq in the early 2000s, drone strikes were considered to be part of the authorized use of military force (Authorization for the Use of Military Force or AUMF legislation passed by congress after the September 11, 2001 terrorist attacks on the United States). For most this seems simple enough, but how this applies when the decision is made to launch a strike at a target outside of traditional war zones is confusing at best and could prove to be hard to explain both diplomatically and domestically. Again, here is where patience in policy development pays off.

The first drone strike outside of a war zone occurred in 2002 when then President George W. Bush ordered a strike in Yemen. Bush's use of drones began with his belief that the direct strikes they offered would spare the innocent, and followed a significant increase in funding for advanced armed drone technology.[10] This is a view taken by every president since and is a view generally held by others despite evidence

[8] Richard Whittle, *Predator: The Secret Origins of the Drone Revolution,* (New York: Henry Holt and Company, 2014), p. 171.

[9] Charles Blanchard, 'This Is Not War by Machine' in *Drone Wars: Transforming Conflict, Law, and Policy,* eds. Peter L. Bergen and Daniel Rothenberg (New York: Cambridge University Press, 2015), p. 122.

[10] John W. Rollins, 'Armed Drones: Evolution as a Counterterrorism Tool', *Congressional Research Service*, 7 November 2023, https://crsreports.congress.gov/product/pdf/IF/IF12342.

that suggests that in some cases, 90% of those killed were unintended targets.[11]

Early in his administration, President Barack Obama appointed Harold Koh, then dean of Yale Law School and a highly regarded expert on international human rights, to the position of State Department Legal Advisor. He quickly began laying the legal framework for defending the administration's drone policy. In 2010, Koh gave a speech outlining the use of drones in attacking terrorists. In this speech, he stated that each strike was analyzed based on the specifics of each case and based on the exercise of the inherent right of self-defense under international law and in particular Article 51 of the UN Charter.[12] In November 2011 the U.S. Department of Justice went further and issued a White Paper titled 'Lawfulness of a Lethal Operation Directed Against a U.S. Citizen Who Is a Senior Operational Leader of Al-Qaeda or an Associated Force.'[13] The importance of this paper here is that it demonstrates at least some level of patience within the Obama administration when dealing with drone strikes. The White Paper sets out the legalities of using lethal force against a U.S. citizen outside of a war zone. In particular, it lays out the legal argument for the conditions required to target a particular person.

Drone strikes outside of war have never violated international law. Rather, they have challenged it because they don't adhere to long-held ideas and meaning of legal classifications.[14] It is precise because drone technology is still—40 years after the current stage of drone operations—in a 'growth and development' phase that laws and treaties have yet to keep up with. Former Secretary of Defense, Robert Gates supported a proposal by the Obama administration to create a court based on the

[11] 'US Deadly Drone Strikes', *Amnesty International*, 18 May 2020, https://www.amnesty.org.uk/thank-you-us-deadly-drones; David Sterman, 'Endless War Challenges Analysis of Drone Strike Effectiveness', *Journal of National Security Law & Policy*, vol. 13:305, 2023, https://jnslp.com/wp-content/uploads/2023/05/Endless_War_Challenges_Analysis_of_Drone_Strike_Effectiveness.pdf.

[12] Harold Koh, 'The Obama administration and international law,' keynote address at the annual meeting of the American Society of International Law, Washington, DC, 25 March 2010. http://www.state.gov/s/l/releases/remarks/139119.htm

[13] US Department of Justice, 'Department of Justice white paper: Lawfulness of a lethal operation directed against a US citizen who is a senior operational leader of Al-Qaida or an associated force,' released February 4, 2013.

[14] Rosa Brooks, 'Drones and the International Rule of Law', *Ethics & International Affairs*, Spring 2014 (28.1), 83.

Foreign Intelligence and Surveillance Act (FISA) for drone use that would place stringent rules on their use to prevent abuse.[15] For clarification, FISA established a federal court to approve surveillance on suspected foreign spies inside the United States. However, congressional Republicans disagreed with Senator Lindsey Graham insisting it would be an intrusion on the president's authority as commander in chief.[16] The idea of a judicial oversight of targeting and killing an enemy, seemed to many, both Democrats and Republicans as unworkable and outside the normal duties of a court. As President Obama worked to ensure the use of drones adhered to standard principles and laws, others seemingly were more worried about the ability to strike quickly and avoid drawn-out discussions of legality.

In the face of fears concerning the creation of 'killer robots' because of AI, legal scholar, and former deputy assistant attorney general in the Office of Legal Counsel John Yoo has argued that despite the unprecedented legal and political regulations needed, a rush to judgment on drones would be a serious mistake due to their great potential to change armed conflict.[17] Indeed, it appears that both the Trump and Biden administrations have continued to use the legal framework laid out by President Obama.[18] While it is important to point out that Trump dropped crucial restrictions contained in Obama's policy, and Biden reinstated them, This patience has both allowed the US to continue to use drones in the War on Terror and wait for international law to catch up to technical innovations.

Just as Wilson's decision to not dive head first into the public and moral debate about the use of submarines in early 1915 was good diplomacy for a nation wanting to remain neutral, Similarly, President Obama's decision to have a legal stance for the use of drone's in targeted killing

[15] 'Gates backs more oversight of drones', CNN, 10 February 2013.

[16] Carlo Muñoz, 'No Court for drone oversight, says GOP', *The Hill*, 13 February 2013, https://thehill.com/policy/defense/142021-no-court-for-drone-oversight-says-gop/.

[17] John Yoo, 'A Legal Framework for the Use of Drones in War', *Strategika*, 14 March 2024, https://www.hoover.org/research/legal-framework-use-drones-war.

[18] Shannon Roddel, 'The Future of US Drone Policy: A Conversation with International Law Professor Mary Ellen O'Connell', 14 October 2022, *Notre Dame News*, https://news.nd.edu/news/the-future-of-us-drone-policy-a-conversation-with-international-law-professor-mary-ellen-oconnell/.

laid out for the public to see is good domestic and foreign policy. By accounting for the decision-making process, the United States would be able to minimize discomfort for targeted killings. The adaptation of drones, in the press at least, has broadened the spectrum of what is considered good policy. Those that want to see Ukraine win have no issue with their use of drones to attach Russian weapons manufacturing deep inside Russia, nor do they have an issue with the use of drones as force multipliers for the Armed Forces of Ukraine. By not placing boundaries on drones as it has done for US made long-range weapons, the Biden administration is following Wilson's lead set more than one hundred years ago—it is good diplomacy.

Lesson Two: Clarity

The first casualty of Germany's use of submarine war occurred on 28 March 1915 when the British ship *Falaba* was sunk off the Irish coast with the loss of 104 crewmembers and passengers including one American. This was followed by an aerial attack on the U.S. ship *Cushing* on 29 April, and then on 1 May the American steamer *Gulflight* was torpedoed. All three attacks were overshadowed in a matter of days with the sinking of the *Lusitania* off the coast of Ireland on 7 May 1915. Then, with a brief period of respite, the British steamer *Arabic* was sunk on 19 August, bringing about a pledge from the Kaiser Wilhelm II himself that no ocean liners should be sunk without full warning. To understand the work done by the diplomats to ease tensions the initial submarine crisis needs to be viewed as a whole and within the framework of the U.S. threat of 'strict accountability.'

As these ships were being sunk, the public (Americans, British, and French citizens), some U.S. leaders, and foreign diplomats held the belief that 'strict accountability' meant the United States would go to war if U.S. ships were attacked or that it would at least 'defend [its] right with force.'[19] It is likely that this theory has grown out of the British belief at the time that 'the logical conclusion to be drawn from this language was

[19] Ross Gregory, *The Origins of American Intervention in the First World War* (New York, 1971), p. 54; Holger Herwig, *Politics of Frustration: The United States in German Naval Planning, 1889–1941* (Boston, 1976), p. 117.

that if an American ship were torpedoed without notice, the American Government would regard this as an act of war.'[20]

But the note to Germany never uses the words force or war, and on 29 April 1915, Lansing wrote that too strong of policy would 'close the door to all compromise.[21] In a letter to Bryan on 5 May 1915, Lansing clearly states that 'Strict Accountability can only mean that the German Government must make full reparation for the act of their naval force and must also repudiate the act, apologize for it, and give ample assurance that it will not be repeated.'[22] This does not mean that Lansing did not see the pitfalls of strong language, for he often wrote to Wilson and Bryan that it is possible that too strong a protest could lead to a break in diplomatic relations and possibly war.[23] But to represent Lansing's position of 'strict accountability' as being directly related to the use of force is a slight misrepresentation of the facts, and this is the problem. It is evident that there was a lack of clarity in both the note sent to Germany and the public' understanding of it.

It has long been assumed that the 'strict accountability' letter of 10 February 1915 set the tone for all future conversations dealing with submarine warfare, but this is not the case. In fact, it was upon being notified that an American citizen had died due to the actions of German submarines that Wilson's advisors, in particular, Lansing and Bryan, began formulating an outline for future U.S. policy—the groundwork for any and all future incidents involving the deaths of U.S. citizens by submarine warfare. Wilson's policy toward submarine warfare did indeed begin with the strict accountability phrase, but this was neither an outline nor a plan of action. At this point, Wilson had yet to define 'strict neutrality,'

[20] Telegram from Spring Rice to Grey, 26 February 1915, PRO 170, FO 800/85 Microfilm, The National Archives of the UK (Hereafter referred to as TNA).

[21] ' Memo, 29 April 1917, The Robert Lansing Papers, 1914–1920, Private Memoranda (Typed Copy), DM 15,347, Reel 1 (Lansing Diary Blue Box 2: Confidential Memoranda and Notes from 15 April 1915 - 30 December 1918, Inclusive), Library of Congress.

[22] Letter from The Counsellor for the Department of State (Lansing) to the Secretary of State, 5 May 1915, The United States Department of State/*FRUS. The Lansing Papers, 1914–1920 (in two volumes),* Vol. 1, p. 384.

[23] Memo, 29 April 1915, Robert Lansing Papers 1914–1920, Private Memoranda (Typed Copy), DM15347, Reel 1 (Lansing Diary Blue Boxed Box 2: Confidential Memoranda and Notes from 15 April 1915–30 December 1918, Inclusive), Library of Congress.

and this was the problem.[24] Leaders of the belligerent countries and the general public had been left to make their own decision as to what the phrase meant.

Lansing admitted that he lacked information about the sinking, but he understood the need for the United States to clarify what it meant by strict accountability.[25] He began by outlining the case against Germany in regards to the *Falaba* noting that if a merchant ship did not abide by the rules of naval warfare, America would have no legal standing to protest the loss[26] of life. Crucially, Lansing's letter stated that if the United States were to make a formal complaint about the sinking, it would have to demand what amounted to reparations for the loss of an American life.[27]

On 5 April 1915, Lansing sent a draft response to Bryan for the President's approval accompanied with a note clearly outlining the issues facing the United States with respect to its response to the sinking of the *Falaba*. The Counsellor pointed out that if the letter to Germany were too soft, there would be no reason for Germany to heed America's warning and drop its submarine tactics, while at the same time acknowledging that if the letter was too harsh, it could be seen as American support for Germany's enemies, leading to German hostility and possibly war.[28] After 2 weeks of discussion, meetings, and rewrites, Lansing drafted a new note that was less harsh, but still firm.[29] The key to this new note was, much

[24] Ernest R. May, *The World War and American Isolation, 1914–1917* (Cambridge, 1963), p. 145; John Milton Cooper, Jr., *Woodrow Wilson: A Biography* (New York, 2009), p. 275.

[25] *Ibid.*, 366.

[26] Letter from The Counsellor for the Department of State (Lansing) to the Secretary of State, 2 April 1915, The United States Department of State/*FRUS. The Lansing Papers, 1914–1920 (in two volumes),* Vol. 1, p. 365.

[27] Ibid., 365.

[28] Letter from The Counsellor for the Department of State (Lansing) to the Secretary of State, 5 April 1915, The United States Department of State/*FRUS. The Lansing Papers, 1914–1920 (in two volumes),* Vol. 1, pp. 369–370.

[29] President Wilson to the Secretary of State, 22 April 1915, The United States Department of State/*FRUS. The Lansing Papers, 1914–1920 (in two volumes),* Vol. 1, pp. 377–378.

like Lansing's original draft, focusing on the rights of neutrals, the violations of international law, and pushing Germany to adhere to the existing laws of visit and search.[30]

Before the United States had a chance to send this letter, reports arrived that on 29 April a German plane had bombed the American steamship *Cushing* with no loss of life. Then on 1 May, the *Gulflight* became the second U.S. flagged ship to be attacked by German forces. While these events caused a stir among some members of Wilson's cabinet, the *Cushing* incident did not involve the loss of life, and in the case of the *Gulflight*, the ship had failed to sink and was traveling with British trawlers making it hard to protest. Regardless of the fact that an American flagged ship had been attacked, the lack of injuries forced Wilson and his advisors to virtually ignore this incident as a case for protest.[31]

While still debating the specifics of the protest letter, and whether or not the incidents of the *Cushing* and *Gulflight* should be included or dealt with separately, news arrived that the British luxury liner *Lusitania* had been sunk with the loss of 1198 lives, 124 of which were American. Based on the yet unclarified meaning of 'strict accountability,' U.S. Ambassador to Great Britain Walter Page and presidential advisor Edward House believed that the incident would usher the United States into the war, and, accordingly, both Page and House—whom the president had sent to Europe for a third time to attempt to bring the warring sides to some type of peace agreement—sent notes to the President declaring that if the United States wanted to maintain the respect of the world it would have to declare war.'[32]

It is at this point that the outline created by Wilson and Lansing on what the 10 February letter and its 'strict accountability' phrase meant

[30] President Wilson to the Secretary of State, 22 April 1915, The United States Department of State/*FRUS. The Lansing Papers, 1914–1920 (in two volumes),* Vol. 1, pp. 377–378.

[31] Letter from The Counsellor for the Department of State (Lansing) to the Secretary of State, 1 May 1915, The United States Department of State/*FRUS. The Lansing Papers, 1914–1920 (in two volumes),* Vol. 1, pp. 381–382.

[32] The Ambassador in Great Britain (Page) to the Secretary of State, 8 May 1915, United States Department of State/*FRUS, 1915, Supplement, The World War*. Part II: Neutral Rights, pp. 385–386.

becomes crucial in understanding the American diplomatic reaction to the German submarine war.

Despite the outrage expressed by Page and House, by the American east coast press, and the British press, the outline created, but not sent, due to the *Falaba* case was the driving force behind Wilson's reaction to the sinking of the *Lusitania*. Leaders in the United States and among the belligerent nations misinterpreted the time it took Wilson to determine the proper response to the sinking of the *Falaba*. Allied governments were reminding American representatives abroad that 'Wilson had not protested the *Falaba* and *Gulflight*.'[33] Back at home, Wilson was being attacked by former President Teddy Roosevelt and Senator Elihu Root for 'failing to respond to Germany's initial attacks' and for issuing 'ultimatums that fail to ultimate.'[34] To many at the time the intense reaction to the sinking of the Lusitania left Wilson with 'no choice but to do something' and that it meant 'America's entry into the war on the side of the Entente had become a serious possibility.'[35]

The idea that the note of 10 February hindered the president's ability to maneuver diplomatically does not take into account the meaning of the 'strict accountability' phrase.[36] As noted, while most people outside of Wilson' s cabinet misunderstood this phrase to mean 'war' as the only option, inside the cabinet the definition had become clearer. At the time of the *Lusitania's* sinking, Wilson, and his advisors had been working for almost three months to create an outline clarifying this point. Events had simply outpaced their ability to put their thoughts into a cohesive, understandable policy.

When Wilson finally emerged from his self-induced sabbatical on 11 May, he shared a draft of a diplomatic note he had written with his cabinet. Wilson's note was firm in its recounting of the earlier note of 10 February, and events between the two countries at sea including the

[33] Ross Gregory, *Walter Hines Page: Ambassador to the Court of St. James's* (Lexington, 1970), p. 97.

[34] Ross A Kennedy, *The Will to Believe: Woodrow Wilson, World War I, and America's Strategy for Peace and Security* (Kent, 2009), p. 111.

[35] Ross Gregory, *The Origins of American Intervention in the First World War* (New York, 1971), p. 60; Reinhard R. Doerries, *Imperial Challenge: Ambassador Count Bernstorff and German-American Relations, 1908–1917* (Chapel Hill, 1989), p. 97.

[36] Ross A. Kennedy, *The Will to Believe: Woodrow Wilson, World War I, and America's Strategy for Peace and Security* (Kent, 2009), p. 82.

attacks on the *Falaba*, *Cushing*, *Gulflight*, as well as the *Lusitania*, and called for Germany to pay reparations, as well as taking steps to ensure that future incidents of this same kind did not occur.[37]

When the final letter was delivered to German Foreign Minister Gottlieb von Jagow on 13 May, it was the culmination of months of thought and policy work by Wilson and his advisors. In the letter, Wilson reiterated his point that the United States did not accept Germany's claim that British measures forced them to bypass the visit and search procedures, internationally recognized as legitimate rules of naval warfare, and as a result, they would be expected to disavow the actions of the submarine commanders at fault.[38] The note continued along the exact phrasing used by Wilson in his draft, calling for the payment of reparations and for the German government to take the actions necessary to ensure that there would be no further incidents of this type.[39] Despite the internal bickering that occurred between Lansing and Bryan, and despite the disagreement over the tone of the letter, the Wilson administration had finally and clearly outlined the policy by which the United States would wage its neutrality against the pressures of submarine warfare; the submarine issue was a diplomatic question, and not a military matter.

With the U.S. decision to use drones as the primary weapon in its war on terror and the lingering questions of legality, clarity is much needed. As noted earlier, the White Paper issued by the Justice Department on 8 November 2011 was seemingly designed to do just that. But as has been pointed out by various legal scholars and critics of the Obama administration, it does not clarify how U.S. policy fits within wider international law.[40] The complication seems to be in the Justice Department's argument for drone strikes on terrorists, not in current war zones. As mentioned earlier, the argument it makes is that Article 51 of

[37] The Secretary of State to the Ambassador in Germany (Gerard), 13 May 1915, United States Department of State/*FRUS, 1915, Supplement, The World War*. Part II: Neutral Rights, pp. 393–396.

[38] *Ibid.*

[39] *Ibid.*

[40] David Kaye, 'International Law Issues in the Department of Justice White Paper on Targeted Killing,' *Insights*, Vol. 17, Issue 8, https://www.asil.org.

the UN Charter authorizes countries to use force as an 'inherent right of self-defense.'[41]

Like Wilson's 'strict accountability' statement, the legal analysis contained in the Justice Department's White Paper leaves much open for questioning. It is limited in how it addresses international law. University of California—Irvine Law Professor David Kaye has pointed out that based on the argument posed by the Justice Department if the U.S. were to cease its war with al-Qaeda, the legality of drone strikes would become less clear.[42] Concurring with this view is the Director of the Institute for National Security and Counterterrorism at Syracuse University, William C. Banks who notes that the defensive use of force has strong legal roots but that the use of lethal force outside a 'traditional battlefield complicates the international legal issues.'[43]

Some of the United States' closest allies, while not publicly reacting to the use of drone strikes, struggle to either support the policy or accuse the US of breaking international law. In a European Council on Foreign Relations policy brief on drones and targeted killing, Anthony Dworkin makes the case that the lack of clarity resulting from the use of new technology gives the European Union an opportunity to take an active role in defining what are acceptable standards for targeted killing in international law.[44] In an opinion piece for the Washington Post in 2011, John B. Bellinger III, a legal advisor to the US Department of State under President Bush, argued that the US should work to 'build a broader international consensus' on drone policy in order to avoid losing support among allies.[45] Bellinger noted that then White House counterterrorism

[41] US Department of Justice, 'Department of Justice white paper: Lawfulness of a lethal operation directed against a US citizen who is a senior operational leader of Al-Qaida or an associated force,' released February 4, 2011, p. 2.

[42] Kaye,' International Law Issues,' *Insight,* p. 2.

[43] William C. Banks, 'Regulating Drones: Are Targeted Killings by Drones Outside Traditional Battlefields Legal?' in *Drone Wars: Transforming Conflict, Law, and Policy,* eds. Peter L. Bergen and Daniel Rothenberg (New York: Cambridge University Press, 2015), p. 144.

[44] Anthony Dworkin, 'Drones and Targeted Killing: Defining a European Position,' *European Council on Foreign Relations*, July 2013, http://www.ecfr.eu/page/-/ECFR84_DRONES_BRIEF.pdf

[45] John B. Bellinger III, 'Will drone strikes become Obama's Guantanamo?', *The Washington Post*, 2 October 2011, https://www.washingtonpost.com/opinions/will-drone-strikes-become-obamas-guantanamo/2011/09/30/gIQA0ReIGL_story.html.

advisor, John Brennen agreed that the success and effectiveness of US counterterrorism strikes were and are dependent on allies.

One aspect of this particular problem stems from U.S. targeted killings of U.S. citizens outside of war zones in Afghanistan and Iraq. The 2011 White Paper does not create a defense for this type of targeted killing within international law, instead, it focused on making the case in U.S. domestic law. Most of the early drone strikes occurred in Pakistan and initially had the consent of the Pakistan government. These strikes fell within the parameters of international law because, although Pakistan was not a war zone, its government had agreed to the use. The 30 September 2011 drone strike in Yemen that killed U.S. citizen Abdulrahman al-Awlaki changed all that. With this new first (that we know of) targeted killing outside of a war zone and without the nation that it occurred in having sanctioned it, the U.S. had stretched the law much like Germany had stretched the law in 1915–1917.

In 2006, the US Department of Defense placed specific and stringent rules on the domestic use of DoD operated unmanned aircraft systems. Even then, DoD drones are only used at the request of local, state, or federal civilian authorities, and in most of these cases the mission is one of support.[46] Even when being used by request, DoD drones must get waivers from the Federal Aviation Administration allowing them to fly pre-coordinated routes within special use airspace. Special use airspace is airspace delegated by the FAA for DoD use above land with express permission of the landowner or government-owned/leased land. The point of this being, that within the US, military drones are not operating outside of the law. Multiple administrations have upheld this notion and to date there have been no known violations by DoD. While a domestic law and not an international law or treaty, its adoption and use do demonstrate US attempts to regulate this technology as it has grown.

Seeking clarity is not always a simple task. In 2014, the Islamic State in Iraq and the Levant (ISIS) began attacking Iraq, Syria, and Libya. The Obama administration claimed ISIS was an offshoot of al Qaeda meaning it fell under the auspices of the AUMF Act passed by Congress after 9/11 and mentioned earlier. Every president since has used the AUMF to provide it with a free hand to kill individuals as well as attack other

[46] Unmanned Aircraft Systems (UAS): DoD Purpose and Operational Use, US Department of Defense, https://dod.defense.gov/UAS/#:~:text=This%20policy%20direction%20is%20set,%2C%20exercises%2C%20and%20testing%20purposes.

nations. Due to the broad nature of the AUMF, presidents have not had to wait for clarity, but have been content to declare clarity where things are more muddy. However, criticism has been leveled at the US because the AUMF does not specifically call for the use of force against any group that is not al Qaeda or the Taliban.[47]

Adding to the convoluted situation are US drone strikes in Pakistan, Yemen, and Somalia and Ukrainian strikes deep in Russia. None of these areas are within an active war zone (though one could easily consider Russia as warzone due to the ongoing Russian invasion of Ukraine), but each strike has been justified because nations have an inherent right to self-defense.[48] In 2022, President Joe Biden signed a new classified policy memo for counterterrorism drone strikes outside of conventional war zones.[49] The aim, it appears, is to reduce the risks to civilians.

There is another area for which clarity is sought, and that is specifically the law. Ethical and moral concerns aside (and to be clear, these are very important, and Peter Lee addresses them in his contribution to this book), international law is ultimately where the legal use of drones will be determined. Regarding the war in Ukraine, Russia has used drones to attack Ukraine's civilian infrastructure and critical national infrastructure (CNI). There is a military logic that can be used for attacking CNI, but deliberate attacks on civilians are forbidden within the principles International Humanitarian Law.[50] But what of the Houthi attacks on shipping in the Red Sea? If the global community does not do more than condemn these attacks does it open the door for the legality of Russia's attacks on Ukraine's CNI, or for drone attacks on CNI in general?

[47] 'The Problems with the AUMF and How To Fix Them', *Human Rights First*, 19 June 2017, https://humanrightsfirst.org/library/the-problems-with-the-aumf-and-how-to-fix-them/.

[48] 'Legality of Drone Warfare', The Bureau of Investigative Journalism, https://www.thebureauinvestigates.com/explainers/legality-of-drone-warfare/.

[49] Oona A. Hathaway, 'Biden's New Counterterrorism Policy Guidance Further Entrenches the Forever War', *Just Security*, 11 October 2022, https://www.justsecurity.org/83487/bidens-new-counterterrorism-policy-guidance-further-entrenches-forever-war/.

[50] Saba Sotoudehfar and Jeremy Julian Sarkin, 'Drones on the Frontline: Charting the Use of Drones in the Russo-Ukrainian Conflict and How Their Use May be violating International Humanitarian Law', *Sciendo*, ICLR, 2023, Vol, 23, No. 2, http://efaidnbmnnnibpcajpcglclefindmkaj/https://novaresearch.unl.pt/files/85371059/Sarkin_Drones-on-the-Frontline.pdf.

Final Thoughts

Part of the problem that faced President Wilson in 1915 and Presidents Obama, Trump, and Biden from 2016 through 2024 (and one would assume future presidents as well) is that the laws governing warfare have not evolved to meet technological advances. Attempting to develop a policy that allows for the use of this new technology while remaining within the parameters of current law has been, and is, a problem U.S. presidents have faced for 100 years now. In the case of the drone, the case is more muddied and could have been (and still can be) clarified. At issue is the idea that drones are no longer new technology and therefore are adequately covered by current rules of war. By defining a drone as an unmanned aerial vehicle, the assertion is that they have been around since at least the First World War.[51] But the drones of the twenty-first century are nothing like the drones of the previous century. If the definition is adjusted from a drone (which can mean anything from a commercial delivery system to a surveillance drone, to a police robot) to a remotely piloted aircraft (RPA), or an unmanned aerial vehicle (UAV), then the discussion is now about a weapon that has been in use for only about 15 years.

When it comes right down to it, the United States in 2024 is asking the same questions of established international law that Germany was asking in 1915—that international law needs to change to meet the needs of modern technology and therefore a modern military. Rosa Brooks was incredibly perceptive when she wrote that the problem with drones is that they 'challenge the international rule of law precisely because they defy straightforward legal categorization.'[52] In other words, there needs to be a clarification of how this new technology fits into the current legal framework. The evolution of drones from surveillance, to weaponization and targeted killing, to mass use on the battlefield, has created a problem for politicians, diplomats, and legal affairs. If the development of submarine use from outlawed to everyday use teaches the world anything, it is that drones are not only here to stay, but that people should get used to their presence.

[51] John Sifton, 'A Brief History of Drones', in *The Nation,* 27 February 2012.

[52] Rosa Books, 'Drones and the International Rule of Law,' 2013, p. 83. Scholarship@Georgetownlaw.edu.

CHAPTER 13

Big Data, Artificial Intelligence, and Autonomous Policy Decision-Making: A Crisis in International Relations Theory?

Ryan David Kiggins

This chapter focuses attention on implications of autonomous information technologies to theories of international relations. As big data, artificial intelligence and, more broadly, fully autonomous information technologies, are increasingly relied upon to make policy determinations, issue policy directives, execute policies, and assess the results of policies executed, international relations theory may face a crisis in explanatory power. International relations theories are, fundamentally, explanations of human decision under certain conditions. When, not if, humans participate less, or no longer conduct policy analysis, make policy decisions, when humans become drones directed by artificial intelligence and autonomous information technologies, theories of international relations may face a crisis of relevance as acute as the production line laborer displaced by autonomous production line robots.

R. D. Kiggins (✉)
Political Science, University of Central Oklahoma, Edmond, USA
e-mail: rkiggins@uco.edu

R. Kiggins (ed.), *The Political Economy of Robots*, International Political Economy Series, https://doi.org/10.1007/978-3-031-95171-8_13

The claim is advanced that international relation theories are, fundamentally, theories for describing and explaining human decision-making under certain conditions. An overview of past present and future uses of information technologies, including artificial intelligence, is then provided to empirically demonstrate that governments have, do, and will continue relying upon information technologies in the policy decision-making process. The last section to this chapter suggests a direction that International Relations scholars may pursue to ensure the continued relevancy, accuracy, and explanative power of international relations theories.[1] On the whole, we may discover that some theoretical traditions in International Relations (IR) are better suited then others in accounting for the advent of information technologies as autonomous or semi-autonomous actors in international relations. The agent/structure ideal type from the social constructivist tradition of IR theory may be best suited to account for the potential agentive capacity to act autonomously inherent in artificial and other information intelligence technologies.

Moving forward, I examine the purpose for theory within IR followed by a brief overview of the increasing reliance on automated decision-making within government foreign-policy making focusing on the United States (US), Great Britain (GB), and Russia in order to show that policy-makers are increasingly reliant on automated decision-making.I conclude by arguing that as reliance on automated decision-making increases apace with the development of more capable artificial intelligence technology, the risk to IR is that some theories on which we rely to explain human decision-making under certain conditions may become obsolete. International Relations must, to remain relevant, adopt a more reflexive posture toward the purpose and analytical application of IR theory by relying on the agent/structure ideal type, and by recognizing that artificial intelligence technology is capable of agentive capacity and must now be treated as an agent capable of effectuating outcomes in international relations.

[1] Following the convention employed by Zehfuss (2002, 1), International Relations herein refers to the academic discipline of IR while international relations refers to its empirical subject matter.

What is IR Theory?

Karl Popper ([1959]2005) noted that all observation is theory-laden and, we may add, international relations is (heavily) laden with theory. One publication offers chapter length overviews of fourteen IR theory traditions. In this section, the argument proceeds by scrutinizing the question, "What is IR theory?" followed by briefly reviewing the current state of IR theory. The section concludes by defending the claim that IR theory is fundamentally concerned with human decision-making.

The question "What is theory", at least within International Relations, remains highly controversial. The reason for the controversy may be a consequence of unsettled epistemological, ontological, and methodological questions continuing to plague the field (Dunne et al. 2013; Jackson 2015; Walker 1993; Wight 2006; Solomon and Steele 2016; Zehuss 2002). The lack of consensus among IR scholars on answers (if there are any) to epistemological, ontological, and methodological questions may be one possible contributor to the plethora of IR theory. Another may be that we lack, as yet, conceptual and empirical tools capable of precisely delineating the complexity that is international relations into a form from which a general theory of IR may arise. Alternatively, the empirical subject matter of international relations may not prove suitable for developing a theory of IR that can stand up to scientific scrutiny. That empirical subject matter is the product of intentional and unintentional human decisions over time in a condition where political authority may or may not serve as a decision constraint. Nevertheless, some IR scholars have argued that IR theory must adhere to certain criteria to be considered a scientific theory eliding the epistemological and ontological sleight of hand pregnant in such a claim (see Jackson 2015).

This view of IR theory holds that theories should describe and explain interactions among social phenomena for the purpose of identifying causal inferences (see King et al. 1994). "Theories indicate what is connected with what and how that connection is made. They convey a sense of how things work, of how they hang together, of what structure of a realm of inquiry may be" (Waltz 1979, 12). Accordingly, the purpose of IR theory is to structure an academic endeavor while offering description, explanation, and, ideally, prediction of the interaction among discreet phenomena. Constructing a theory is more often than not, a long process of trial and error, with theory predictions, hypotheses, and underlying assumptions being confirmed through iterative testing and peer-review

(Hempel 1966; Kuhn 1996). Some hold that IR scholars should ensure the subject intended to be studied can be done so utilizing "the analytical method of classical physics—examining the attributes and interactions of two variables while others are kept constant" (Waltz 1979, 12).

Of course, this idealized view of science and the role of IR scholars in knowledge production elides, "that the scientific process is not driven solely by the ideals of impartial and measured dialogue, drawing on empirical and rational support. Rather, it would appear that presuppositions or priors, aggressive rhetoric, economic and legal muscle, and authority all have a role to play in securing scientific knowledge" (Moses and Knutsen 2007, 2). Kuhn's investigation of scientific revolutions offers tales in which ego, rapacious competition, and greed often spurred scientific progress (1996). And let us not refrain from neglecting that American political science especially, has pursued research in support of and response to US foreign policy objectives (Oren 2003). Hardly a rational empiricist approach to science. The point is that what counts as theory is inherently political with tradeoffs being made as to what counts epistemologically and ontologically. Recognizing that theory, especially international relations theory, is inherently political should not be read as dismissive of IR scholarship and theory. Rather, it should be read as an acknowledgment of the innate difficulties of actualizing idealized research design norms within IR scholarly practice, drawn from another academic discipline that may reflect an underlying problem of incommensurability for International Relations are investigating the effects of human decision-making under certain conditions with all that entails as far as human biases, assumptions, and foibles.

International relations as scholarly endeavor was, allegedly, born out of concern to understand the causes of war in an effort to limit its recurrence (Angell 1913; Burchill and Linklater 2013; Carr [1939]2001; Guzzini 2013; Hoffman 1977; Smith 1989). Normative commitments served as motivator for the beginning of International Relations, specifically the normative belief in the perniciousness of war and hope that, through scholarly scrutiny, the causes of war could be identified and remedies proposed and implemented to prevent war. Such origin narratives or, better yet, disciplinary histories, may be fairly viewed as bound up in presentist concerns related to substantiating a claim to ontological and epistemological legitimacy within an academic discipline (Buzan and Lawson 2013; Schmidt 1998; Smith 2000; Turton 2015; Waever 1998). Regardless of how International Relations scholars sociologically, politically, and economically view their discipline, an important feature of International Relations remains unchanged: to investigate, describe, and explain human decision-making under certain conditions through postulating explanative theory. The cornucopia of international relations theories available reflects this commitment to explaining why, how, when, where, and what political decisions humans take under certain conditions.

Burchill et al. (2013) offer substantive overviews for 12 traditions of IR theory all with disparate epistemologies, ontologies, and multiple variants contributing to a condition of paradigmatic instability. International Relations scholars view this paradigmatic instability either positively or negatively. According to Lake (2013; 2011), the existence of multiple traditions of IR theory has encouraged eclecticism within IR and this eclecticism is a hallmark of vibrancy and theoretical creativity (also see Ferguson 2015). Alternatively, this eclecticism has given rise to consternation among some IR scholars concerned with the apparent lack of progress toward the formation of a "grand theory" that explains international relations (Keohane 1984; Mearsheimer and Walt 2013; Waltz 1979). For, the development of an IR grand theory is viewed as essential for a scholarly endeavor that fancies itself a science *qua* science. Skinner (1990) notes that C. Wright Mills coined the term "grand theory" for use in critiquing different strategies for progress in the social sciences. Grand theory may be understood as a "belief that the primary goal of the social disciplines should be that of seeking to construct a systematic theory of the nature of man and society" (Skinner 1990, 3). On this account, IR grand theory would be a systematic theory of the nature of war and cooperation among actors operating under certain conditions. According to Skinner's account of Mills, grand theory was an "impediment to the progress of the social sciences" (1990, 3). And yet, as Solomon and Steele note that within IR, "there seems to be a default admonition to scholars, and students, to re-embrace grand theory" (2016, 2). Grand theory is viewed by some IR scholars as quintessential proof that IR is a scientific endeavor and should be taken and treated as such by practitioners, policymakers, and academic peers (see Baldwin 1993; Keohane 1984; Mearsheimer 2001; Mearsheimer and Walt 2013; Waltz 1979). Indeed, some scholars continue to advocate that IR remain focused on producing grand theory (Harrison and Mitchell 2013; Ikenberry 2012; Mearsheimer and Walt 2013; Snyder 2013). Like good little positivists, some IR scholars seem especially fretful about facing the prospect that IR may be, or become, a degenerative research program unworthy of the social and pecuniary legitimacy conferred by the appellation "science." Yet some philosophers of science, notably Kuhn (1996) and Winch (1990), share Mill's skepticism regarding the development of grand theory in the social sciences. We would do well to share that skepticism. Fortunately, there is emerging in International Relations a condition of détente among disparate theoretical traditions.

A 2013 special issue in the *European Journal of International Relations* speaks to this détente being largely sanguine about the future prospects for IR and its theory. The current state of international relations theory is labeled as pluralist being characterized by theory testing (Dunne et al. 2013). Lake (2011) similarly advocates that IR scholars focus attention on middle-theorizing rather than on resorting to producing research that negatively animates the politics of IR theory—the incessant narcissism inherent in defending one's theoretical tradition while attacking another's. Solomon and Steele (2016) also note IR theory's progress toward a richer understanding of international relations being enabled through scrutinizing micropolitics (everyday individual activities) to enrich our theorizing about macropolitics. Overall, scholarly attention to the state of IR theory, its usefulness, and, especially, its purpose is energetically optimistic (see Buzan and Lawson 2013; Ferguson 2015; Tickner and Blaney 2013). Indeed, there appears to be an emerging consensus that IR scholars must be conversant in a number of IR theories given that each IR theory emphasizes certain features of international relations while (intentionally) de-emphasizing other features. Waltz (1979) suggests that de-emphasizing certain features of international relations must occur in order for theory to possess any sort of explanatory power. Otherwise, the shear complexity of international relations would prove especially overwhelming. Of course, such normative commitments drawn from philosophy of science are but one way to think about IR theory (Burchill et al. 2013; Sylvester 2013; Walker 1993; Wight 2006).

What does any of this have to do with the purpose of IR theory? As much as we fancy IR theory as reflecting certain virtues we characterize as scientific because scientific has social, political, and economic cache, the purpose of IR theory is to explain why independent and interdependent human decisions congeal into the outbreak of war, conflict, and exploitation. It is to explain and describe why humans can be so reprehensibly cruel to other humans. It is to explain, in short, us. Whether in using dehumanized and, ironically, anthropomorphized, terms such as "actors", "states", "institutions", or any number of other terms to be found in IR theory, as IR scholars we are attempting to identify connections between discreet phenomena that tell us something about how human decisions result in certain outcomes given certain conditions. That the outcome may be war, poverty, democratization, globalization, gender subjugation, or dozens of other topics matters not to why we investigate international relations or the purpose for constructing International Relations theory. It's all about us. IR theory is fundamentally concerned with describing and explaining human decision-making under certain conditions.

The condition most commonly incorporated into international relations theory is anarchy, the notion that no central authority exists capable of imposing limits to the interests and actions of independent actors. Milner (1991) takes to task those in IR that assert anarchy as an empirical fact by effectively showing that IR scholars haphazardly employ the inherently (intentionally?) ambiguous concept risking, from a positivist standpoint, a degenerative research program. Milner's critique pushed further open the door for other IR scholars, notably Lake (2009; 1996) and, more recently, Pouliot (2016), to investigate instances of hierarchy, what Carr ([1939]2001) would have simply referred to as international order or, more abstractly, as utopia. Utopia in the sense that implicit in hierarchy are certain norms, ideas, interests, rules, decision-making procedures, and identities that exert, independently or interdependently, influence on human decision-making under certain conditions. Wendt's (1999; 1992) claim that "anarchy is what states make of it", essentially means that anarchy serves as a blank canvas painted by humans acting through any number and variety of institutions and in accordance with a plethora of normative frameworks they themselves have created for managing transnational interactions with each other. Institutions may include states, international intergovernmental organizations, international non-governmental organizations, multinational corporations, transnational terrorist networks, other types of transnational networks, and norms such as sovereignty or territoriality; all of which are utilized by humans to effect outcomes in global politics. All of which are subjects and objects of IR theory that may assist in the parsing of international relations but, at the cost of de-humanizing IR theory, of ignoring that at its core IR theory is fundamentally an attempt to describe and explain human decision-making under certain conditions—call it anarchy, hierarchy, or some other concept.

Those decisions under anarchy made by humans, on which International Relations focus, are existentially consequential to humans, the planet on which we live, and other species also native to this planet; and are taken by individuals that we might describe as political decision-makers, who occupy positions of authority within governing and commercial institutions. In attempting to make decisions that may be existentially consequential for humans, political decision-makers may choose to rely on a number of sources and tools. Since the invention of the internet, information available to political decision-makers has increased exponentially and with that increase the need to parse, analyze, and correlate that information for the purpose enhancing political decision-making. We now turn to investigating the increasing use of automated information technology in support of political decision-making observed in public and private settings. The purpose is to note that humans are rapidly being replaced as political decision-makers.

Governments, States, and Computer Automated Decision-Making

Alan Turing (1912–1954) is widely viewed as the inventor of computing. Computing may be understood as a set of (re)programmable steps specifying which information inputs are required to produce certain information outputs through a machine. Turing's formidable intellect was put to use breaking German naval codes by the British government during World War II. Relying on concepts and a machine design he published as an academic paper in 1936, Turing designed and built an electro-mechanical machine that successfully predicted the daily encryption key for German naval codes. Breaking the German naval codes enabled the British and allies to gain intelligence on German war strategy that, it has been estimated, shortened World War II by two to four years (Hodges 2003). The success of Turing's electro-mechanical machine during World War II, marked a turning point for the use of computing machines in support of government decision-making. In this section, we provide an all-to-brief overview on the use of computing machines to automate government decision-making in the post-World War II era. The purpose is to provide empirical support to the claim that automated policy decision-making is not a new trend and, with the rapid development of information technologies, including (potentially) artificial intelligence, is fast moving beyond human input and control. The section concludes by advancing the claim that the advent of automated autonomous policymaking may present a crisis of explanatory power to International Relations theory.

In his study of the growth in English government during the period 1688–1783, Brewer remarked that, "The power of governments has always been and always will be in large part dependent upon their capacity to order and manipulate different sorts of information" (1989, 221). Computing was intentionally designed to accomplish the task on which government power relies: the ordering, manipulation, and storage of information. Computing power, understood as the capacity to order and manipulate information to advance political goals, strategies, and policies, is the foundation of modern government power. However, government risks losing that power to the tsunami of big data now deluging government, enterprise, and private computing systems. Big data remains conceptually contested lacking consensus regarding a precise definition (De Mauro et al. 2015; HU et al. 2014; Ward and Barker 2013). Nevertheless, big data may be understood as, "*Information assets characterized by such a High Volume, Velocity and Variety to require specific Technology and Analytical Methods for its transformation into Value*" (De Mauro et al. 2015, 103; emphasis original). What precisely are information assets and what value do they offer to government?

An answer to this question must begin with some understanding of digital universe size. One approach to measuring the size of the digital universe is in terms of information (data) storage capacity. Gantz and Reinsel (2012) note that the digital universe, measured in terms of data available, between 2005 and 2020, will grow by a factor of 300 from 130 exabytes to over 40,000 exabytes.[2] To put that in perspective, the Library of Congress stores the equivalent of 432 terabytes of information or the equivalent of 432,000 gigabytes compared to my tablet that can store up to 128 gigabytes of information. Where does all this data originate? To answer, let's start with your smartphone, reasonable to assume that you own one. Your use of a smartphone may include voice calls, recording or viewing video, texting or emailing, web browsing, social networking, gaming, and any economic transactions associated with those activities. Each activity, including economic transactions (purchasing of consumables while playing mobile games, products on retail websites, ordering food), are data, digital kernels of information about you, specifically your life choices and behavior, that you freely and unwittingly volunteer to

[2] A Terabyte is equivalent to 1,000 Gigabytes. A Petabyte is 1,000 Terabytes; an Exabyte is 1,000 Petabytes; A Zettabyte is 1,000 Exabytes; A Yottabyte is equivalent to 1,000 Zettabytes.

cyberspace with each mouse click, thumb swipe, or thumb press. As our dependence on networked computing has increased so has the volume, velocity, and variety of data volunteered by us available for collection by any enterprising entity whether government, commercial, or private. We both generate data and consume data—simultaneously. These data are information assets. How is value extracted from these assets? This depends on whether the imperative underlying the objective for extracting value is economic, political, or both in addition to being a technical question regarding computing power capability.

Since the digital revolution exploded, some companies specialize in collecting, cataloging, and selling access to terabytes-sized database files containing the digital kernels—the data—derived from recording your online behavior; what big data scientists, government officials, and entrepreneurs refer to as "transactions." For example, Acxiom Corporation, has as its primary information asset and the foundation of its profit-making model, a database of approximately 500 million active consumers worldwide each with 1,500 data points per person (Peled 2014). On an annual basis, Acxiom processes 50 trillion data transactions. This is made possible by your use of your smartphone combined with other internet connected devices you use daily. Acxiom collects, stores, and develops methods that structure your digital data for later analyses and correlations from which value propositions may be derived for business profit. That is to say, your digital data are analyzed to discern how best to extract more money from your wallet through increased economic consumption. To this end, other companies such as Facebook stores 500 terabytes of data while Google stores 10 exabytes of data.[3] Storing data is but one feature of extracting value. Another necessary feature is technical: accessing and analyzing that data to produce novel insights into consumer or human behavior. Machine learning is a field within computer science that focuses on building computing systems that autonomously improve through experience (Mitchell 2006). "[A] machine learns with respect to a particular task T, performance metric P, and type of experience E, if the system reliably improves its performance P at task T, following experience E" (2006, 1). In effect, machine learning employs a model of probabilistic reasoning to make predictions about novel relationships that may exist in

[3] Daniel Price; "Surprising facts and stats about the big data industry." *Cloud Tweaks* (17 Mar 2015). Accessed Jul 2016. http://cloudtweaks.com/2015/03/surprising-facts-and-stats-about-the-big-data-industry/

big data sets. Google, for example, uses such technology to anticipate the text string one is entering into the search field at google.com. Self-driving cars are employing machine learning algorithms to accomplish the task of safely conveying passengers from one point to another. Netflix employs machine learning to develop recommendations for users and Facebook relies on this technology to predict products and content Facebook users may desire. These companies have successfully monetized your data to generate economic value for their respective profit models. Yet, the size of commercial databases are dwarfed by government databases.

Government has an advantage over commercial entities in the collection of your data: government may compel you to provide data you otherwise would not provide. Often, government is able to invoke extraordinary authority to gain access to your data especially during exigent crises. The political imperative in such exigent crises is security—preventing future terrorist attacks through preventive policing and military action supported through intelligence gleaned from big data (Schneier 2015). Such was the case within the United States in the aftermath of the September 11, 2001 terrorist attacks. During the 2-year period following those attacks, the National Security Agency of the US government (NSA) collected and stored over two trillion domestic or international phone calls (Peled 2014). NSA By the end of 2011, the NSA disclosed that four times a day it had collected as much data as the Library of Congress (LOC) has been storing as printed material since its 1800 founding (Peled 2014). The NSA recently completed construction of a new data storage facility in Utah. While the exact storage capacity figure is classified, estimates in the popular press range from 3 exabytes to 1 yottabyte of information. Peled suggests 5 zettabytes that is the capacity to process 69 times more data than Google did in 2013 (2015, 28). Why so much data? The logic is simple. Given the uncertainty of who will commit an act of terror in the future, maintaining data on people enables government to scrutinize patterns that correlate with potential terrorist activities leading up to an attack, enabling quicker preventive action (Mayer-Schonberger and Cukier 2013). Such clairvoyance is deemed necessary to ensure security. This security imperative drives government efforts to acquire, develop, and grow computing power to process stored big data (Schneier 2015). For, without the capacity to make sense of all that data information, to identify patterns of behavior that correlate with terrorist or threatening activity, there is no big data value add proposition for government where national or global security is concerned. Enter artificial intelligence, machine learning, and other information technologies as a (potential) means for making sense of big data and with that entry, increased US government funding of research and, more broadly, investment in information technology.

US government funding of technologies capable of processing and analyzing data to effect outcomes to the advantage of US policy began to take shape during the Vietnam War. During the conflict, US military battle tested emergent automated technologies leading US Army Chief of Staff, General William Westmoreland (1914–2005), to claim that battlefields of the future will be fully automated, "enemy forces will be located, tracked, and targeted almost instantaneously through the use of data-links, computer assisted intelligence evaluation, and automated fire-control…the need for large forces to fix the opposition physically will be less important…the future portends a more automated battlefield" (as quoted in Nikutta 1987, 101). Automated battle in the sense that human decision-making is abrogated in favor of computing power through the interaction of software, hardware, and big data to find, fix, and fire on opponents. The United States was not the only nation to employ such technologies. A recently declassified US intelligence review of the Able Archer-83 military exercise, prepared during the George H. W. Bush administration (1989–1993), sheds some light on the use of computing power by Union of Soviet Socialist Republic (USSR) senior policymakers. The intelligence review was written by the *President's Foreign Intelligence Advisory Board* (PFIAB 1990) and dated February 15, 1990. Able Archer-83 was the code name given to an annual Cold War era NATO exercise that practiced nuclear war procedures. The "83" indicates the last two numbers of the year in which an Able Archer exercise took place. Thus, Able Archer-83 occurred in November 1983 (Fischer 2006). According to the intelligence review, Soviet leadership apparently took seriously the possibility that Able Archer-83 was a prelude to a nuclear first strike by NATO. How this occurred was a primary question addressed in the PFIAB (1990) review.

The PFIAB review recounts that in 1979, the USSR foreign and domestic intelligence service, known in the West by its initials KGB, developed a computing machine that was part of a vast intelligence gathering network known as VRYAN on which USSR policymakers relied to assess the risk for nuclear war with the US. The machine relied on thousands of security and economic data inputs to determine the correlation or, relative security position, of the USSR to the US by assigning a value of 100 to the US (PFIAB 1990). So long as the Soviet Union was at sixty or seventy percent of the US, Soviet leaders were confident that nuclear war was avoidable. However, shortly before Able Archer-83, the VRYAN machine calculated the correlation with the US at forty-five percent sending Soviet leaders into a heightened state of alarm fearing that the USSR had entered a position of vulnerability due to a decrease in its nuclear deterrence capability. In this state of alarm, Soviet leaders mobilized the KGB to determine whether or not the US and NATO were making preparations for a nuclear attack, put Soviet strategic nuclear forces in Germany and Poland on alert, and waited anxiously for any indication that the US and or NATO was making a nuclear move. Fortunately, the British foreign intelligence service had recruited a KGB agent, Oleg Gordiyevskiy, then serving as second in command at the KGB London residency, who confirmed that Soviet leadership were convinced Able Archer-83 was a prelude to nuclear war (PFIAB 1990). The US and NATO allies accordingly scaled back Able Archer-83 to signal to the USSR that nuclear war was not imminent (Fischer 2006). In an effort to receive "objective" analysis of US and USSR relations during the Cold War, the Soviet Union had developed a computing system which, in hindsight inaccurately, predicted a worsening of relative security with the US. Soviet leaders relied on automated digital technology to assess the risk for nuclear war effectively abrogating human reasoning and in doing so, nearly came to a fateful decision that would have radically altered our world. The trend toward computer-driven decision-making and away from human-driven decision-making is increasing as new digital technologies enable the sort of automated battlespace Westmoreland predicted and as computing power is further developed to access, scrutinize, and identify correlations in big data to predict policy. The combination of big data and semi-autonomous and autonomous militarized machines presages a future for national and global security in which humans have an increasingly decreased role in decision-making, war fighting, and relevance as agents in international relations theory.

The use of unmanned aerial vehicles (UAVs or, more commonly drones), by the United States, other countries, and non-state actors such as Hezbollah and ISIS, all portend the further automation of the battlefield where automated, semi-autonomous, or autonomous machines are put in harm's way with humans being far from actual kinetic fire and, thus, potential death and or injury. Singer (2009) demonstrates that the push to automate US national security decision-making has, in the twenty-first century, taken on a more urgent tone as technological advances in computing power, computer networked communications, and robotics are leading to increased distance of humans from the battlefield. The political risk of committing US military forces to advance US diplomatic and national security objectives is, accordingly, plummeting. If fewer US military personnel are at risk of injury or death due to combat operations, US political leaders may be more apt to employ US military force to advance US interests leading to a more unstable global security environment. Of course, the form of that military force matters: kinetic or digital? Consider US efforts to slow the advance in nuclear weapons technology acquisition and development by Iran. Options included a full-scale air campaign that would have presented considerable political risk should US pilots be killed, captured, and paraded for global media consumption. In the end, the US government opted for a strategy that presented less political risk: the use of a computer virus, the first digital warhead. The Stuxnet computer virus was allegedly designed by the national security agency, provided to its Israeli counterpart which, then modified the virus before successfully planting the computer virus on the computer network used at Iranian nuclear facilities (Zetter 2014). Significantly, and unlike kinetic munitions, computer viruses and other forms of malware are reusable by friend and foe. Making the use of such weapons perhaps more risky for in utilizing digital weapons, one has disclosed and effectively given a capability to an adversary that can, in turn, be used against you. Furthermore, computer viruses and other malware become autonomous agents as such digital weapons perform the functions for which it was programmed. Digital weapons simply make decisions based on preprogrammed instructions. In the case the stuxnet virus, it was designed to collect data, transmit that data home, and conduct disruptive operations to the production of highly enriched uranium for use in Iranian nuclear weapons. All tasks that it reportedly did well (Zetter 2014).

Others have noted the long-term use of technology to advance US national security interests through demonstrating US technological prowess and capacity for automated (quicker) decision-making during crisis (Adas 2006; Headrick 1991; Winseck and Pike 2007). For example, in 1983, the US Defense Advanced Research Projects Agency (DARPA), funded a \$600 million 5-year program, known as the Strategic Computing Initiative (SCI), to develop artificial intelligence and related technologies to support US defense and national security decision-making

(Arkersten 1987; Bonvillian 2006; Din 1987; Fong 2001; Roland and Shiman 2002). The initiative was renewed for another 5 years in 1988, ending in 1993 and estimated total expenditures by DARPA during the period on SCI at $1 billion (Grudin 2009). Why invest vast sums into nascent computing technology if computing power mattered little to government? Fast forward to the twenty-first century, according to Figliola and Fischer (2015), US government spending on information technology (IT) has increased from $46 billion in FY2001 to $81 billion by FY2013 averaging an annual increase of 6%. The increase in IT spending parallels US government security concerns associated with preventing terrorism through big data-fueled analysis leading to more focused and precise actions.

Additionally, a major domestic policy initiative of the Obama administration (2009–2017) has been to migrate US government legacy database systems to cloud-based database systems in the hope of lowering database maintenance costs while supporting data fusion accessible to US government intelligence, local law enforcement, and military agencies. Data fusion is a concept latched onto by the Bush administration (2001–2009) in response to intelligence failures associated with the September 11, 2001 terrorist attacks. Failure by the US government to prevent the attacks was in part attributed to excessive compartmentalization of information among federal agencies that prevented sharing of disparate pieces of information that, taken together, may have created a mosaic of actionable intelligence potentially leading to the prevention of the attacks. The Department of Homeland Security has accordingly been given funding and responsibility for fielding data fusion centers in all fifty states for the purpose of remedying past intelligence shortcomings to prevent future terrorist attacks (Monahan and Palmer 2009; Rollins 2008). These data fusion centers leverage big data for the purpose of developing intelligence necessary for preventive policing and military action before a terrorist event occurs (Guzik 2009; Monahan 2010; Newkirk 2010). Data fusion centers employ a wide array of data mining methods developed to plumb digital information stored in government databases for the purpose of correlating patterns of behavior with models of terrorist or criminal behavior. Data mining methods are algorithms—set of steps to solve a problem—designed to identify, prioritize, and correlate information stored in a database (Kodratoff 2014). Essentially the tasks that the field agent, intelligence analyst, and experts perform in decision-making processes. Instead of relying on trained espionage agents in the field making decisions about what information is relevant and should be collected; instead of relying on intelligence analysts who receive the "raw" intelligence from field agents and make determinations as to the weight that should be given such information by decision-makers; instead of relying on decision-makers who cogitate upon intelligence estimates, expert analysis, and their own knowledge and experience to develop government plans in response; the field agent and intelligence analyst is replaced with algorithms that automatically collect data which, is then analyzed by other algorithms which, then spew out a series of ranked probabilistic statements describing options from which government decision-makers choose. At least, in this scenario, humans remain in the decision loop albeit as final decision-makers who may simply choose the option with the best probability of succeeding as predicted by algorithms. As Thiele (this volume) notes, there is a potential loss in human cognition and political judgment where human reasoning is replaced with automation through algorithm. Nonetheless, the US government presses forward with developing computing power for myriad uses.

As users and devices connected to the Internet have grown exponentially, so has available data from which value propositions may be developed in support of government objectives. It is those very value propositions on which the future of global security rest. The future is now given the proliferation of remotely piloted armed drones, computer viruses that damage nuclear centrifuges in Iran, and proto-artificial intelligence data mining systems such as *Palantir* that enable automated decision-making leading to direct use of extraordinary government authority in the form of drone launched hellfire missiles at targets deemed a threat (Harris 2014; Owen 2015; Singer and Friedman 2014). But, as Owen (2015) notes, data-based decision-making is only as good as the data, "Palantir's algorithms…carry the biases and errors of the people who wrote them" (Owen 2015, 171). To what extent can trust be placed in automated decision-making if the underlying technology is itself as biased as human decision making? Regardless, the US government presses forward with steely determination to develop ever larger databases, better data mining technologies, and big data-based decision-making. If computing power is a core component of government power, then whichever government first develops artificial intelligence, quantum computing, and related information technologies will be in a superior position relative to other state and non-state actors, able to "process and understand vast stores of real-time surveillance data" enabling the sort of predictive policymaking that renders human decision-making irrelevant (Owen 2015, 172; also see Howard 2015). Humans are fast being automated out of the decision-making process. As Thiele (this volume) notes, the risk of automating decision-making is to lose the capacity to make consequential and wise political decisions. I add that as humans are automated out of political decision-making, what use are theories of international relations as models of political decision-making taken by humans? International Relations theory faces a crisis of relevancy associated with a lack of explanatory power regarding the contribution to international relations of emergent computer autonomous agents such as algorithms, big data, robots, and, soon, artificial intelligence.

We are long past the point where a shift in ontology is required that will allow for flattening the hierarchy extant among the object/subjects we investigate, theorize about, and argue over. This flattening out must also include a broadening of perspectives and views as to what and who matters materially to international relations outcomes. For, as we have, albeit all too briefly, reviewed in this section, emergent computer-automated information technologies do effect outcomes in international relations. The use of the Stuxnet computer virus that autonomously disrupted Iranian nuclear research initiatives, VRYAN to assess the correlation of forces between NATO and the Warsaw Pact during the Cold War, and data fusion centers all support the claim that political decision-making is moving away from human input and control. What is to be done?

Toward a Reflexive International Relations

In support of the claim that theories of international relations are rapidly approaching a crisis of explanative power, I have provided an all too brief overview of information technology use in government. My argument is that as governments come to rely ever more on information technologies for policy analysis, recommendation, and execution up to and including the use of deadly and or kinetic force, theories of international relations will need to account for new actors of consequence in the form of those very same information technologies. Robots, artificial intelligence, and other information technologies to which have been delegated decision-making capability in the policy-making process, must be treated with the same ontological and epistemological status as individuals, states, and other actors. Doing so requires the adoption and development of reflexivity.

An initial step to adopting a reflexive posture as an International Relations scholar is to accept that the objects under scrutiny reflect anthropomorphisms of "thinking human beings who continually reflect on themselves and the social world surrounding them. This self-reflexivity changes how humans understand themselves and their world and hence how they act in the world" (Oren 2006, 79; also see Jackson 2010, 157). Humans have the capacity to become critically aware of their biases and assumptions, make adjustments as conditions dictate, and press forward generating new social, political, and economic outcomes. The application of reflexivity to international relations theories follows the same path: acknowledging the epistemological and ontological biases and assumptions underpinning each IR theory tradition leading to understanding of IR theory's contextual limits (Barkin 2009; Guzzini 2013; Oren 2009; Walker 1993; Wight 2006). The suggestion to adopt reflexivity as a (temporary?) Archimedean point from which to assess the applicability of IR theory to research questions, echoes an ongoing conversation about the status of International Relations as a science (Brown 2013; Jackson 2010; Lake 2013; Mearsheimer and Walt 2013; Oren 2003; Sylvester 2013). This conversational is meta-theoretical resting on the claim best expressed by Jackson (2010) that any decision regarding whichever philosophy of science tradition International Relations adopts as foundational, is a wager, an act of faith. The implication to IR theories is acknowledgment of their limits as explanative frameworks. I want to suggest that working with(in) these limits is as much a matter of coming to a rough consensus on being a post-foundational science; as it is a matter of developing sufficient empirical information, data, and concepts from which one may theorize about international relations. Fortunately (or not), International Relations has a long and inglorious history of developing such for research purposes.

International Relations has evolved or adapted a number of useful heuristic devices to assist in the explanation of its empirical subject matter. These heuristic devices include state, nation, national interest, identity, anarchy, hierarchy, balance of power, distribution of power, security dilemma, structure, agent, actor/s, to highlight a few. These heuristic devices are necessary for discovery, illumination, and elucidation of international relations for they assist in parsing complexity into manageable relatively intelligible forms from which IR scholars may theorize about human decision-making under certain conditions. As humans increasingly become displaced in the policy decision-making process by super intelligent learning and, soon, artificial intelligence-enabled machines, how will IR theory theorize about computer-automated international relations? Our focus must broaden beyond human-based policy decision-making by taking seriously the proposition that computer-automated decision-making whether by big data-based algorithm, quantum computing, robots, or artificial intelligence are consequential actors, actors with agentive capacity to effect international relations outcomes of equal ontological and epistemological weight as the humans and human anthropomorphisms that pepper contemporary IR theory.

The notion agentive capacity is meant to capture interventions by inanimate material bodies, in social milieu, that provoke "effects dramatic and subtle" (Bennett 2010, 6). Thinking this way about international relations is to move away from subject/object dichotomization characterizing the epistemological and ontological corpus of International Relations. It is, within International Relations, to move toward privileging ontology over epistemology by viewing the world as being "composed of objects, that these objects are varied and include entities as diverse as mind, language, cultural and social entities, and objects independent of humans such as galaxies, stones, quarks, tardigrades and so on" (Bryant 2011, 18). Recognizing the capacity of information computing technologies to effect international relations outcomes clears ontological space for theorizing about computer, algorithm, Internet, big data, information, computer virus, and malware. Yet, how do International Relations scholars theorize about these consequential objects in order to operationalize our findings across various temporally distinct outcomes? "ICKY!" One might exclaim, a closet positivist has come out!! Let's not throw the proverbial baby out with the bathwater. We do have some responsibility for adhering to certain scientific norms without being dogmatic about our commitments a la King, Keohane, and Verba (1994). We are, after all, engaged in an activity to investigate and empirically demonstrate why international relations outcomes occur.

To this end, a useful heuristic to International Relations scholars, consistent with a reflexive orientation that treats all entities investigated by International Relations as object-subjects, may be that of agent. A sufficiently ambiguous term broad enough to account for state, non-state, and inanimate material objects while simultaneously communicating sufficient specificity regarding the capacity for affecting, effecting, and participating in decision-making under certain conditions. Precisely what International Relations scholars seek to understand, describe, and explain. Of course, this means subsuming state-centric, structure-centric, or identity-centric theory to that of agent-centric theory where agents may be anything with a demonstrable material effect on international relations outcomes. The challenge will be identifying precisely what and who constitutes an agent and justifying such a designation by empirically demonstrating the consequentiality of actions taken by "the what" and "the who" under certain conditions that effect international relations outcomes. This scholarly exercise must be done reflexively without prejudice as to whether or not an object is animate or inanimate; for, what matters is the consequentiality of an object to international relations outcomes: the capacity to alter the articulations, decisions, and interests of other objects—and on this score information and computing technology is consequential. Adopting a reflexive posture consistent with the recognition that material objects matter as subjects to international relations potentially offers an approach to account for new consequential agents in IR theory. In doing so, International Relations may avoid a crisis of explanatory power brought on by the advent of artificial intelligence-enhanced machines making policy decisions effecting outcomes in global politics and consequences to human lives.

Addendum

Reflecting on technological innovations occurring since publication of this chapter, offers the rare opportunity to assess whether claims made in 2018, remain relevant years later. Baldy speaking, claims staked in this chapter remain relevant. Recent technological innovations and efforts by state leaders to adopt in policy decision-making lend further support. The overarching argument in this chapter is International Relations theory will experience a crisis of explanative power as autonomous technologies develop sufficient agentive capacity to take on increasing cognitive load and decision-making authority in state policy decision-making processes. The proposed remedy is an ontological shift within International Relations from a state-centric interpretation of global politics to an agent-centric interpretation of global politics. In what follows I offer brief reflections supporting the claims advanced in this chapter.

Reflection #1: What is meant by agentive capacity?

Agentive capacity is influenced by *New Materialism* (see Bennet 2009, Bryant 2011, or Verbeek 2005), which brings attention to ways in which things or "artifacts" have direct affects. That things "set constraints and create possibilities which work in favour of the genesis of certain meanings and courses of action rather than others" (Caronia and Mortari 2015, 403). Work within the New Materialist approach also demonstrated how artifacts do create, reflect, and alter extant relational and institutional structures. In the original contribution to this volume, I endeavored to illustrate in the limited space available, how autonomous technologies utilized in policy decision-making processes *do* set constraints and *create* possibilities for certain courses of action over alternatives. Autonomous technologies it follows do possess agentive capacity, and on that basis, merit ontological treatment as agents in international relations. Three questions come to mind.

First, do present autonomous technologies possess sufficient agentive capacity to overtake humans in policy decision-making processes, such that International Relations theory *should* ontologically shift to an agent-centric interpretation of international relations? Second, as an empirical question, how does one measure or observe agentive capacity consistent with discipline and field scientific norms? Third, will a shift to AI decision-making lead to more or less ethical policy decision-making?

As to the first question, autonomous technologies appear to have not yet demonstrated sufficient and consistent agentive capacity that overtakes humans in policy decision-making processes. Additional research is necessary on this front. The development of LLMs, however, is a strong indicator that sufficient agentive capacity will develop sooner rather than later. Confronting International Relations theory with a problem of explanative power, for which IR would do well to prepare. Regarding the second question, measuring or observing agentive capacity of autonomous technologies utilized in policy decision-making processes, is a matter of assessing cognitive load shared between humans and AI, deference of humans to AI predictions, and degree of autonomy granted to AI to execute or, at least, direct execution of strategies and actions to achieve policy objectives. One could utilize an analytical framework of strategic, tactical, and operational decision-making to assist with concretely identifying shared cognitive load, human deference to AI predictions, and degree of AI autonomy. A key barrier is the "black box" nature of AI analysis and prediction. Researchers developing AI systems are presently unable to explain how an AI system crafted analysis and, on that basis, offered predictions. Work by AI scientists and system developers is proceeding to create tools for use creating transparency of AI analysis and predictions. If successful, such tools will overcome this key barrier creating conditions for increased human trust leading to deference to AI systems in policy decision-making processes.

An interesting issue concerns the ethics of policy decisions. Admittedly questions about the ethics of a policy decision turn on a number issues including the decision to utilize an ethical framework (over alternatives) to evaluate the ethics of a policy decision, the socio-historical period in which the policy decision is taken, the ethical framework relied upon by policy decision-makers during the decision-making process, existing information asymmetries, available state power capabilities, the interests, ethics, and actions of other actors. As to whether an AI policy decision would be more or less ethical, the answer lies in the question of whose ethics were incorporated into the algorithms comprising the AI agent and for what purpose? Whose ethics are incorporated into an AI agent is a political question that may be answered quite differently across state actors operating in a condition of anarchy. AI agents developed by historically, culturally, economically, and politically distinct sovereign actors are more likely to reflect ethical policy decision-making consistent with ethical traditions observed within states yet influenced by foreign policy imperatives defined by state leaders. The upshot for IPE/IR scholars is a range of ethics observed in AI agent policy decision-making is probable in the absence of global agreement on a common set of ethics all AI agents should abide in policy decision-making. Assuming such an agreement is desirable, possible, and enforceable among state actors.

Reflection #2: Does the development of AI-LLMs lend support (weaken) our claim?

In terms of technology innovations, the development of AI Large Language Models (LLMs) lends direct support to the notion of agentive capacity—a key cog in my argument. To the extent that LLMs enhance the agentive capacity of autonomous technologies including AI and robotics systems. While LLMs appear to be a long way from agentive capacity meriting human deference to AI predictions or directing execution of strategies, LLMs do support cognitive load sharing between human and machine. Indeed, this may be the most significant value-add of LLMs to policy decision-making processes, over the short- to medium-term. Think in terms of analytic productivity gains for intelligence analysts, defense analysts, or more broadly, policy analysts. The fusion of human and LLMs offers the prospect of compressing the time cycle necessary to produce novel intelligence, defense, or policy insights and recommendations (products) contributing to productivity gain. Analytical productivity supported by LLMs will contribute to compressed policy decision-making cycles allowing for more nimble and quicker policy response. Trust in AI analysis and predictions is likely to grow as is the capability of learning algorithms to take on additional cognitive load. Both developments in turn support increasing human deference to AI in policy decision-making processes. Sufficient agentive capacity is on the horizon.

Reflection #3: Still a science of human decision-making under certain conditions?

As it presently stands, International Relations remains a state-centric scientific endeavor. The "state" it is proposed remains the most consequential actor in a condition of anarchy.[4] Global political economy is characterized by state actors pursuing prosperity and security through enacting policies of cooperation or competition, constrained by available state power, factor endowments, degree of economic development, and the interests and capabilities of other state actors. Within this milieu, the crux of the matter is how AI and more generally, autonomous technologies, alter patters of cooperation and competition among state actors in global political economy and international relations more broadly? Addressing this question begins with settling questions about ontology. The purpose of this chapter has been to argue as AI and associated automated technologies develop agentive capacity, International Relations ontology must be revised to account for the agentive capacity of AI and

[4] Anarchy may be understood as a political condition in which the absence of a central authority capable of settling disputes and enforcing agreements is experienced by actors.

AI-directed technologies. The innovation of LLMs are harbingers of the need for IR scholars to initiate a dialogue about how to theorize international relations in the emerging AI-dominated world. Else AI-LLMs will do so for them. Wait! It already does.

Reflection #4: They're already better than us.

On 4 September 2024, I prompted ChatGPT 4o as follows, "how will chatgpt impact global politics." I have the response on file which was crafted by ChatGPT 4o in about a minute or less. Below is the response to the prompt from ChatGPT 4o:

> *The impact of AI models like ChatGPT on global politics could be significant, influencing various aspects of political processes, governance, and international relations. Here are a few key areas where AI could make an impact:*
>
> 1. ***Information Dissemination and Propaganda***
>
> – ***Rapid Spread of Information**: ChatGPT can quickly generate content, which could be used to disseminate information or misinformation on a massive scale. This can influence public opinion, potentially affecting elections, policy decisions, and international relations.*
> – ***Deepfakes and Disinformation**: The ability to generate highly convincing text can be used to create fake news, deepfakes, or misleading narratives, potentially destabilizing governments or creating conflicts between nations.*
>
> 2. ***Diplomacy and Negotiation***
>
> – ***Automated Diplomacy**: AI can assist in diplomatic communication, drafting proposals, or analyzing complex international agreements. It might be used to simulate negotiation scenarios, helping diplomats make more informed decisions.*
> – ***Risk of Miscommunication**: On the downside, reliance on AI for sensitive diplomatic communications might lead to misinterpretations or errors, potentially escalating conflicts.*

3. **Surveillance and Security**

- ***Enhanced Surveillance**: Governments might use AI to analyze vast amounts of data for surveillance, tracking political dissent, or monitoring international developments. This could lead to more authoritarian control in some regions.*
- ***Cybersecurity Threats**: AI can both strengthen and weaken cybersecurity. While it can be used to detect and prevent cyber-attacks, it can also be employed to develop more sophisticated cyber-attacks, leading to new forms of digital warfare.*

4. **Policy Making and Governance**

- ***Data-Driven Policy Making**: AI can analyze complex data sets to provide insights into economic trends, social issues, or environmental concerns, potentially leading to more informed policy decisions.*
- ***Ethical and Bias Concerns**: The use of AI in governance raises ethical questions about bias, transparency, and accountability. If not carefully managed, AI could reinforce existing inequalities or be used to justify harmful policies.*

5. **Global Power Dynamics**

- ***AI Arms Race**: As countries invest in AI technology, it could become a new frontier for global competition, similar to nuclear or space technologies. Nations with advanced AI capabilities may gain significant geopolitical advantages.*
- ***Shifting Alliances**: AI could also influence global alliances, as countries might collaborate on AI development or form blocs to counterbalance the influence of AI superpowers.*

6. **Public Engagement and Advocacy**

- ***Increased Public Participation**: AI could be used to engage citizens in political processes, through tools that analyze public opinion, suggest policy alternatives, or simulate the outcomes of different decisions.*
- **Manipulation of Public Discourse**: However, the same tools could be used to manipulate or suppress dissenting opinions, influencing the democratic process in subtle ways.

> *In summary, while AI models like ChatGPT offer opportunities to enhance global politics through improved communication, analysis, and governance, they also pose risks related to misinformation, surveillance, and geopolitical competition. The impact will largely depend on how these technologies are developed, deployed, and regulated on a global scale.*

Few could craft a better response in as short of time. One that is cogent, concise, and clear noting numerous ways AI and associated autonomous technologies are impacting global politics. The productivity gains are astonishing! From a policy decision-making view, the ChatGPT response above offers an executive summary from which a more detailed issue by issue policy analysis report may be composed, supported by LLM capacity to share cognitive load. With that report utilized by decision-makers to craft appropriate policy responses, with additional AI & LLM analytical support. The current value-add of LLMs is contributing to increased productivity in activities supporting policy decision-making. The pace of analysis supporting decision-making is increasing and with that the pace of policymaker responsiveness. Ironically, humans are losing pace vis-à-vis AI and autonomous technologies. If we are not vigilant, we may miss the point at which we are overtaken, losing opportunity to develop ontologies better suited for theorizing about the emerging AI dominated world.

References

Adas, M. (2006). *Dominance by design: Technological imperatives and America's civilizing mission*. Harvard University Press.

ÅKERSTEN, S. I. (1987). The Strategic Computing Program. *Arms and Artificial Intelligence: Weapon and Arms Control Applications of Advanced Computing*. A. M. Din. Stockholm, Sweden, SIPRI: 87–99.

Angell, N. (1913). *The great illusion: A study of the relation of military power to national advantage*. GP Putnam's sons.

Baldwin, D. A. (1993). *Neorealism and neoliberalism: the contemporary debate*. Columbia University Press.

Barkin, S. (2009). "Realism, Prediction, and Foreign Policy." *Foreign Policy Analysis*, 5(3): 233–246.

Bennett, J. (2009). *Vibrant matter: A political ecology of things*. Duke University Press.

Bennett, J. (2010). Vibrant Matter: A Political Ecology of Things. In *Vibrant Matter*. Duke University Press.

Bonvillian, W. (2006). "Power play." *The American Interest,* 2: 39–48.

Brewer, J. (1989). *The sinews of power: War, money, and the English state, 1688–1783.* Harvard University Press.

Brown, C. (2013). "The poverty of grand theory." *European Journal of International Relations* 19(3): 483–497.

Bryant, L. R. (2011). *The democracy of objects.* Open Humanities Press.

Burchill, S. & Linklater, A. (2013). "Introduction," in Burchill, S., Paterson, M., Reus-Smit, C., Linklater, A., Devetak, R., True, J., & Donnelly, J. (2013). *Theories of international relations* (pp. 1–31). Palgrave Macmillan, 5th Edition.

Burchill, S., Paterson, M., Reus-Smit, C., Linklater, A., Devetak, R., True, J., & Donnelly, J. (2013). *Theories of international relations.* Palgrave Macmillan. Fifth Edition.

Buzan, B., and G. Lawson (2013). The global transformation: The nineteenth century and the making of modern international relations. *International studies quarterly*, 57(3), 620–634.

Buzan, B., and G. Lawson (2016). Theory, history, and the global transformation. *International Theory*, 8(3), 502–522.

Caronia, L., & Mortari, L. (2015). "The agency of things: how spaces and artefacts organize the moral order of an intensive care unit." *Social Semiotics*, *25*(4), 401–422.

Carr, E. H. (2001). *The twenty years' crisis, 1919–1939: an introduction to the study of international relations.* Basingstoke, Palgrave.

Din, A. M. (1987). *Arms and artificial intelligence: weapon and arms control applications of advanced computing.* Stockholm International Peace Research Institute.

Dunne, T., et al. (2013). "The end of International Relations theory?" *European Journal of International Relations,* 19(3): 405–425.

Ferguson, Y. H. (2015). "Diversity in IR Theory: Pluralism as an Opportunity for Understanding Global Politics." *International Studies Perspectives,* 16(1): 3–12.

Figliola, P. M. and E. A. Fischer (2015). *Overview and issues for implementation of the federal cloud computing initiative: Implications for federal information technology reform management.* C. R. Service. Washington, D.C., US Congressional Research Service (CRS). 1.

Fischer, B. B. (2006). "The Soviet–American War Scare of the 1980s." *International Journal of Intelligence and Counterintelligence,* 19(3): 480–518.

Fong, G. R. (2001). "ARPA does Windows: The defense underpinning of the PC revolution." *Business and Politics,* 3(3): 213–237.

Gantz, J. and D. Reinsel (2012). "The digital universe in 2020: Big data, bigger digital shadows, and biggest growth in the far east." Available at: https://www.emc.com/collateral/analyst-reports/idc-digital-universe-united-states.pdf. Last accessed, 2 September 2016.
Grudin, J. (2009). "AI and HCI: Two fields divided by a common focus." *AI Magazine* 30(4): 48.
Guzik, K. (2009). "Discrimination by Design: predictive data mining as security practice in the United States''war on terrorism'." *Surveillance & Society*, 7(1): 3–20.
Guzzini, S. (2013). "The ends of International Relations theory: Stages of reflexivity and modes of theorizing." *European Journal of International Relations*, 19(3): 521–541.
Harris, S. (2014). *@ War: The Rise of the Military-Internet Complex*. Houghton Mifflin Harcourt.
Harrison, E. and S. M. Mitchell (2013). *The Triumph of Democracy and the Eclipse of the West*. Springer.
Headrick, D. R. (1991). *The invisible weapon: Telecommunications and international politics, 1851–1945*. Oxford University Press.
Hempel, C. G. (1966). *Philosophy of natural science*.
Hodges, A. (2003). The military use of Alan Turing. In *Mathematics and war* (pp. 312–325). Basel: Birkhäuser Basel.
Hoffmann, S. (1977). "An American social science: international relations." *Daedalus*: 41–60.
Howard, P. N. (2015). *Pax Technica: How the Internet of things may set us free or lock us up*. Yale University Press.
Hu, H., et al. (2014). "Toward scalable systems for big data analytics: A technology tutorial." *IEEE Access* 2: 652–687.
Ikenberry, G. J. (2012). *Liberal Leviathan: The origins, crisis, and transformation of the American world order*, Princeton University Press.
Jackson, P. T. (2010). *The conduct of inquiry in international relations: philosophy of science and its implications for the study of world politics*. Routledge.
Jackson, P. T. (2015). "Must international studies be a science?" *Millennium-Journal of International Studies* 43(3): 942–965.
Keohane, R. O. (1984). *After hegemony: cooperation and discord in the world political economy*. Princeton, N.J. Princeton University Press.
King, G., Keohane, R.O., and Verba, S. (1994). *Designing social inquiry scientific inference in qualitative research*. Princeton, N.J., Princeton University Press: xi, 245 p.
Kodratoff, Y. (2014). *Introduction to machine learning* Morgan Kaufmann.
Kuhn, T. S. (1996). *The structure of scientific revolutions*, University of Chicago press.

Lake, D. A. (1996). "Anarchy, Hierarchy, and the Variety of International Relations." *International organization* 50(1): 1–33.

Lake, D. A. (2009). *Hierarchy in international relations*, Cornell University Press.

Lake, D. A. (2011). "Why "isms" Are Evil: Theory, Epistemology, and Academic Sects as Impediments to Understanding and Progress." *International Studies Quarterly*, 55(2): 465–480.

Lake, D. A. (2013). "Theory is dead, long live theory: The end of the Great Debates and the rise of eclecticism in International Relations." *European Journal of International Relations* 19(3): 567–587.

De Mauro, A., et al. (2015). *What is big data? A consensual definition and a review of key research topics*. AIP Conference Proceedings.

Mayer-Schönberger, V. and K. Cukier (2013). *Big data: A revolution that will transform how we live, work, and think*. Houghton Mifflin Harcourt.

Mearsheimer, J. J. (2001). *The tragedy of great power politics*. New York; London, W.W. Norton & Company.

Mearsheimer, J. J. and S. M. Walt (2013). "Leaving theory behind: Why simplistic hypothesis testing is bad for International Relations." *European Journal of International Relations*, 19(3): 427–457.

Milner, H. (1991). "The Assumption of Anarchy in International Relations Theory: A Critique." *Review of International Studies*, 17(1): 67–85.

Mitchell, T. M. (2006). *The discipline of machine learning*, Carnegie Mellon University, School of Computer Science, Machine Learning Department.

Monahan, T. and N. A. Palmer (2009). "The emerging politics of DHS fusion centers." *Security Dialogue*, 40(6): 617–636.

Monahan, T. (2010). "The future of security? Surveillance operations at homeland security fusion centers." *Social Justice* 37(2/3 (120–121): 84–98.

Moses, J. and T. Knutsen (2007). *Ways of knowing: Competing methodologies in social and political research*. Palgrave Macmillan.

Newkirk, A. B. (2010). "The rise of the fusion-intelligence complex: A critique of political surveillance after 9/11." *Surveillance & Society*, 8(1): 43–60.

Nikutta, R. (1987). "Artificial intelligence and the automated tactical battlefield." *Arms and Artificial Intelligence: Weapons and Arms Control Applications of Advanced Computing*. A. M. Din. Stockholm, Sweden, SIPRI: 100–134.

Oren, I. (2003). *Our enemies and US: America's rivalries and the making of political science*, Cornell University Press.

Oren, I. (2006). "Can political science emulate the natural sciences? The problem of self-disconfirming analysis." *Polity* 38(1): 72–100.

Oren, I. (2009). "The Unrealism of Contemporary Realism: The Tension between Realist Theory and Realists' Practice." *Perspectives on Politics*, 7(02): 283–301.

Owen, T. (2015). *Disruptive power: The crisis of the state in the digital age*, Oxford University Press, USA.

PFIAB. (1990). President's Foreign Intelligence Advisory Board, "The Soviet 'War Scare,'" February 15, 1990. Available at https://nsarchive.gwu.edu/document/21038-4-pfiab-report-2012-0238-mr. Last accessed 1 August 2025.

Peled, A. (2014). *Traversing digital Babel: information, e-government, and exchange*, MIT Press.

Popper, K. (2005). *The logic of scientific discovery*, Routledge.

Pouliot, V. (2016). *International Pecking Orders: The Politics and Practice of Multilateral Diplomacy*, Cambridge University Press.

Roland, A. and P. Shiman (2002). *Strategic computing: DARPA and the quest for machine intelligence, 1983–1993*, MIT Press.

Rollins, J. (2008). "Fusion centers: Issues and options for Congress," DTIC Document.

Schmidt, B. C. (1998). *The political discourse of anarchy: a disciplinary history of international relations*, SUNY Press.

Schneier, B. (2015). *Data and Goliath: The hidden battles to collect your data and control your world*, WW Norton & Company.

Singer, P. W. (2009). *Wired for war: The robotics revolution and conflict in the 21st century*, Penguin.

Singer, P. W. and A. Friedman (2014). *Cybersecurity: What Everyone Needs to Know*, Oxford University Press.

Skinner, Q. (1990). *The return of grand theory in the human sciences*, Cambridge University Press.

Smith, S. (1989). "Paradigm dominance in international relations: The development of international relations as a social science." *The Study of International Relations*, Springer: 3–27.

Smith, S. (2000). "The discipline of international relations: still an American social science?" *The British Journal of Politics & International Relations*, 2(3): 374–402.

Snyder, Q. Z. (2013). "Taking the System Seriously: Another Liberal Theory of International Politics." *International Studies Review*, 15(4): 539–561.

Solomon, T. and B. J. Steele (2016). "Micro-moves in International Relations theory." *European Journal of International Relations*, July: 1–25. https://doi.org/10.1177/1354066116634442.

Sylvester, C. (2013). "Experiencing the end and afterlives of International Relations/theory." *European Journal of International Relations*, 19(3): 609626.

Tickner, A. B. and D. L. Blaney (2013). *Thinking international relations differently*, Routledge.

Turton, H. L. (2015). *International Relations and American Dominance: A Diverse Discipline*, Routledge.

Verbeek, P. P. (2005). *What things do: Philosophical reflections on technology, agency, and design*. Penn State Press.

Waever, O. (1998). "The sociology of a not so international discipline: American and European developments in international relations." *International organization*, 52(04): 687–727.

Walker, R. B. (1993). *Inside/outside: international relations as political theory*. Cambridge [England]; New York, Cambridge University Press.

Waltz, K. N. (1979). *Theory of international politics*. Reading, Mass., Addison-Wesley Pub. Co.

Ward, J. S. and A. Barker (2013). "Undefined by data: a survey of big data definitions." Available at: https://arxiv.org/abs/1309.5821. Last accessed 2 September 2016.

Wendt, A. (1992). "Anarchy is what states make of it: the social construction of power politics." *International organization*, 46(2): 391–425.

Wendt, A. (1999). *Social theory of international politics*. Cambridge; New York, Cambridge University Press.

Wight, C. (2006). *Agents, structures and international relations: politics as ontology*, Cambridge University Press.

Winch, P. (1990). *The idea of a social science and its relation to philosophy*, Routledge.

Winseck, D. R. and R. M. Pike (2007). *Communication and empire: Media, markets, and globalization, 1860–1930*. Duke University Press.

Zehfuss, M. (2002). *Constructivism in international relations: the politics of reality*. Cambridge, UK; New York, Cambridge University Press.

Zetter, K. (2014). *Countdown to zero day: Stuxnet and the launch of the world's first digital weapon*. Broadway Books.

CHAPTER 14

Comrades in Arms Revisited: Shifting Ground in the Era of AI and Robotics in Warfighting

Mary Manjikian

Military planners have long asked, "What will the military of the twenty-first century look like? What unique challenges might it face, and what capabilities will individual warfighters and their units need to possess to prevail in future conflicts?" As Cronin notes, one constant in warfare has been the impact of technological change. Here, she argues that in the twentieth and twenty-first centuries, technological innovations have often eventually led to a "democratization" of warfare, allowing for the creation of cheaper, more accessible weapons and providing more individuals and minor non-state actors with access to deadly force. (Cronin 2020) That is, while states continually seek to maintain a monopoly on violence, new technologies frequently fall into the hands of new and unexpected players, disrupting the status quo of traditional warfare.

M. Manjikian (✉)
College of Arts and Sciences, Regent University, Virginia Beach, US
e-mail: mmanjikian@regent.edu

R. Kiggins (ed.), *The Political Economy of Robots*, International Political Economy Series, https://doi.org/10.1007/978-3-031-95171-8_14

While military organizations strive to develop technologies that enhance safety and minimize casualties, they are also confronted with the urgent challenge of adapting to the rapid proliferation of emerging technologies. The introduction of new technologies, such as AI-driven weapons and autonomous drones, raises questions of control, effectiveness, and integration into existing systems. These innovations have the potential to radically reshape traditional notions of military leadership, unit cohesion, and combat effectiveness, underscoring the need for an immediate and thorough reevaluation of our leadership strategies.

For this reason, today, some analysts argue that AI will inevitably lead to a so-called Revolution in Military Affairs or RMA. Here, the belief is that artificial intelligence and its integration into robotic systems at both strategic and tactical levels represents introducing a so-called emerging technology into warfighting. Emerging technology refers to new and innovative technologies currently in development or have just begun to impact society and industry, with the potential to significantly alter existing systems and practices in a myriad of ways, including politically, organizationally, economically, and socially. Emerging technologies are also, by their nature, unpredictable. It often needs to be clarified in the earliest stages what precisely the new technology represents, and it is equally challenging to predict the effects that it might have. In military affairs, emerging technologies might include advancements such as artificial intelligence (AI), autonomous systems, cyber capabilities, and hypersonic weapons. These technologies can potentially reshape military strategy, tactics, and operations, providing advantages in speed, precision, and decision-making, offering hope for the future of warfare. (Singer 2009).

As we evaluate the potentially revolutionary effect of integrating AI systems into our modern military, a key issue is the balance between technological determinism and human agency. Will new systems "seek their own ends" growing organically to occupy a particular niche in war planning (as a technological materialist might suggest), or will how these systems evolve ultimately be steered by human planners who encourage certain developments while actively discouraging or forbidding others? Furthermore, as Verbeek (2005) argues, technologies do more than augment human abilities—they actively shape human actions and outcomes. Utilizing AI as a military planner may unwittingly build in specific values (such as a preference for efficiency and stable, predictable plans of action) while precluding other values that are often valuable in a military context (such as the opportunity to see an opportunity and quickly shift gears to take advantage of it—what Clausewitz defines as a component of military leadership.) For this reason, policy planners will want to think carefully about which levels and stages of military leadership and planning should be delegated to autonomous systems, and which will still require a human in the loop. However, adopting such a stance rests on assumptions that this technology is ultimately controllable by humans and that humans can make decisions about the adoption of this technology independently, uninfluenced by data that autonomous systems might already provide.

Top-Down or Bottom-Up? Technological Determinism, Human Agency, and AI in Military Decision-Making

In discussions about AI's role in U.S. military decision-making, questions surrounding technological determinism and human agency loom. As explored in my previous work, technology does not simply impose itself on human systems in a deterministic manner; its eventual role is shaped by how it is adopted, integrated, and used by human actors. This leads to a critical inquiry: Will AI's integration into the military be primarily driven by top-down leadership decisions, or will bottom-up forces, such as the adaptation by individual warfighters and units, shape its final form?

Technological determinism posits that once a technology is introduced, its trajectory is inevitable, mainly shaping organizational, political, and social outcomes regardless of human intentions. In military contexts, this determinism can be seen in historical examples where new technologies like aviation or nuclear power fundamentally reshaped military strategies and organizational structures. (Feickert 2006). However, as many scholars of military innovation, including Freedman (Freedman 2003) and Krepinovich (Krepinevich 1994), argue that new military technologies are often introduced without fully formed doctrines to guide their use. This doctrinal lag leaves space for human agency—individual warfighters, commanders, and even the broader public—to shape how and where these technologies are deployed, sometimes in unanticipated ways.

In the case of AI, the top-down vision for its use is apparent in recent documents like the *2023 National Defense Strategy (NDS)* and the

National Security Strategy (NSS), which emphasize AI as a transformative tool. These strategies present AI as key to maintaining U.S. strategic superiority, particularly in decision-making, command, and control. The *NDS* calls for integrating AI to enhance decision-making speed and precision, implicitly presenting AI as a solution to the growing complexity of global threats (Department of Defense 2023). However, as the historical example of the Future Combat Systems (FCS) program illustrates, these top-down initiatives often face practical challenges and resistance when new technologies are introduced without sufficient grounding in existing military structures and culture (Feickert 2006; Kosiak 2007).

This raises the question of whether AI will follow a similar pattern, with its use and potential risks being shaped by bottom-up forces. Indeed, in earlier technological shifts, we have seen doctrine develop reactively, sometimes after the technology has already started changing operational practices. For example, early air power doctrine was developed only after aviation technology had already altered the dynamics of World War I (Freedman 2003). AI's introduction into the U.S. military may follow a similar trajectory, where individual units and warfighters adapt AI to their needs before formal doctrine fully catches up.

Defining Robotics in the Context of AI and Military Innovation

In discussing the integration of AI and autonomous systems in warfighting, it is essential to clarify the concept of "robotics." A robot refers to any machine capable of performing tasks autonomously or semi-autonomously, often using sensors, actuators, and decision-making algorithms to navigate and interact with its environment (Scharre 2018). While the term "robotic warfighters" may conjure images of humanoid machines, such as those developed by Boston Dynamics, robotics in the military encompasses a wide range of automated systems, including unmanned aerial vehicles (UAVs), ground-based robots, and maritime systems.

Robotic warfighters are typically integrated with advanced AI to perform tasks ranging from surveillance and reconnaissance to logistical support and combat operations. For example, Boston Dynamics has been at the forefront of developing highly mobile robotic systems, such as their "Spot" robots, tested in military exercises for applications like mapping dangerous environments and performing search-and-rescue operations

(Boston Dynamics, n.d.). However, in military terms, "robotics" often refers to any system with automated or semi-automated functionality closely related to AI. (Waltzman 2020).

When discussing AI in military applications, we refer to the software that powers these robotic systems, enabling them to make decisions autonomously or assist human operators in real time. Thus, while "AI" and "robotics" are sometimes used interchangeably in defense discussions, they refer to different components of an integrated system. AI provides the cognitive capabilities, while robotics provides the physical platform to carry out tasks autonomously (Scharre 2018; Singer 2010).

Robotics and AI are transforming how state actors conduct war and how smaller states and non-state actors participate in conflicts. As a prominent type of robotic warfighter, drones exemplify how new technology democratizes warfare. Initially developed by technologically advanced states, drones are now accessible to smaller nations and militant groups. The use of drones in Israel, Palestine, and Ukraine provides real-world examples of how AI-driven technologies are reshaping the battlefield. (Bennett 2020) (Kofman 2021).

In Israel and Palestine, drones have been central to both state and non-state military strategies. The Israeli Defense Forces (IDF) employ drones for surveillance, targeted strikes, and border security, while militant groups in Gaza use them for reconnaissance and attacks, often utilizing commercially available drones. (Bennett 2020) In Ukraine, both Ukrainian and Russian forces have heavily relied on drones for reconnaissance, tactical strikes, and intelligence-gathering, underscoring how lower-cost robotic systems are increasingly central to modern conflict. (Kofman 2021).

This accessibility points to a critical trend: robotics and AI have the potential to democratize warfare by lowering the entry barriers for more minor actors. Unlike traditional kinetic weapons, drones can be inexpensive, and innovation often comes from the bottom up. Non-state actors can acquire and repurpose commercially available drones or develop rudimentary systems at a fraction of the cost of conventional military systems. This democratization of warfare allows smaller groups to challenge more significant, technologically advanced military powers (Cronin 2020). The idea that technology empowers more minor actors disrupts traditional hierarchies in warfare. The democratization of warfare through drones also poses challenges for conventional military forces, which must rapidly adapt to counter the increasing use of low-cost, highly effective UAVs by non-state actors and smaller military forces.

In addition to AI and robotics, emerging technologies such as **hypersonic weapons** and **quantum computing** are reshaping the future of warfare. Hypersonic weapons, capable of traveling at speeds greater than Mach 5, are designed to evade current missile defense systems using unpredictable flight paths. Both Russia and China have advanced their hypersonic missile programs, raising concerns about the U.S.'s ability to defend against these threats. Hypersonic weapons offer speed and increased precision, which could significantly impact strategic military planning by compressing decision-making timelines for military leaders. (Trevithick 2020).

Moreover, **quantum computing** is another technology on the horizon with the potential to revolutionize military operations, particularly in cryptography, intelligence analysis, and logistics. Quantum computers could vastly accelerate data processing and optimization tasks, enabling more precise targeting, faster decryption of encrypted communications, and more efficient resource management. While the U.S. invests heavily in these areas, adversaries like China have also made significant strides, positioning these technologies as critical components in the next phase of global military competition. (Kania 2019) Together, AI, robotics, hypersonics, and quantum computing represent the next frontier in technological advancements that could fundamentally alter the nature of war. This would push military planners to rethink conventional strategies and prepare for an era of rapid, automated warfare.

Revolutionary Change or Incremental Improvement?

One of the central debates in military AI and robotics thus concerns the extent to which it will bring revolutionary changes to the nature of warfare. A *Revolution in Military Affairs* (RMA) refers to a fundamental transformation in how wars are fought, driven by innovations in military technology, doctrine, and organization, which leads to new operational strategies and renders previous methods of warfighting obsolete (Krepinevich 1994). Some argue that AI represents a new RMA, necessitating a complete overhaul of traditional military strategies, doctrines, and leadership structures. This idea is not limited to American military thought; Russian military analysts, too, have used the concept of RMA to grapple with how new technologies like AI might fundamentally change warfighting, with some Russian strategists suggesting that AI has the potential to revolutionize combat in the same way that nuclear weapons did in the twentieth century. (Adamsky). Chinese analysts, similarly, have debated whether AI constitutes an RMA, mainly focusing on its ability to alter strategic calculations and power dynamics (Kania 2018). However, others maintain that AI's impact will be more incremental, improving operational efficiency but leaving core military structures largely intact (Schneider and MacDonald 2024) This perspective divergence reflects the broader global uncertainty about whether AI will revolutionize warfare or refine existing practices.

Historically, new technologies have often sparked RMAs, transforming military operations and broader societal structures (Booth1990). For example, introducing nuclear weapons and jet propulsion in the twentieth century reshaped global power dynamics and necessitated new strategic doctrines (Adamsky, n.d.). However, if history provides any lessons, then we can see that not all technological innovations necessarily lead to Revolutions in Military Affairs (RMAs), and indeed, military history is filled with examples of failed innovations that began with the great promise of revolutionary transformation but where either the transformation failed to materialize or where it turned out to be less significant than initially anticipated. The introduction of nuclear-powered bombers, for instance, was expected to be a game-changer during the Cold War but ultimately faced insurmountable technical challenges. Similarly, the rise of battleships as the dominant naval force was short-lived once air power and submarines became integral to naval warfare during World War II. Even the Future Combat Systems (FCS) program, initiated by the U.S. Army in 2003, sought to develop networked systems of manned and unmanned technologies but faced significant technical hurdles and was eventually canceled in 2009 due to costs and the rapidly changing nature of asymmetric warfare. (Feickert 2006) (Kosiak 2007).

In considering whether AI represents a *Revolution in Military Affairs* (RMA), several criteria can be used to assess its transformative potential. Scholars like Krepinevich (1994) and Knox Murray (Knox 2001) have suggested that for a technology to qualify as an RMA, it must introduce fundamental changes in several areas: technological innovation, organizational adaptation, doctrinal change, strategic impact, and operational advantage. These criteria provide a framework for evaluating the role of AI in the military. We must ask whether AI introduces groundbreaking technologies, such as autonomous systems, that fundamentally alter how wars are fought. Are military organizations restructuring to integrate AI, and are doctrines evolving to reflect AI's role in decision-making and operations? Moreover, does AI shift the global balance of power, offering clear operational advantages like enhanced precision, speed, and lethality? These questions guide the ongoing debate about whether AI will bring revolutionary or incremental improvements to warfare.

As Table 1.1 indicates, the introduction of AI into military operations can be assessed against a rubric of critical criteria commonly used to define a *Revolution in Military Affairs* (RMA). The table below outlines these criteria, poses relevant questions for AI's role in military operations, and lists key authors who have contributed to these discussions.

The questions proposed here also allow us to connect changes to military organization and values created by the advent of AI to earlier changes occasioned by similar technological advances. The challenges that military planners and decision-makers face today are not unique. Military planners and analysts have long grappled with whether a new technology represents revolutionary change or incremental improvement. Similarly, the advent of AI raises questions about whether we are on the cusp of another such revolution.

Does AI Constitute a True RMA?

The Revolution in Military Affairs (RMA) concept has been subject to much debate. Some analysts argue that RMAs represent profound and sudden changes that fundamentally alter the nature of warfare, while others see these shifts as more incremental and evolutionary. This debate extends to contemporary technologies like AI and drones, where opinions differ on whether these innovations represent a new RMA or extend existing military capabilities. Scholars such as Krepinevich (1994) and Knox and Murray (2001) suggest that an RMA must introduce fundamental changes across multiple areas: technological innovation, organizational adaptation, doctrinal evolution, strategic impact, and operational advantage. The question remains whether AI fulfills these criteria or faces challenges similar to previous technologies, such as the U.S. Army's Future Combat Systems (FCS) program.

Historical attempts at revolutionary innovation can serve as cautionary tales, demonstrating that not all advances lead to paradigm shifts and that some technological innovations we expected would be revolutionary did not prove so. That is, not every military technological innovation has catalyzed an RMA. Battleships, once seen as the pinnacle of naval power, became obsolete with the rise of air power and submarines during World War II. Likewise, nuclear-powered bombers, anticipated to revolutionize air combat during the Cold War, faced technical challenges and were eclipsed by advancements in missile technology. Military history also provides examples of innovations that appeared dominant initially but failed to transform warfare, such as tanks in urban warfare, which proved vulnerable to anti-tank ambushes, and early twentieth-century hopes for zeppelins in strategic bombing, which were dashed by their vulnerability to anti-aircraft fire (Feickert 2006; Kosiak 2007).

Even the Maginot Line, a formidable fortification built by the French to deter German invasions, was circumvented by Germany during World War II. Similarly, older military technologies like the bayonet became outdated with modern firearms and artillery. These examples underscore that technological advancements do not always lead to fundamental changes in military strategy or operations, especially when they face practical, logistical, or strategic limitations. The Future Combat Systems (FCS) program, launched by the U.S. Army in 2003, is another example of how ambitious military innovations can struggle to meet expectations (O'Hanlon 2009).

More recently, the United States Future Combat Systems Program (FCS), launched in 2003, was an ambitious initiative to integrate advanced unmanned systems, tanks, and combat vehicles into a cohesive network. However, the program faced technical complexities and struggled with interoperability issues, cost overruns, and evolving operational contexts such as asymmetric warfare in Iraq and Afghanistan. Ultimately, the program was canceled in 2009 due to these challenges. (Feickert 2006) (Kosiak 2007) (O'Hanlon 2009). The FCS example highlights the difficulty of achieving revolutionary technological advancements in military settings, especially when facing financial and strategic constraints. These challenges suggest that caution is necessary when predicting the transformative potential of technologies like AI.

Thus, it is perhaps too early to declare that AI is a revolutionary technology or that its implementation would cause or necessitate a Revolution in Military Affairs. Moreover, considering the criteria necessary for a technology to constitute an RMA, there are strong arguments on both sides. AI enhances operational capabilities, from autonomous drones to predictive analytics, allowing for more precise and responsive military actions.

However, whether these developments constitute an RMA or a gradual evolution of existing capabilities remains unclear. As Fig. 14.1 illustrates in the Rubric for Assessing AI's Revolutionary Potential, the U.S. military has begun integrating AI technologies in decision-making, operational planning, and command structures. However, the level of fundamental organizational change required to declare an RMA remains a subject of debate. Others maintain that AI's impact will be more incremental, improving operational efficiency but leaving core military structures intact (Schneider and MacDonald 2024).

Incorporating AI into military operations raises similar concerns. AI systems require significant infrastructure, including computational power, data management capabilities, and cybersecurity safeguards. There are also issues with integrating AI across legacy platforms, ensuring reliability in unpredictable environments, and addressing ethical concerns regarding

RMA Criteria	Questions for AI	Supporting Authors
Technological Innovation	Does AI introduce fundamental changes to military technology, such as autonomous systems or real-time decision-making?	Adamsky, Kania
Organizational Adaptation	Are military organizations restructuring to integrate AI and adopting new command structures or training?	Scharre, Schneider & MacDonald
Doctrinal Change	Are doctrines evolving to incorporate AI's role in decision-making and operations?	Krepinevich, Farrell et al.
Strategic Impact	Does AI shift the global balance of power or provide strategic advantages?	Kania, Adamsky
Operational Advantage	Does AI offer clear operational benefits, such as increased precision, speed, or lethality?	Schneider & MacDonald, Scharre

Fig. 14.1 RMA criteria applied to AI

decision-making in autonomous operations. Like FCS, these practical considerations may limit AI's potential despite its promise as a revolutionary technology. As Fig. 14.2 illustrates, introducing military AI is not a decision that can merely be made unilaterally by one highly-placed official. Instead, AI must be integrated within preexisting legal, organizational, and safety structures. These constraints may look slightly different depending on a nation's unique institutions, values, and history. In particular, the U.S. may face challenges related to its democratic structures of accountability and transparency, giving its more autocratic adversaries, like Russia and China, an advantage when introducing AI and robotics into its military.

Category	Preexisting Constraints	Considerations for AI Integration
Legal	- Strict data privacy regulations (e.g., GDPR) that complicate data sharing and usage. - International laws governing autonomous weapons and lethal force use.	- Ensuring compliance with emerging data protection laws. - Developing clear legal frameworks for autonomous systems.
Organizational Leadership	- Military doctrines and command structures not designed for AI integration. - Resistance to change within traditional military hierarchies.	- Developing new doctrines that incorporate AI capabilities. - Training programs for AI use.
Economic	- High costs of upgrading infrastructure to support AI. - Competition for funding with other priorities. - Financial risk associated with large-scale AI investment.	- Allocating sufficient resources for AI development and integration. - Balancing AI investment with other military expenditures.
Technological	- Legacy systems not designed for integration with AI. - Limited computational power and data processing in current hardware. - Vulnerabilities in cybersecurity.	- Upgrading hardware/software across platforms. - Ensuring AI systems are secure and interoperable with legacy systems.
Ethical	- Compliance with the Law of Armed Conflict (LOAC) and principles of Just War. - Ensuring accountability for autonomous decision-making.	- Developing ethical guidelines and frameworks for AI use in combat. - Ensuring AI systems adhere to LOAC and Just War principles.

Fig. 14.2 Constraints related to the introduction of AI technology into U.S. warfighting doctrine and operations utilizing a system of systems approach

The Role of Strategic Culture in AI Integration

That is, the introduction of AI into military decision-making cannot be fully understood without considering the role of strategic culture, as Adamsky (n.d.) and Farrell et al. (Farrell, Payne, & Kartchner 2024)) argue that each nation's strategic culture—its traditions, values, and attitudes toward using force—plays a critical role in shaping how new technologies are adopted and integrated. The U.S., China, and Russia each present distinct strategic cultures that influence their approach to AI, highlighting tensions between ethical considerations, risk tolerance, and the pace of technological adoption.

In the U.S., military planners prioritize ethical concerns and operational safety, which can slow the adoption of AI technologies. The 2018 *Department of Defense Artificial Intelligence Strategy* emphasized that the U.S. would deliver "AI-enabled capabilities" in a way that ensures "safety, security, and ethical behavior," reflecting a cautious approach (Department of Defense 2018). Furthermore, the *National Security Commission on Artificial Intelligence* (NSCAI) Final Report (2021) outlined the need for "continuous, iterative innovation," balancing speed with ensuring AI systems are trustworthy and reliable (NSCAI 2021). This risk-averse culture concerns AI's potential unintended consequences, accountability, and the possibility of civilian casualties (Horowitz and Mahoney 2018). However, as adversaries like China and Russia make rapid strides in military AI, the U.S. faces increasing pressure to keep pace. Thus, the U.S. military faces a nuanced strategic dilemma in balancing its cautious approach to AI integration with the pressure to keep pace with adversaries such as China and Russia. The U.S. Department of Defense's AI strategy emphasizes ethical AI development and accountability, recognizing the need to mitigate risks associated with autonomous systems. However, the rapid pace of AI advancements by other global powers has created an AI arms race where the U.S. risks falling behind due to its more stringent ethical and operational constraints. (Lee 2018) (Work 2015) In addition, the U.S. faces unique challenges stemming from its democratic system. Ethical concerns, public opinion, and congressional oversight shape the pace at which AI technologies are integrated into U.S. military operations. The need to navigate legal and ethical frameworks, such as the Law of Armed Conflict (LOAC), further slows the introduction of autonomous systems into combat operations. This contrast between the rapid development strategies of Russia and China and the more measured U.S. approach may give rise to strategic vulnerabilities. (Allen 2017).

The AI arms race highlights the strategic dilemma faced by the U.S. military. Although U.S. warfighters may prefer to proceed cautiously, they are compelled to move faster due to the advancements of adversaries like China and Russia (Kania 2019). Indeed, China's *New Generation Artificial Intelligence Development Plan* sets a clear target of becoming the global leader in AI by 2030, driven by deep integration of AI with military, economic, and industrial sectors (State Council of the People's Republic of China 2017). And autocracies may enjoy an advantage in implementing technological changes in the military sector. Both China and Russia operate under centralized governments that can streamline AI integration without the same ethical or legal constraints faced by the U.S. Autocratic regimes may also think differently about risks and accountability. For instance, Russia's doctrine accepts higher levels of operational risk to quickly deploy AI capabilities in the military domain (Gerasimov, 201). China's *Military-Civil Fusion* strategy promotes the rapid adoption of AI across national defense, emphasizing speed over caution (State Administration for Science, Technology, and Industry for National Defense 2018). Similarly, Russia's 2019 *Foundations of the Russian Federation State Policy in the Field of AI* identifies AI as crucial for national defense and promotes the development of autonomous military systems, even at the expense of short-term risks (Government of the Russian Federation 2019). Russian military doctrine also reflects this aggressive stance, prioritizing AI to enhance military readiness and strategic parity with adversaries (Government of the Russian Federation 2019).

. In contrast, the U.S. is more concerned with responsible AI deployment, balancing the need for speed with democratic accountability and international collaboration (Department of Defense 2023). This tension creates a complex strategic challenge: moving too slowly risks falling behind in the AI arms race, but moving too quickly could introduce operational vulnerabilities and ethical breaches (Scharre 2018).

Technological Determinism and Human Agency: Top-Down vs. Bottom-Up AI Integration

Thus far, in considering whether the introduction of AI into military operations would constitute an RMA, we have proceeded from the assumption that the introduction of this technology would proceed in a top-down fashion, ordered by and implemented by the military's leadership—as we see in China and Russia. However, as noted earlier in this essay, we must also consider the tension between top-down and bottom-up processes when considering how AI will transform military operations. In a recent analysis, business school professor Ethan Mollick (2024) has suggested that current trends point to the uneven adoption of AI technology in the civilian workplace—with "early adopters" scattered throughout all levels of an organization. He argues that it may be difficult for organizational leaders to fully control the adoption of a technology that is in some ways so compelling—offering fast and efficient ways of automating one's workflow and transforming one into a more efficient and productive employee. In such a scenario, early adopters within an organization might "go rogue," implementing changes at their operational levels without waiting for (or even informing) their superiors. (Mollick 2024).

Indeed, such a pattern of early adopters "going rogue" with new technologies has precedents in military innovations. During World War I, for instance, aviators like Manfred von Richthofen pioneered new tactics for aerial combat long before formal air power doctrines had been established. Similarly, in World War II, U.S. Navy pilots developed dive-bombing techniques independently, which later became vital to American naval success at the Battle of Midway. (Symonds 2011). These "gonzo" pilots, as they were sometimes called, operated independently of formal military doctrines, demonstrating how bottom-up innovations can reshape military operations.

Mollick's insights suggest that a similar dynamic may unfold with AI. While top-down strategies, as outlined in the *2023 National Defense Strategy* (Department of Defense 2023), envision AI as a tool for enhancing situational awareness and decision-making, the reality on the ground may differ. Skilled individuals within the military might use AI unanticipated ways, potentially disrupting traditional hierarchies and introducing new challenges for leadership. As Verbeek (2005) reminds us, technologies actively shape human actions, and how AI is used may differ significantly from the intentions of military planners.

Leadership Challenges and Generational Differences in AI Adoption

Finally, we return to the matter of military leadership and unit cohesion and the effects AI might have on these battlefield readiness issues. One of the more concerning aspects of AI's integration into the military is the potential disruption to unit cohesion and leadership structures. As noted earlier, Mollick (2024) argues that those most skilled in working with AI may not necessarily be in leadership positions, creating tensions between traditional leaders and "AI specialists" operating within units. This dynamic could challenge existing norms of military hierarchy, where leadership is traditionally based on rank, experience, and character.

Moreover, the increasing reliance on AI for decision-making could undermine trust between leaders and subordinates. If AI systems are perceived as making decisions independently of human oversight, subordinates may question the legitimacy of their leaders' authority. This erosion of trust could have profound implications for unit cohesion, particularly in high-stakes combat situations where trust and camaraderie are essential for success.

Another challenge already identified by military leaders relates to the generational gap in comfort and trust toward AI. Research indicates that younger service members, particularly Millennials and Gen Z, are more comfortable with AI technologies than their older counterparts. (Miller 2019) (Chu and Smith 2023). Younger generations, raised in a world saturated with AI-driven technologies, such as GPS navigation and algorithmic recommendations, are more likely to trust AI's decision-making capabilities. In contrast, older generations approach AI cautiously, concerned about its reliability and potential unintended consequences. (Faverio 2022). This generational divide can affect how AI is adopted within military units. Younger warfighters, more accustomed to digital tools and AI, may be more open to trusting AI-driven systems for decision-making, surveillance, and combat operations. In contrast, older leaders, influenced by lessons from past wars, may be slower to embrace AI, relying instead on traditional command-and-control structures (Chu and Smith 2023).

However, this reliance on AI carries significant risks. Overtrust in AI-driven systems may lead to complacency, with younger soldiers less likely to scrutinize the outputs of AI technologies. Studies indicate that the more familiar people are with AI, the less likely they are to question its conclusions. This trend could have dire consequences in military contexts where adversaries actively seek to exploit system vulnerabilities. (Techflo 2023) In addition, it is possible that. Gen Z's exposure to algorithm-driven content on platforms like TikTok makes them more vulnerable to disinformation campaigns that can manipulate AI systems and mislead young soldiers (Pew Research Center 2023). In contrast, older military leaders, including Baby Boomers and Gen X, are generally more cautious about integrating AI into warfighting. Surveys indicate that older adults are more likely to view advancements like driverless cars—and, by extension, AI—as "bad for society" and are more reluctant to rely on AI systems (Chu and Smith 2023). This generational divide creates friction in military decision-making processes, with younger soldiers eager to adopt AI technologies and older leaders advocating a more measured approach.

The skepticism of older generations can be attributed not only to their unfamiliarity with AI but also to legitimate concerns about the ethical implications of AI systems in military operations. As Scharre (2018) and Horowitz and Mahoney (2018) argue, these older leaders are often more focused on the potential risks of AI systems, such as unintended civilian casualties, bias, and ethical breaches. Moreover, AI systems can reproduce societal biases, including ageist attitudes, potentially undermining the contributions of older military personnel (Chu and Smith 2023). Military leadership must navigate this generational divide, ensuring that younger soldiers are empowered to leverage AI's advantages while maintaining critical oversight to mitigate risks.

As the U.S. military integrates AI into its operations, leaders must balance encouraging the innovative use of AI by younger generations and instilling a healthy level of caution. This balance will be critical in maintaining operational effectiveness, as overreliance on AI could lead to errors that adversaries could exploit. At the same time, the cautious approach favored by older generations must not stifle the military's ability to innovate and keep pace with global adversaries, particularly in the ongoing AI arms race with China and Russia (Kania 2019).

Human Agency and Strategic Culture

As AI becomes more integrated into U.S. military operations, the uneven distribution of AI literacy and skills within organizations could lead to significant leadership challenges. This dynamic could profoundly affect unit cohesion and leadership in a military context. If AI-savvy individuals are scattered throughout units or command structures without recognition, their potential contributions to decision-making processes may be underutilized. Alternatively, these individuals might operate in silos, using AI without fully integrating their insights into broader strategic goals, thereby disrupting the unity and coherence of decision-making. Mollick's ideas highlight the possibility that new forms of leadership might emerge, where those who master AI tools can wield disproportionate influence, potentially creating tensions between traditional military leaders and those perceived as "AI specialists."

Furthermore, this hidden reliance on AI could create challenges for military leaders attempting to maintain control over increasingly automated systems. As Mollick suggests, the reluctance to acknowledge the role of AI in decision-making may lead to gaps in leadership transparency, where decisions are influenced by AI tools that remain outside the formal chain of command. This tension could undermine traditional leadership structures and challenge the cohesion of military units, as different actors may hold varying levels of trust in AI-generated decisions.

Mollick's insights tie directly into the broader conversation about technological determinism and human agency. While military leadership may seek to control AI's role through formal doctrine, skilled individuals' informal and hidden uses of AI could shape its deployment in unpredictable ways, further complicating efforts to maintain traditional command structures.

Conclusion: Navigating the Challenges of AI in Modern Warfare

As AI continues to be integrated into U.S. military operations, balancing the promise of technological advancement with the need for careful, ethical, and strategic oversight will be essential. The generational divide in trust toward AI and the strategic pressures of keeping pace with adversaries like China and Russia underscores the complex landscape that military leaders must navigate. Whether AI will lead to an actual Revolution in Military Affairs remains to be seen. Still, its potential to disrupt traditional military structures, decision-making processes, and leadership models is undeniable. By addressing these challenges thoughtfully, the U.S. military can harness AI's power to enhance its operational capabilities while maintaining the values and ethical frameworks that have long defined its strategic culture.

Bibliography

Adamsky, D. (n.d.). *Russian MIlitary Innovation in the 21t Century: The Case of Artificial Intelligence.*

Allen, G. A. (2017). *Artificial Intelligence and National Security.* Cambridge, MA: Belfer Center for Science and International Affairs, Harvard Kennedy School.

Bennett, S. (2020). *Russian Military Robotics: Recent Trends.* Annapolis, MD: Center for Naval Analyses.

Booth, K. (1990). The Concept of Strategic Culture. In C. (. Jacobsen, *Strategic Power: USA/USSR.* London: Palgrave.

Chu, A., & Smith, D. A. (2023). Generational Differences in AI Adoption in the Armed Forces: A Cross-Sectional Study. *Military Science Review, 14*(1), 112–134.

Cronin, A. K. (2020). *Power to the People: How New Technologies Are Reshaping Armed Conflict.* New York: Oxford University Press.

Government of the Russian Federation. (2019, October). Decree of the President of the Russian Federation on the development of artificial intelligence in the Russian Federation [National strategy document]. Retrieved August 1, 2025, from https://cset.georgetown.edu/wp-content/uploads/Decree-of-the-President-of-the-Russian-Federation-on-the-Development-of-Artificial-Intelligence-in-the-Russian-Federation-.pdf

Farrell, T., Payne, K., & Kartchner, K. (2024). Artificial Intelligence and Strategic Culture. In *Routledge Handbook of Strategic Culture* (pp. 47–484). New York: Routledge.

Faverio, M. (2022). *Older Americans are more worried than Younger Adults about the Prospect of Driverless Cars on the Road.* Philadelphia, PA: Pew Research Center.

Feickert, A. (2006). *The Army's Future Combat Systems (FCS): Background and Issues for Congress (CRS Report for Congress, RL32888).* Washington, DC: Congressional Research Service.

Freedman, L. (2003). *The Evolution of Nuclear Strategy, 3rd ed.* New York: Palgrave MacMilllan.

Horowitz, M. C., & Mahoney, C. (2018, December 25). Artificial intelligence and the military: Technology is only half the battle. War on the Rocks. Retrieved August 1, 2025, from https://warontherocks.com/2018/12/artificial-intelligence-and-the-military-technology-is-only-half-the-battle/

Kania, E. (2019). *Battlefield Singularity: Artificial Intelligence, Military Revolution, and China's Future Military Power.* Washington, DC: Center for A New American Security.

Knox, M. A. (2001). *The Dynamics of Military Revolution, 1300–2050.* Cambridge, UK: Cambridge University Press.

Kofman, M. (2021). *Russian AI Strategy: Military Applications and Implications.* Washington, DC: Carnegie Endowment for International Peace.

Kosiak, S. (2007). *Analysis of the FY 2008 Defense Budget Request (Working Paper).* Washington, DC: Center for Strategic and Budgetary Assessments.

Krepinevich, A. (1994). . Cavalry to Computer: The Pattern of Military Revolutions. *The National Interest.*

Lee, K. (2018). *AI Superpowers: China, Silicon Valley, and the New World Order.* New York: Houghton Mifflin Harcourt.

Miller, C. (2019). The Impact of AI on Military Leadership: Bridging the Generational Divide. Defense Innovation Journal. *Defense Innovation Journal, 7*(2), 45–61.

Mollick, E. (2024). *Co-Intelligence: Living and Working with AI.* New York: Penguin, Penguin Random House. Retrieved August 1, 2025, from https://www.penguinrandomhouse.com/books/741805/co-intelligence-by-ethan-mollick/PewResearchCenter2023

O'Hanlon, M. E. (2009). *The Science of War: Defense Budgeting, Military Technology, Logistics, and Combat Outcomes.* Princeton, NJ: Princeton University Press.

Pew Research Center. (2023, August 28). U.S. Public Concern Grows Over Role of Artificial Intelligence in Daily Life. Pew Research Center. Retrieved August 1, 2025, from https://www.pewresearch.org/short-reads/2023/08/28/growing-public-concern-about-the-role-of-artificial-intelligence-in-daily-life/

Scharre, P. (2018). *Army of None: Autonomous Weapons and the Future of War.* New York: Norton.

Schneider, J., & MacDonald, J. (2024). Looking Back to Look Forward: Autonomous Systems, Military Revolutions and the Importance of Cost. *Journal of Strategic Studies, 47*(2), 162–184.

Singer, P. W. (2009). *Wired for War: The Robotics Revolution and Conflict in the 21st Century.* New York: Penguin Press.

State Administration for Science, Technology and Industry for National Defense (SASTIND). (2018). Guidelines for development of the defense technology industry at the grassroots level: Basic research and cutting-edge technology projects [Policy document]. Retrieved August 1, 2025, from https://cset.georgetown.edu/publication/project-to-strengthen-development-of-the-defense-technology-industry-at-the-grassroots-level-guidelines-for-basic-research-and-cutting-edge-technology-projects-2018/

State Council of the People's Republic of China. (2017, July 8). A New Generation Artificial Intelligence Development Plan [White Paper]. Retrieved August 1, 2025, from https://digichina.stanford.edu/work/full-translation-chinas-new-generation-artificial-intelligence-development-plan-2017/

Symonds, C. (2011). *The Battle of Midway.* New York: Oxford University Press.

Techflo. (2023). *AI in Defense: Risks, Vulnerabilities, and Overtrust.* No Location: Techflo Analytics.

Trevithick, J. (2020, March 16). "Hypersonic Missiles: How the U.S., Russia, and China Are Developing New Weapons That Will Revolutionize Warfare. *The War Zone.*

Verbeek, P. P. (2005). *What Things Do: Philosophical Reflections on Technology, Agency and Design.* State College, PA: Pennsylvania State University Press.

Waltzman, R. (2020). *Cognitive Security: The Impact of AI on Decision Making.* Los Angeles: Rand Corporation.

Work, R. (2015). *Remarks by Deputy Secretary Work on Third Offset Strategy. U.S. Department of Defense.* Washington, DC: United States Department of Defense.

CHAPTER 15

Robots Writing Chinese and Fighting Underwater

Greg Austin

As of 2024, China attaches even greater importance to military robotics than it did when the original chapter was published in the first edition of this book. Prominent Chinese news outlets, such as the government-linked *Global Times* and independent *The Paper*, were quick to report on remarks by the former Chairman of the US Joint Chiefs of Staff, General Mark Milley, who said that within ten to 15 years, "one-third of the U.S. military, or one-quarter to one-third, will be robots" (Ceder 2024). He linked this perspective to the powerful role played by AI in deployed military robots, which as a combination would give the country possessing them "a very significant—and potentially decisive—advantage in an armed conflict" (Demarest 2024). The related conclusion of the original chapter is pertinent here: "China cannot join the robotics RMA until it is much further along the road of the information revolution" (Austin 2017, 286).

G. Austin (✉)
East West Institute, Bengaluru, India
e-mail: gaustin@ewi.info

R. Kiggins (ed.), *The Political Economy of Robots*, International Political Economy Series, https://doi.org/10.1007/978-3-031-95171-8_15

Milley's remarks did not make a distinction between drones and robots. This short note, like the original chapter, prefers Singer's definition of a robot which sees it as a machine that can sense (detect), think (process information) and act (affect its environment according to pre-programmed or self-generated instructions) (Singer 27). One could add to those qualities the ability to manipulate its physical form for a variety of tasks, from basic to quite complex. While the boundary between robots and unmanned vehicles will be blurred, the original chapter suggested that discussion of robots should not uncritically include discussion of all unmanned systems (Austin 2017, 272). Chinese sources regularly discuss autonomous unmanned military systems and military robots as varieties of a single class of objects. Statistics on national robotics development can be misleading if they do not provide clear definitions of the types of robots covered by the data (for example both single function, fixed-station industrial robots, and more sophisticated machines).

State of Play

Public media carry regular reports of Chinese breakthroughs in military robotics. For example, in May 2024, the Chinese army demonstrated two AI robot dogs equipped with automatic rifles in a military exercise with Cambodia (BBC News Chinese 2024). This was reported to be the first time that China had publicly used military AI robots in a combat setting. Several Chinese scholars with military ties have offered a nuanced assessment of the current state of play. The use of unmanned equipment on the battlefield has become widespread, stimulating rapid growth in research on unmanned autonomous technology, especially unmanned combat systems. They observe that "military robots with higher intelligence, more flexible movements, and faster reactions are moving from behind the scenes to the front stage of war" (Zhai et al. 2023, 24). Nevertheless they conclude that while the "application of robotics technology in military warfare is not very widespread ... robots will play an irreplaceable and significant role in the future battlefield".

On the other hand, these same authors assess that the "China's robotics industry has not yet formed" but given technological breakthroughs, its creations will be "able to perform dangerous operations, planet exploration, and ocean operations" (p. 27). One of the constraints raised by another specialist (Dou Qiaoqi, the head of Wuhan Gelanruo Precision Technology Company), speaking on the occasion of a visit by China's leader, Xi Jinping, is that there appears to be a shortage of appropriate Chinese talent for robotics work (Hubei Daily 2024).

Another Chinese analysis suggested areas of improvement of their national military robot sector: theoretical research and test verification, formulating strategic planning and standards, deepening civil-military integration and technology joint research, and strengthening technical security and information confidentiality (Song et al. 2024, 789). A European assessment sees the US as likely to maintain its dominance in this field, based on the authors assessment of its superiority in "defence innovation, doctrinal evolutions, and fielding of military robotics" (Vestner and Lusenti 2023, 81).

In 2023, China set in train new measures to promote robotics development, including the Implementation Plan for the New Industry Standardization Pilot Project (2023–2035) and the Guiding Opinions on the Innovation and Development of Humanoid Robots. These directives have noted the importance of integration of advanced technologies (AI, high-end manufacturing and new materials (Yu 2024). In January 2024, the Implementation Opinions on Promoting the Innovation and Development of Future Industries promoted a special focus on R&D for humanoid robots in intelligent manufacturing, home services, and special environment operations. These policy imperatives followed almost a decade of implementation of other similar initiatives which did not enable China to achieve breakthroughs on the level it had hoped.

An analysis published by the Shanghai government's Science and Technology Commission concluded in respect of humanoid robots, the world's leading research is "mainly concentrated in the United States, Japan and some European countries" (Yu 2024). It reported that the core technology of humanoid robots in China is "far behind that of the U.S. and Japan, and key components are in urgent need of breakthroughs". It also observed importantly that global landscape is characterized by diversification, regionalisation, and cross-border integration.

The global market is a reasonable proxy indicator of aspects of the current state of play in robotic development. A Canadian market analysis firm has estimated that the global market for military robots will double in the ten years to 2034 (from US$23.92 billion to US$44.23 billion) (Precedence 2024a) while a similar study form China saw a much faster market growth to reach US$64.13 billion by 2032 (Fortune 2024). The industrial base of military robotics in China is still highly dependent on foreign skills, imported components and investment (Atkinson 2024).

According to this study, while some indicators of competitive advantage (patents and published articles) suggest China is doing as well as or better than the US, a closer look reveals that the US retains quite an edge. It produces more robotics innovations and has a much more highly mobilized private sector R&D. Moreover, the report notes that the leading Japanese manufacturer outperforms its Chinese peers. As in other areas of advanced technology, this report lays out a policy agenda for US to do more to retain its edge over China.

AI as the Key

In June 2024, China announced that it had created its first AI military commander, albeit housed in and confined to a laboratory (Chen 2024a). In the following November, China disclosed for the first time research that directly applies large language models to weapons systems to command military drones to attack enemy radar systems (Chen 2024b). The revelation in a scientific journal noted that work is still experimental.

At the same time, military-associated specialists in China have doubts that China can catch up easily with the US leadership position in AI and related technologies, especially high-end sensors, and in managing the risks (Bresnick 2024a). A survey of their published attitudes between 2022 and 2022 found that they have concerns (similar to those in the US) about network and cyber security, military data collection and analysis. They also see difficulties in building trustworthy AI systems to the point where commanders can rely on them in combat.

This survey mentioned the difficulty of understanding in much detail where in practice the PLA sist on these issues (p. 1). A study published by Brookings concluded that China's progress in military application of AI is "remains uncertain" and "cannot be assessed with high confidence at this point" (Kania 2020, 1). Gains will depend on the country's capacity to "operationalize emerging weapons systems" by overcoming fundamental current technological and organizational challenges. A 2024 testimony to the US Congress observed that "open-source information about the PLA fielding specific military AI systems remains sparse" (Stokes 2024, 3). One reason is that the activities are undertaken in software systems which are largely unrevealed. This report concluded that China "will have to overcome multiple obstacles to fulfil its objectives" (p.5). A European commentary concluded that China remains dependent on foreign semiconductors needed for high-end AI applications and lacks the appropriately qualified workforce (Bresnick 2024b).

Taiwan Scenarios

The relative strengths of the industrial base or scientific research in China and the US will not determine war outcomes under the influence of military robots in a war over Taiwan. The main determinants will relate to fielded units, their tactical or strategic employment and counter-measures deployed by the other side. In respect of military drones, according to a Canadian analysis, the US has the larger share (around one-third) of the global market (Precedence 2024b). This assessment delivers nuance to an American claim that "Chinese companies have cornered the commercial drone market, and few companies can match the performance and cost of Chinese drones due to the scale of their market share" (Pettyjohn et al. 2024, 19). If the US leads the world in deployment and development of military drones, the authors' claim that China is better placed than the US to exploit drones for military purposes in a war over Taiwan (p. 6) is one that would benefit from some critical analysis.

Their conclusion is premised on several judgements, notably that "Mass production of an affordable mix of drones" would be needed for a likely protracted conflict and that geography would significantly disadvantages the US by imposing a requirement for long range drones with significant "endurance" (p.1). Their position stands in contrast to an expectation in the Pentagon that it can subject Chinese forces to a drone "hellscape" in the Taiwan Strait. The Defense view is that the US can deploy unmanned vehicles (undersea, surface and air) in the thousands to degrade Chinese capabilities to allow time for the US and its allies to respond more forcefully (Rogin 2024). According to the report, this plan depends on classified capabilities, so they are not on the market, a position which negates the effect of China's lead in export of military drones (cited—unconvincingly—as a possible indicator of Chinese drone superiority).

In such a scenario, the drones would not operate independently of other US capabilities, especially cyber, intelligence, and command and control. The US is also on record, via Admiral Davidson, while serving as the head of US Pacific Command, as planning a cyber blitzkrieg that would aim to penetrate an adversary's networks, neutralize their decision-making and thereby blunting their kinetic capabilities: "penetration and then disintegration of an adversary's systems and decision-making, thereby defeating their offensive capabilities" (Davidson 2020). It will be the connectivity of US drones to the country's far superior cyber and intelligence capabilities, and command and control that makes the "hellscape" in the Strait a credible outcome. Clear US superiority in military space capabilities, especially for intelligence and surveillance (Austin et al. 2022), help underpin this potential.

Chinese Breakthroughs

We cannot discount the likelihood of Chinese technological and prototype breakthroughs that see it field superior robotics or robotic drones that can deliver battlefield advantage over the US and its allies. One source of this type of disruption would be a significant weakening of US alliances, either in NATO or with Japan. Another source might be a dramatic change in the strategic reorientation of investment funds from wealthy Arab states toward China's ICT sector in a way that positively transforms current research, workforce, and production patterns. On the other hand, such changes would have to occur in relatively pronounced forms within this decade to disrupt what we see now as the likely trajectory of robotics power in that time frame. Much will depend on changes in the organizational and technology policies in the armed forces and security agencies of the major technology powers.

References

Atkinson, R. D. 2024. How Innovative is China in the Robotics Industry? Information Technology and Innovation Foundation. Washington DC. https://itif.org/publications/2024/03/11/how-innovative-is-china-in-the-robotics-industry/

Austin, G. 2017. Robots Writing Chinese and Fighting Underwater. In Kiggins, R. (ed). 2017. The Political Economy of Robots: Prospects for Stability and Peace in the Automated 21st Century. Springer International Publishing. 271–290

Austin, G., Rajagopalan, R.P., & Wright, T. (2022). Military Ambitions and Competition in Space: The Role of Alliances. The International Institute for Strategic Studies. https://www.iiss.org/research-paper/2022/02/military-ambitions-and-competition-in-space-the-role-of-alliances/

BBC News Chinese. 2024. Zhōngguó AI jīqìrén fǎngshēng "wēi biǎoqíng" jīxiè gǒu shǒu xiàn jūn yǎn [China's AI robot mimics "micro-expressions"; robot dog makes its first appearance in a military exercise]. June 20. https://www.bbc.com/zhongwen/simp/science-69125361

Bresnick, S. 2024a. China's Military AI Roadblocks PRC Perspectives on Technological Challenges to Intelligentized Warfare. Centre for Security and Emerging Technologies, Georgetown University. Washington DC. https://cset.georgetown.edu/wp-content/uploads/CSET-Chinas-Military-AI-Roadblocks-1.pdf

Bresnick, S. 2024b. China Bets Big on Military AI. Centre for European Policy Analysis. April 3. https://cepa.org/article/china-bets-big-on-military-ai/

Ceder, Riley. 2024. "One-third of US military could be robotic by 2039: Milley." Military Times, July 14.

Chen, S. 2024a. Chinese scientists create and cage world's first AI commander in a PLA laboratory. South China Morning Post. June 16. https://www.scmp.com/news/china/science/article/3266444/chinese-scientists-create-and-cage-worlds-first-ai-commander-pla-laboratory

Chen, S. 2024b. J-20 designer arms electronic warfare drones with generative AI. South China Morning Post. November 1. https://www.scmp.com/news/china/science/article/3284588/words-can-kill-j-20-designer-arms-electronic-warfare-drones-generative-ai

Davidson, P. 2020. Transforming the Joint Force: A Warfighting Concept for Great Power Competition. Speech. West San Diego. March 3. https://www.pacom.mil/Media/Speeches-Testimony/Article/2101115/transforming-the-joint-force-a-warfighting-concept-for-great-power-competition/

Demarest, C. 2024. One-third of U.S. military could be robotic, Milley predicts. Axios. July 11. https://www.axios.com/2024/07/11/military-robots-technology

Fortune. 2024. Military Robot Market. https://www.fortunebusinessinsights.com/zh/military-robots-market-104663

Hubei Daily. 2024. Xí Jìnpíng zǒng shūjì zài wǔhàn chǎnyè chuàngxīn fāzhǎn yán jiù yuàn kǎochá huífǎng jì [A return visit to General Secretary Xi Jinping at the Wuhan Institute of Industrial Innovation and Development] November 7. https://www.hubei.gov.cn/zwgk/hbyw/hbywqb/202411/t20241107_5403933.shtml

Kania, E., 2020. "AI Weapons" In China's Military Innovation. Brookings Institution in cooperation with the Centre for Security and Emerging Technology, Georgetown University. https://www.brookings.edu/wp-content/uploads/2020/04/FP_20200427_ai_weapons_kania_v2.pdf

Nan B. and Gao S. *Měiguó jūn fāng dàlì fāzhǎn wú rén zuòzhàn xìtǒng, fàngyán "15 niánnèi měijūn sān fēn zhī yī shì jīqìrén.* [The U.S. military is vigorously developing unmanned combat systems, boasting that "one-third of the

U.S. military will be robots within 15 years.] The Paper. Shanghai. July 16. https://m.thepaper.cn/newsDetail_forward_28087567

Pettyjohn S., Dennis H. and Campbell, M. 2024. Swarms over the Taiwan Strait. Center for New American Security. https://www.cnas.org/publications/reports/swarms-over-the-strait

Precedence. 2024a. Military Robot Market Size, Share and Trends 2024 to 2034. September. https://www.precedenceresearch.com/military-robot-market

Precedence. 2024b. Military Drones Market Size, Share, and Trends 2024 to 2034. November. https://www.precedenceresearch.com/military-drones-market

Rogin, J. 2024. The U.S. military plans a 'Hellscape' to deter China from attacking Taiwan. The Washington Post (Online) June 10. https://www.proquest.com/blogs-podcasts-websites/u-s-military-plans-hellscape-deter-china/docview/3066163516/se-2

Singer, P. W. 2009. *Wired for War: The robotics Revolution and Conflict in the 21st Century*, Penguin, New York

Song, H., Jia, H. and Wang, S., 2024, February. Development of unmanned intelligent support equipment of military powers. In *International Conference on Mechatronic Engineering and Artificial Intelligence (MEAI 2023)* (Vol. 13071, pp. 789–794). SPIE.

Stokes, J., 2024. Military Artificial Intelligence, the People's Liberation Army, and U.S.-China Strategic Competition. Testimony before the U.S.-China Economic and Security Review Commission. Hearing on Current and Emerging Technologies in U.S.-China Economic and National Security Competition. February 1. https://s3.us-east-1.amazonaws.com/files.cnas.org/documents/Jacob_Stokes_Testimony.pdf

Vestner, T. and Lusenti, A. 2023. Great Powers' Military Robotics. *stratos Militärwissenschaftliche Zeitschrift der Schweizer Armee Revue scientifique militaire de l'Armée suisse*. 79–86. https://deftech.ch/wp-content/uploads/2023/07/stratos_1-23.pdf#page=81

Yu Q. 2024. Humanoid Robots: Future Development Trends and Strategic Recommendations for Shanghai. 2 July. Science and Technology Commission of Shanghai. 2 July. https://stcsm.sh.gov.cn/cxyj/jsyj/shxjcg/20240702/cddb1d60802c4de4ad981aa835354aa9.html

Zhai C., Zhang P., Xu H, Yuan X., Zhou L. and Wu R. 2023. The Application and Inspiration of Robots in the US Military. *2023 8th Asia-Pacific Conference on Intelligent Robot Systems (ACIRS)*, Xi'an, China, 2023, pp. 24–28, https://doi.org/10.1109/ACIRS58671.2023.10239710.

CHAPTER 16

Armed Drones: Automation, Autonomy, and Ethical Decision-Making

Peter Lee

In this chapter I explored the ethical, moral, political, and technological challenges posed by armed drones in modern warfare. The chapter focused on the growing automation and autonomy of drones and how this evolution impacts ethical decision-making in military contexts.

I argued that the increasing reliance on drones raised profound ethical questions, particularly concerning the human role in the decision to kill. I differentiated, probably too much, between automation—where humans are involved in operational decisions—and autonomy, where drones may act independently of human control using 'rule based behavious' (Cummings 2017, 5). As military technology advances, the boundary between automated and autonomous systems has blurred into a continuum, heightening concerns about accountability, responsibility, and moral justification in decisions to take life in military operations.

I also examined how, by distancing controller-combatants from the battlefield, thereby reducing the political risk that accompanies the deaths

P. Lee (✉)
University of Portsmouth, Portsmouth, UK
e-mail: peter.lee@port.ac.uk

R. Kiggins (ed.), *The Political Economy of Robots*, International Political Economy Series, https://doi.org/10.1007/978-3-031-95171-8_16

of soldiers in far-off wars, drones shift the ethical calculus, potentially lowering the threshold for initiating conflicts or for making lethal decisions. This chapter also considered whether existing international laws and conventions adequately regulated the use of autonomous weapon systems. I suggested that without robust ethical frameworks and regulation, the deployment of autonomous armed drones could erode moral accountability in warfare, probably with a degree of hope that some level of international agreement would be reached. On that last point, I was somewhat hopeful but the reality has been sobering. In response to issues that have arisen in drone warfare since I originally wrote the chapter, I will now reflect on key developments since: technological advances in artificial intelligence (AI) and autonomy; the realpolitik that is shaping autonomous drone development and use; and how levels of autonomy shift the risk and ethical calculus involved.

Increasing Autonomy

Since 2018, increasingly autonomous drones have significantly impacted military operations in conventional and non-conventional applications, by state and non-state groups. The pursuit of increasing autonomy in drones is primarily driven by developments in artificial intelligence (AI), allowing them to conduct increasingly complex tasks with less and less human intervention. Autonomous drones can perform reconnaissance, target identification, and precision strikes, adapting to a changing battlespace in real time. This advancement reduces the need for direct human control, thereby accelerating decision-making processes during critical missions. However, reducing human input also raises ethical questions about the extent to which an autonomous drone system should be able to kill human targets.

Brown (2023, 735ff) sets out three clear ethical arguments: realist (concerned with power), deontological (duty to follow rules), and teleological (focuses on the aim or intended outcome). In doing so he recognises a political reality—that there are some states, such as Russia, for whom operational imperative is the sole consideration when it comes to using autonomous drones, while the US, UK, and other allies are more constrained by democratic process and the need for ethical justification. The US position has developed somewhat since former US Deputy Secretary of Defense Robert Work stated that the US, 'will not delegate lethal authority to a machine to make a decision' when it comes to the use of military force (Work 2016). The DoD position in 2023 is set out:

> Persons who authorize the use of, direct the use of, or operate autonomous and semiautonomous weapon systems will do so with appropriate care and in accordance with the law of war, applicable treaties, weapon system safety rules, and applicable rules of engagement (ROE). The use of AI capabilities in autonomous or semi-autonomous weapons systems will be consistent with the DoD AI Ethical Principles. (US DoD: 2023, 4)

Political Reality

A commitment to the use of AI ethical principles in the use of autonomous weapons contrasts with Russia's approach to drone use in Ukraine. At the time of writing in 2024, Russia is using waves of so-called 'kamikaze' drones to try to overwhelm Ukraine's air defences (Plichta 2024). These drones are a low-tech form of autonomous drone and should perhaps more rightly be referred to as automated drones. There is no attempt to avoid civilians—the drones are frequently and deliberately aimed towards population centres to cause as much damage and harm as possible. These might not, yet, be drones that are AI-enabled with machine learning and decision-making, but Russia's battlefield conduct indicates what can be expected when more sophisticated variants are eventually produced. At a less technologically sophisticated level, non-state groups like Hamas, Hezbollah and ISIS have used small, usually off-the-shelf adapted drones with increasing sophistication over the past decade. The *modus operandi* of these groups includes the targeting and terrorising of civilian populations. The drones they have used do not come close to the sophistication of Western military grade, aircraft-sized drones. However, as Blount and Sammut argued in 2016, drones help such groups create a narrative of air power sophistication for their supporters that may well be more powerful than the weapon capability itself (Blount and Sammut 2016). Successful drone use by Hamas against Israel in 2023 and 2024, and Hezbollah's 2024drone incursions into Israel—including its attack on the home of Prime Minister Netanyahu (Bennett and Lukiv 2024)—is measured more by political and psychological impact than tonnage of explosives. Such propaganda victories ensure that drone technology will continue to be embraced by these and other groups and, on this evidence, more capable, future autonomous versions will be similarly deployed without hesitation. This, in turn, will further fuel the ongoing defensive, counter-drone arms race which, inevitably, will prompt further innovation and drone destructiveness. Thereby rapidly advancing the need for states who are committed (for now) to not delegating life-taking decisions to autonomous weapon systems, for defensive, AI-enabled autonomous systems to counter those that will be used offensively without ethical concern.

Levels of Autonomy, Risk, and Ethical Calculus

In the original chapter, the terms automation and autonomous were used in opposition throughout, a reflection of several years of public discourse around the meaning of autonomy in Lethal Autonomous Weapon Systems (LAWS). The debate became somewhat dualistic: something was either autonomous or it was not, and if it was not 'fully' autonomous then had to be automated. This binary approach to terminology and the conceptualisation of autonomous weapon systems was extended to three distinct levels, related to John Boyd's decision-making OODA loop (observe-orient-decide-act): '[Autonomous weapon systems] may require human input during the execution of tasks ("human in the loop"), operate independently but under the supervision of a human who can intervene ("human on the loop"), or be fully autonomous ("human out of the loop")' (Caves and Carus 2021). Practically, however, this conceptualisation is not as useful as the six levels of autonomy classified by the Society of Automobile Engineers (SAE) that are rapidly becoming ubiquitous, used by the European Union (EU) and the US National Highway Traffic Safety Administration (NHTSA), for example (Bhila et al. 2024, 15).

The'loop' classification for military applications and the SAE autonomy levels for vehicles can be seen juxtaposed in Table 16.1:

The significance of different levels of autonomy in risk calculations and ethical considerations lies in the balance between human oversight and system independence. As the autonomy level of a weapon system increases, the risks it poses also grow and the ethical considerations are greater. At lower levels of autonomy, such as SAE Levels 0–2, humans maintain control or oversight of weapon systems, reducing risks because human intervention is possible at critical moments. However, as systems move towards higher levels of autonomy (3–5), where human input is

Table 16.1 Risk Matrix: Autonomous vehicles & Autonomous weapon systems (Peter Lee 2023 in Bhila et al. 2024, 30)

RISK MATRIX: AUTONOMOUS VEHICLES & AUTONOMOUS WEAPON SYSTEMS

Peter Lee, 2023

Autonomy Level	**SAE Definition (Civilian)**	**Military Applications**	**AI system functions and degrees of autonomy**				
0	No Driving Automation	**Human in loop.** Actively controlling with different levels of AI input	No risk from system				
1	Driver Assistance				Low		None
2	Partial Driving Automation				Moderate		
3	Conditional Driving Automation	**Human on loop.** System can request human input			Medium		
4	High Driving Automation	**Human out of loop.** No human input necessary.			High		
5	Full Driving Automation				Extreme	Extreme	Critical
			Routine Admin-istrative tasks.	System AI-informed routine decisions.	AI-based complex decision support to human operators in benign environments.	Autonomous system uses AI based decision-making in high risk environments with human input available.	Autonomous system uses AI based decision-making in high risk environments and lethal environments with no human input available.

limited or eliminated, risks increase due to potential system malfunctions or unforeseen circumstances.

Autonomous systems in high-risk scenarios, particularly in military, life-critical applications, raise serious ethical concerns. The greater the autonomy, especially with a machine learning element in the AI, the greater the potential for unintended consequences: crucial, if lethal force is involved. Fully autonomous systems, capable of making decisions without human intervention, challenge traditional concepts of human accountability and responsibility. Ethical dilemmas arise around International Humanitarian Law compliance and the principles of proportionality and distinction in combat, because of concerns about whether these systems can make sound decisions. Since they are not human, I take the view that autonomous systems cannot—in themselves—take moral decisions. That is the purview of human beings.

Ultimately, rapid developments in AI-driven, autonomous weapon systems necessitates robust regulatory frameworks and ethical guidelines to manage the associated risks and ensure that their deployment, especially in military contexts, aligns with international legal standards and societal values. There is still much to be done by states who take such challenges seriously.

References

Bennett, Tom, and Lukiv. 2024. 'Netanyahu Undeterred after Reported Drone Attack on His Home'. BBC News. 19 October 2024. https://www.bbc.com/news/articles/cwyl4e7w2e7o.

Bhila, Ishmael, Peter Lee, and Alison Wakefield. 2024. 'Autonomous Vehicles: Threats, Risks, and Opportunities'. ASIS Foundation. https://store.asisonline.org/autonomous-vehicles-threats-risks-and-opportunities.html.

Blount, Clive, and Charlie Sammut. 2016. '"A Gift to Our People": The Use of Drone Technology by Islamist Insurgents'. *Air Power Review* 19 (1): 8–25.

Brown, Andrew. 2023. "Ethics, Autonomy, and Killer Drones: Can Machines Do Right?" *Comparative Strategy* 42 (6): 731–46.

Caves, John P Jr, and W S Carus. 2021. 'The Future of Weapons of Mass Destruction: An Update'. Washington DC: National Intelligence University. https://apps.dtic.mil/sti/trecms/pdf/AD1147523.pdf.

Cummings, Mary L. 2017. "Artificial Intelligence and the Future of Warfare." *Chatham House Research Paper*. https://www.chathamhouse.org/sites/default/files/publications/research/2017-01-26-artificial-intelligence-futurewarfare-cummings-final.pdf

Plichta, Marcel. 2024. 'Russia's Growing Kamikaze Drone Fleet Tests Ukraine's Limited Air Defenses'. *Atlantic Council*. 14 May 2024. https://www.atlanticcouncil.org/blogs/ukrainealert/russias-growing-kamikaze-drone-fleet-tests-ukraines-limited-air-defenses/. Accessed 16 October 2024.

US DoD. 2023. 'DoD Directive 3000.9 "Autonomy in Weapon Systems"'. Department of Defense. https://www.esd.whs.mil/portals/54/documents/dd/issuances/dodd/300009p.pdf

Work, Robert. 2016. Quote from an interview. 'David Ignatius and Pentagon's Robert Work Talk About New Technologies to Deter War'. *Washington Post*. 30 March 2016. https://www.washingtonpost.com/blogs/post-live/wp/2016/02/29/securing-tomorrow-with-davidignatius-whats-at-stake-for-the-world-in-2016-and-beyond/. Accessed 16 October 2024.

CHAPTER 17

Lethal Autonomous Systems and the Plight of the Non-combatant

Ronald Arkin

It seems a safe assumption, unfortunately, that humanity will persist in conducting warfare, as evidenced over all recorded history. New technology has historically made killing more efficient, for example with the invention of the longbow, artillery, armored vehicles, aircraft carriers, or nuclear weapons. Many view that each of these new technologies has produced a Revolution in Military Affairs (RMA), as they have fundamentally changed the ways in which war is waged. Many now consider robotics technology a potentially new RMA, especially as we move towards more and more autonomous[1] systems in the battlefield.

[1] We do not use autonomy in the sense that a philosopher does, i.e., possessing free will and moral agency. Rather we use in this context a roboticists' definition: the ability to designate and engage a target without additional human intervention after having been tasked to do so.

R. Arkin (✉)
School of Interactive Computing, College of Computing, Georgia Institute of Technology, Atlanta, GA, USA
e-mail: arkin@cc.gatech.edu

R. Kiggins (ed.), *The Political Economy of Robots*, International Political Economy Series, https://doi.org/10.1007/978-3-031-95171-8_17

Robotic systems are now widely present in the modern battlefield, providing intelligence gathering, surveillance, reconnaissance, target acquisition, designation, and engagement capabilities. Limited autonomy is also present or under development in many systems as well, ranging from the Phalanx system "capable of autonomously performing its own search, detect, evaluation, track, engage and kill assessment functions",[2] fire-and-forget munitions, loitering torpedoes, and intelligent antisubmarine or anti-tank mines among numerous other examples. Continued advances in autonomy will result in changes involving tactics, precision, and just perhaps, if done correctly, a reduction in atrocities as outlined in research conducted at the Georgia Tech Mobile Robot Laboratory (GT-MRL).[3] This paper asserts that it may be possible to ultimately create intelligent autonomous robotic military systems that are capable of reducing civilian casualties and property damage when compared to the performance of human warfighters. Thus, it is a contention that calling for an outright ban on this technology is premature, as some groups already are doing.[4] Nonetheless, if this technology is to be deployed, then restricted, careful, and graded introduction into the battlefield of lethal autonomous systems must be standard policy as opposed to haphazard deployments, which I believe is consistent with existing International Humanitarian Law (IHL).

Multiple potential benefits of intelligent war machines have already been declared by the military, including: a reduction in friendly casualties; force multiplication; expanding the battlespace; extending the warfighter's reach; the ability to respond faster given the pressure of an ever-increasing battlefield tempo; and greater precision due to persistent stare [constant video surveillance that enables more time for decision making and more eyes on target]. This argues for the inevitability of development and deployment of lethal autonomous systems from a military efficiency and economic standpoint, unless limited by IHL.

It must be noted that past and present trends in human behavior in the battlefield regarding adhering to legal and ethical requirements are questionable at best. Unfortunately, humanity has a rather dismal record

[2] U.S. Navy, "Phalanx Close-in Weapons Systems", United States Navy Factfile, http://www.navy.mil/navydata/fact_display.asp?cid=2100&tid=487&ct=2, accessed 7/23/2013.

[3] R.C. Arkin, Governing Lethal Behavior in Autonomous Robots, Chapman-Hall, 2009.

[4] Notably Human Rights Watch, International Committee on Robot Arms Control (ICRAC) and Article 36.

in ethical behavior in the battlefield. Potential explanations for the persistence of war crimes include[5]: high friendly losses leading to a tendency to seek revenge; high turnover in the chain of command leading to weakened leadership; dehumanization of the enemy through the use of derogatory names and epithets; poorly trained or inexperienced troops; no clearly defined enemy; unclear orders where intent of the order may be interpreted incorrectly as unlawful; youth and immaturity of troops; external pressure, e.g., for a need to produce a high body count of the enemy; and pleasure from power of killing or an overwhelming sense of frustration. There is clear room for improvement and autonomous systems may help address some of these problems.

Robotics technology, suitably deployed may assist with the plight of the innocent non-combatant caught in the battlefield. If used without suitable precautions, however, it could potentially exacerbate the already existing violations by human soldiers. While I have the utmost respect for our young men and women warfighters, they are placed into conditions in modern warfare under which no human being was ever designed to function. In such a context, expecting a strict adherence to the Laws of War (LOW) seems unreasonable and unattainable by a significant number of soldiers.[6] Battlefield atrocities have been present since the beginnings of warfare, and despite the introduction of International Humanitarian Law (IHL) over the last 150 years or so, these tendencies persist and are well documented,[7] even more so in the days of CNN and the Internet. "Armies, armed groups, political and religious movements have been

[5] Bill, B. (Ed.), Law of War Workshop Deskbook, International and Operational Law Department, Judge Advocate General's School, June 2000; Danyluk, S., "Prevent- ing Atrocities", Marine Corps Gazette, Vol. 8, No. 4, pp. 36–38, Jun 2000; Parks, W.H., "Crimes in Hostilities. Part I", Marine Corps Gazette, August 1976; Parks, W.H., "Crimes in Hostilities. Conclusion", Marine Corps Gazette, September 1976; Slim, H., Killing Civilians: Method, Madness, and Morality in War, Columbia University Press, New York, 2008.

[6] Surgeon General's Office, Mental Health Advisory Team (MHAT) IV Operation Iraqi Freedom 05–07, Final Report, Nov. 17, 2006.

[7] For a more detailed description of these abhorrent tendencies of humanity discussed in this context, see Arkin, R.C., "The Case for Ethical Autonomy in Unmanned Systems", Journal of Military Ethics, 9:4, pp. 332–341, 2010.

killing civilians since time immemorial".[8] "Atrocity… is the most repulsive aspect of war, and that which resides within man and permits him to perform these acts is the most repulsive aspect of mankind".[9] The dangers of abuse of unmanned robotic systems in war, such as the Predator and Reaper drones, are well documented; they occur even when a human operator is directly in charge.[10]

Given this, questions then arise regarding if and how these new robotic systems can conform as well as, or better than, our soldiers with respect to adherence to the existing IHL. If achievable, this would result in a reduction in collateral damage, i.e., noncombatant casualties and damage to civilian property, which translates into saving innocent lives. If achievable this could result in a moral requirement necessitating the use of these systems. Research conducted in our laboratory[11] focuses on this issue directly from a design perspective. No claim is made that our research provides a fieldable solution to the problem, far from it. Rather these are baby steps towards achieving such a goal, including the development of a prototype proof-of-concept system tested in simulation. Indeed, there may be far better approaches than the one we currently employ, if the research community can focus on the plight of the noncombatant and how technology may possibly ameliorate the situation.

As robots are already faster, stronger, and in certain cases (e.g., Deep Blue, Watson, [12]) smarter than humans, is it really that difficult to believe they will be able to ultimately treat us more humanely in the battlefield than we do each other, given the persistent existence of atrocious behaviors by a significant subset of human warfighters?

[8] Slim, H., Killing Civilians: Method, Madness, and Morality in War, Columbia University Press, New York, 2008, p. 3.

[9] Grossman, D., On Killing: The Psychological Cost of Learning to Kill in War and Society, Little, Brown and Company, Boston, 1995, p. 229.

[10] Adams. J., "US defends unmanned drone attacks after harsh UN Report", Christian Science Monitor, June 5, 2010; Filkins, D., "Operators of Drones are Faulted in Afghan Deaths", New York Times, May 29, 2010; Sullivan, R., "Drone Crew Blamed in Afghan Civilian Deaths", Associated Press, May 5, 2010.

[11] For more information see Arkin, R.C., Governing Lethal Behavior in Autonomous Systems, Taylor and Francis, 2009.

[12] http://en.wikipedia.org/wiki/Deep_Blue_chess_computer, http://en.wikipedia.org/wiki/Watson_computer.

Why Technology Can Lead to a Reduction in Casualties on the Battlefield

Is there any cause for optimism that this form of technology can lead to a reduction in non-combatant deaths and casualties? I believe so, for the following reasons.

- The ability to act conservatively: i.e., they do not need to protect themselves in cases of low certainty of target identification. Autonomous armed robotic vehicles do not need to have self-preservation as a foremost drive, if at all. They can be used in a self-sacrificing manner if needed and appropriate without reservation by a commanding officer. There is no need for a "shoot first, ask-questions later" approach, but rather a "first-do-no-harm" strategy can be utilized instead. They can truly assume risk on behalf of the noncombatant, something that soldiers are schooled in, but which some have difficulty achieving in practice.

  The eventual development and use of a broad range of robotic sensors better equipped for battlefield observations than humans currently possess. This includes ongoing technological advances in electro-optics, synthetic aperture or wall penetrating radars, acoustics, and seismic sensing, to name but a few. There is reason to believe in the future that robotic systems will be able to pierce the fog of war more effectively than humans ever could.
- Unmanned robotic systems can be designed without emotions that cloud their judgment or result in anger and frustration with ongoing battlefield events. In addition, "Fear and hysteria are always latent in combat, often real, and they press us toward fearful measures and criminal behavior".[13] Autonomous agents need not suffer similarly.
- Avoidance of the human psychological problem of "scenario fulfillment" is possible. This phenomenon leads to distortion or neglect of contradictory information in stressful situations, where humans use new incoming information in ways that only fit their pre-existing belief patterns. Robots need not be vulnerable to such patterns of

[13] Walzer, M., Just and Unjust Wars, 4th ed., Basic Books, 1977.

premature cognitive closure. Such failings are believed to have led to the downing of an Iranian airliner by the USS Vincennes in 1988.[14]
- Intelligent electronic systems can integrate more information from more sources far faster before responding with lethal force than a human possibly could in real-time. These data can arise from multiple remote sensors and intelligence (including human) sources, as part of the Army's network-centric warfare concept and the concurrent development of the Global Information Grid. "Military systems (including weapons) now on the horizon will be too fast, too small, too numerous and will create an environment too complex for humans to direct".[15]
- When working in a team of combined human soldiers and autonomous systems as an organic asset, they have the potential capability of independently and objectively monitoring ethical behavior in the battlefield by all parties, providing evidence, and reporting infractions that might be observed. This presence alone might possibly lead to a reduction in human ethical infractions.

Addressing Some of the Counter-Arguments

But there are many counterarguments as well. These include the challenge of establishing responsibility for war crimes involving autonomous weaponry, the potential lowering of the threshold for entry into war, the military's possible reluctance to give robots the right to refuse an order, proliferation, effects on squad cohesion, the winning of hearts and minds, cybersecurity, proliferation, and mission creep.

There are good answers to these concerns I believe, and are discussed elsewhere in my writings.[16] If the baseline criteria become outperforming humans in the battlefield with respect to adherence to IHL (without mission performance erosion), I consider this to be ultimately attainable, especially under situational conditions where bounded morality [narrow,

[14] Sagan, S., "Rules of Engagement", in Avoiding War: Problems of Crisis Management (Ed. A. George), Westview Press, 1991.

[15] Adams, T., "Future Warfare and the Decline of Human Decisionmaking", in Parameters, U.S. Army War College Quarterly, Winter 2001–02, pp. 57–71.

[16] E.g., Arkin, R.C, op. cit., 2009.

highly situation-specific conditions] applies,[17] but not soon and not easily. The full moral faculties of humans need not be reproduced to attain to this standard. There are profound technological challenges to be resolved, such as effective in situ target discrimination and recognition of the status of those otherwise hors de combat, among many others. But if a warfighting robot can eventually exceed human performance with respect to IHL adherence, that then equates to a saving of noncombatant lives, and thus is a humanitarian effort. Indeed, if this is achievable, there may even exist a moral imperative for its use, due to a resulting reduction in collateral damage, similar to the moral imperative Human Rights Watch has stated with respect to precision-guided munitions when used in urban settings.[18] This seems contradictory to their call for an outright ban on lethal autonomous robots[19] before determining via research if indeed better protection for non-combatants could be afforded.

Let us not stifle research in the area or accede to the fears that Hollywood and science fiction in general foist upon us. By merely stating these systems cannot be created to perform properly and ethically does not make it true. If that were so, we would not have supersonic aircraft, space stations, submarines, self-driving cars, and the like. I see no fundamental scientific barriers to the creation of intelligent robotic systems that can outperform humans with respect to moral behavior. The use and deployment of ethical autonomous robotic systems is not a short-term goal for use in current conflict, typically counterinsurgency operations, but rather will take considerable time and effort to realize in the context of interstate warfare and situational context involving bounded morality.

A Plea for the Non-combatant

How can we meaningfully reduce human atrocities on the modern battlefield? Why is there persistent failure and perennial commission of war crimes despite efforts to eliminate them through legislation and advances in training? Can technology help solve this problem? I believe that simply

[17] Wallach, W. and Allen, C., Moral Machines: Teaching Robots Right from Wrong, Oxford University Press, 2010.

[18] Human Rights Watch, "International Humanitarian Law Issues in the Possible U.S. Invasion of Iraq", Lancet, Feb. 20, 2003.

[19] Human Rights Watch, "Losing Humanity: The Case Against Killer Robots", Nov. 19, 2012.

being human is the weakest point in the kill chain, i.e., our biology works against us in complying with IHL. Also, the oft-repeated statement that "war is an inherently human endeavor" misses the point, as then atrocities are also an inherently human endeavor, and to eliminate them we need to perhaps look to other forms of intelligent autonomous decision-making in the conduct of war. Battlefield tempo is now outpacing the warfighter's ability to be able to make sound rational decisions in the heat of combat. Nonetheless, I must make clear the obvious statement that peace is unequivocally preferable to warfare in all cases, so this argument only applies when human restraint fails once again, leading us back to the battlefield.

While we must not let fear and ignorance rule our decisions regarding policy towards these new weapons systems, we nonetheless must proceed cautiously and judiciously. It is true that this emerging technology can lead us into many different futures, some dystopian. It is crucially important that we not rush headlong into the design, development, and deployment of these systems without thoroughly examining their consequences on all parties: friendly forces, enemy combatants, civilians, and society in general. This can only be done through reasoned discussion of the issues associated with this new technology. Toward that end, I support the call for a moratorium to ensure that such technology meets international standards before being considered for deployment as exemplified by the recent report from the United Nations Special Rapporteur on Extrajudicial, Summary, or Arbitrary Executions.[20] In addition, the United States Department of Defense has recently issued a directive[21] restricting the development and deployment of certain classes of lethal robots, which appears tantamount to a quasi-moratorium.

Is it not our responsibility as scientists and citizens to look for effective ways to reduce man's inhumanity to man through technology? Where is this more evident than in the battlefield? Research in ethical military robotics can and should be applied toward achieving this end. The advent of these systems, if done properly, could possibly yield a greater adherence to the laws of war by robotic systems than from using soldiers of flesh and blood alone. While I am not averse to the outright banning of

[20] Christof Heyns, Report of the Special Rapporteur on Extrajudicial, Summary, and Arbitrary Execution, United Nations Human Rights Council, 23rd Session, April 9, 2013.

[21] United States Department of Defense Directive Number 3000.09, Subject: Autonomy in Weapons Systems, November 21, 2012.

lethal autonomous systems in the battlefield, If these systems were properly inculcated with a moral ability to adhere to the laws of war and rules of engagement, while ensuring that they are used in narrow bounded military situations as adjuncts to human warfighters, I believe they could outperform human soldiers with respect to conformance to IHL. The end product then could be, despite the fact that these systems could not ever be expected to be perfectly ethical, a saving of noncombatant lives and property when compared to human warfighters' behavior.

This is obviously a controversial assertion, and I have often stated that the discussion my research engenders on this subject is as important as the research itself. We must continue to examine the development and deployment of lethal autonomous systems in forums such as the United Nations and the International Committee of the Red Cross to ensure that the internationally agreed upon standards regarding the way in which war is waged are adhered to as this technology proceeds forward. If we ignore this, we do so at our own peril.

The Way Forward?

It clearly appears that the use of lethality by autonomous systems is inevitable, perhaps unless outlawed by international law—but even then enforcement seems challenging. But as stated earlier, these systems already exist: the Patriot missile system, the Phalanx system on Aegis class cruisers, anti-tank mines, and fire-and-forget loitering munitions all serve as examples. A call for a ban on these autonomous systems may have as much success as trying to ban artillery, cruise missiles, or aircraft bombing and other forms of standoff weaponry (even the crossbow was banned by Pope Innocent II in 1139[22]). A better strategy perhaps is to try and control its uses and deployments, which existing IHL appears at least at first glance to adequately cover, rather than a call for an outright ban, which seems unenforceable even if enacted.

The horse is out of the barn. Under current IHL, these systems cannot be developed or used until they can demonstrate the capability of adequate distinction, proportionality, and shown that they do

[22] Royal United Services Institute for Defence and Security Studies, "The Ethics & Legal Implications of Unmanned Vehicles for Defence and Security Purposes", Workshop webpage, held Feb. 27, 2008, http://www.rusi.org/events/ref:E47385996DA7D3, accessed 5/12/2013.

not produce unnecessary suffering, and must only be used given military necessity. Outside those bounds, any individuals responsible should be held accountable for violations of International Humanitarian Law, whether they are scientists, industrialists, policymakers, commanders, or soldiers. As these systems do not possess moral agency, the question of responsibility becomes equated to other classes of weapon systems, and a human must always ultimately bear responsibility for their use.[23] Until it can be shown that the existing IHL is inadequate to cover this RMA, only then should such action be taken to restructure or expand the law. This may be the case, but unfounded pathosdriven arguments based on horror and Hollywood in the face of potential reductions of civilian casualties seems at best counterproductive. These systems counterintuitively could make warfare safer in the long run to the innocents in the battlespace, if coupled with the use of bounded morality, narrow situational use, and careful graded introduction.

Let it be restated that I am not opposed to the removal of lethal autonomous systems from the battlefield, if international society so deems it fit, but I think that this technology can actually foster humanitarian treatment of noncombatants if done correctly. I have argued to those that call for a ban, they would be better served by a call for a moratorium, but that is even hard to envision occurring, unless these systems can be shown to be in clear violation of the LOW. It's not clear how one can bring the necessary people to the table for discussion starting from a position for a ban derived from pure fear and pathos.

For those familiar with the Martens clause[24] in IHL, a case could be made that these robotic systems potentially 'violate the dictates of the public conscience'. But until IHL lawyers agree on what that means, this seems a difficult course. I do believe, however, that we can aid the plight of non-combatants through the judicious deployment of these robotic

[23] Cf. Arkin, R.C., "The Robot Didn't Do it.", Position Paper for the Workshop on Anticipatory Ethics, Responsibility, and Artificial Agents, Charlottesville, VA., January 2013.

[24] The clause reads "Until a more complete code of the laws of war is issued, the High Contracting Parties think it right to declare that in cases not included in the Regulations adopted by them, populations and belligerents remain under the protection and empire of the principles of international law, as they result from the usages established between civilized nations, from the laws of humanity and the requirements of the public conscience." (Available at the ICRC website, http://www.icrc.org/eng/resources/documents/misc/57jnhy.htm, last visited on 30 April 2013).

systems, if done carefully and thoughtfully, particularly in those combat situations where warfighters have a greater tendency or opportunity to stray outside International Humanitarian Law. But what must be stated is that a careful examination of the use of these systems must be undertaken now to guide their development and deployment, which many of us believe is inevitable given the ever-increasing tempo of the battlefield as a result of ongoing technological advances. It is unacceptable to be "one war behind" in the formulation of law and policy regarding this revolution in military affairs that is already well underway. The status quo with respect to human battlefield atrocities is unacceptable and emerging technology in its manifold forms must be used to ameliorate the plight of the noncombatant.

CHAPTER 18

The Political Economy of Bots: Theory and Method in the Study of Social Automation

Samuel Woolley

Introduction

Automation forms a foundational part of the device network known as the Internet. Computerized algorithms determine what users see on their Facebook news feed or the Uber map. Near instantaneous communication occurs via pre-arranged technical networks of nodes and edges. The rise of the Internet of Things (IoT) means that automated connection and organization will only grow, and the effects upon society are manifold. How, though, are particular—even racist, classist, or sexist—values programmed in at the algorithmic level of the Internet? How might digitally enabled automation be used for user manipulation? Could the devices in the IoT be used to eavesdrop on publics or predict their behavior? Could automated online personas—social bots—be used to alter public opinion?

S. Woolley (✉)
University of Washington, Seattle, WA DC, USA
e-mail: samwooll@uw.edu

R. Kiggins (ed.), *The Political Economy of Robots*, International Political Economy Series, https://doi.org/10.1007/978-3-031-95171-8_18

Questions like these, and the pervasive nature of automation in our daily lives, lie at the center of this chapter. Specifically, I focus on the rise of political bots—software-driven social actors that can be used to spread propaganda and obstruct activism. I draw together in-depth analyses of literature on political communication, science and technology studies, qualitative research methods, and bot history in order to develop a multi-faceted understanding of the political economy of the political bot. Mapping continuities between political communication and the current literature on technology and politics illuminates how uses of social bot technology interact with ideas related to communication, power, and structures of agency. Analysis of literature from science and technology studies, especially from the emerging sub-field of ethnography of information, explains how qualitative methods and historical context can help us to understand the phenomenon of the political bot.

In order to understand the ways in which historical context, political communication, and qualitative methods can help build an understanding of bots, it is first important to understand what a bot is. In the simplest sense, bots are pieces of automated software used to do tasks online on behalf of human coders. They can crawl the Internet for information on a particular topic or perform repetitive tasks that would otherwise take a human day. Social bots are a particular kind of bot, unique in that they have a front-facing function wherein they communicate with real human users. Social bots are closely related to chat bots, programs build to talk to people across numerous parts of the Internet, but are different in that they only exist on social networking services (SNS) like WeChat, Twitter, Weibo, VK, and Facebook.

The ability to build bots, whether as automated fetching software or as chat capable personas, is bound up in the digital divide—while anyone can hypothetically learn to code from open educational resources, the cost of equipment, the free time necessary, and the background knowledge required places limits upon the sorts of people who can develop such skills. In other words, to create bots Internet users must have a good understanding of computer code and the time and money to do so. Furthermore, a high degree of Internet literacy is required to simply identify an automated social media account—in the form of a political bot—on a site like Twitter. Politics exists in the very ability to build a social bot, let alone launch one for political means.

It's not surprising, then, that the powerful—here those with resources required to build, deploy, and sustain bots—can use bots in attempts to control those who don't have a sophisticated knowledge of coding and the Internet. Social bots have been harnessed by technologically adept marketers to send blatant spam in the form of automatically messaged or recycled advertising content since the beginning of Twitter and on the social platforms that came before (Chu et al. 2010). Politicians have, in the last several years, taken note of and emulated celebrity Twitter users' tactics of purchasing massive numbers of bots to significantly boost follower numbers (Chu et al. 2012). Militaries, state-contracted firms, and elected officials worldwide now use political bots to invasively spread various forms of propaganda and flood newsfeeds with political spam (Cook et al. 2014; Forelle et al. 2015).

Recent research reveals the pervasive breadth of global political bot use across online social networks (Boshmaf et al. 2011). Automated Political bots have been the main tools for online astroturf—or fake grassroots—smear campaigns during political moments worldwide: the US midterm elections of 2010 (Metaxas and Mustafaraj 2012), the ongoing crisis in Syria (Qtiesh 2011), and the 2014–2015 disputes over Crimea (Alexander 2015).

Politically oriented bots are emerging phenomena and are amongst the most important recent innovations in political strategy and communication technology. Bots are prevalent, and active, in social media conversations—and their presence in these spaces continues to grow. The noise, spam, and manipulation inherent in many bot deployment techniques threaten to disrupt civic conversations and organization worldwide—jeopardizing equal participation in the public sphere and democracy more generally.

A Brief History of Bots

An understanding of bot history enables insight into how political bots came to be. Bots have existed online as long as the Internet has been public. In the 1990s, they played a prominent role in Internet Relay Chat (IRC), an early chat system online. Bots on IRC appear as a normal user to others on a channel. In other words, they function as chat bots—though they are mainly used to perform automated functions on behalf of other users. IRC bots perform several useful tasks: keeping the IRC from being overtaken by malicious or foreign users, preventing other types of IRC warfare such as malicious file sharing, keeping the IRC open while no other users are online, and controlling who is able to join the channel.

Robey Pointer is credited by some as having developed the first bot—"Eggdrop"—for IRC regulation in 1993. This program is, to date, the oldest IRC bot still in active development.

Throughout the 1990s and early 2000s, the number of bots active across the Internet grew exponentially. According to Stone-Gross et al. (2009), over 6 months of 2004 the number of new bots launched online each day rose from 2,000 to 30,000. Search engines like Google built their platform around the ability to use bots in a connective way; the Googlebot is a Internet-crawling bot that builds out the platform's database of Internet sites. Advertisers, marketers, and scammers also began to use bots in high numbers. Automated emails and messages containing sales links on chat sites are only one way these groups have used bots to spread their content.

Newer online social platforms have mimicked IRC by also building bots into their core functions. Twitter, for instance, has an open application programming interface (API) that allows coders to easily build and launch bots on the platform. Bots can act as useful tools on a SNS and social bots have been constructed that draw attention to social issues, make jokes, and send out news. The last 5 years, however, have seen social bots deployed for purposes of political manipulation.

There are unique benefits of socially oriented automation, and the more general use of bots and automated technology for aggressive means is also not new. The development of malicious uses for bots came alongside that of more strictly utilitarian bots like Eggdrop. Both denial of service attacks (DoS) and distributed denial of service (DDoS) attacks function via the use of bots and botnets—networked collections of bots.

It's important to draw a distinction between botnets and social bots. Both are software driven and automated, but they diverge in the interactive arena. Whereas botnets are designed to simultaneously attack a particular Internet site or group of sites, social bots tend to be independently functioning agents designed to have front-end interactions with human users. The "bot" in botnets refers to an infected computer used to log on to a site in conjunction with other bot computers; the bot in social bot refers to an automated persona—though one that might exist in a network with other stand-alone personas.

Today, both the number of bots online and their diversity is astounding. *The New Yorker* sums up this trend well: "bots, whose DNA can be written in nearly any modern programming language, live on cloud servers, which never go dark and grow cheaper by the day" (Dubbin

2013, para. 1). Filippo Menczer, a professor at the Indiana University School of Informatics and Computing and one of the early researchers of politicized social bot technology, makes a similar point on the rise of the social bot: “Bots are getting smarter and easier to create, and people are more susceptible to being fooled by them” (Urbina 2013).

A Bot Timeline

- 1950: Alan Turing publishes “Computing Machinery and Intelligence.”
- 1954: The “Georgetown Experiment” successfully converts over 60 Russian sentences to English automatically.
- 1964–1970: SHRDLU, an early natural language processing (NLP) program that uses pre-programmed handwritten NLP rules, is created.
- 1967–1969: ELIZA, another early natural language processing program, is created.
- 1972: PARRY, an early interactive chat bot, is created.
- 1975–1981: Several “conceptual ontologies,” which translate embodied information into computational data, are created.
- 1984: Racter, an early automatic prose generator, is created.
- 1988–1989: Machine learning algorithms begin to revolutionize the field of natural language processing.
- 1989: IBM research spearheads the use of statistical models in machine translation.
- 1993: The Eggdrop IRC bot is created.
- 1994: A.L.I.C.E. (Artificial Linguistic Internet Computer Entity) is created.
- 1997: Deep Blue, a chess-playing computer, beats Grandmaster Garry Kasparov.
- 1997: Cleverbot, a bot that learns from human input and interaction, is launched.
- 2006: IBM runs initial tests for its question-answering system Watson.
- 2011: Supporters of Republican Senate candidate Scott Brown use Twitter bots to attack, and spread propaganda, about opponent Martha Coakley (Metaxas and Mustafaraj 2012).
- 2011: @Horse_ebooks, allegedly a Twitter bot, becomes an Internet phenomenon.

- 2012: Journalists at Gawker reveal that @Horse_ebooks is human-driven.
- 2012: Google Now, a digital personal assistant, is launched.
- 2012–2015: Mexican government officials use Twitter bots to silence dissent. Opponents respond with their own bot-driven attacks (Orcutt 2012).
- 2013: US Presidential candidate Mitt Romney is accused of using bots to boost his Twitter following (Coldewey 2012).
- 2014: Bot Eugene Goostman passes the Turing test.
- 2014: Humor bot @oliviataters is launched by comedian Ron Dubbin.
- 2014: A Twitter Securities and Exchange report (SEC) reveals that 12% of the accounts on the platform are bots (Motti 2014).
- 2014: Turkish Prime Minister Recep Tayyip Erdogan allegedly authorizes thousands of bots to support his party and attack opponents (Medeiros 2014).
- 2015: Reddit Moderator Bots on the /technology subreddit allegedly delete comments related to the Snowden revelations and the US National Security Administration (Alfonso 2014).
- 2015: Coca Cola's bot-driven Twitter account tweets sections of "Mein Kampf."
- 2016: Facebook launches a bot platform for Messenger.
- 2016: CEO of Messenger app Kik predicts a "bot goldrush."
- 2016: Microsoft's Twitter bot "Tay," billed as an AI chatbot, is fooled into publicly tweeting racist, misogynist, and generally offensive content.
- 2016: US Presidential candidate Donald Trump is accused of using bots to attack opponents and boost his online credibility (Woolley & Howard, 2016).

Automated Actors and Political Communication

According to Andrew Chadwick (2013), any study of media—whether digital, broadcast, or print—and politics should "foreground complexity, interdependence, and transition" (p. 4). Chadwick argues that it is the broader evolutions in the technical and social spheres that have led to the construction of a "hybrid" media system that has altered the dynamic of political communication. Socio-technical innovation has led to ever-changing political affordances of this multi-media landscape, including the uses of the Internet, networks, algorithms, online data caches, and more (Karpf 2012, 7). Political bots represent the most recent trend in the "hybrid organization" that is central to the study and understanding of media and politics today. The automation and computational power behind political bots afford their users the power to massively magnify their message: whether for good or ill.

The rise of hybridized technology and the "networked society" has not only affected the way political conversations occur but has also altered the way political campaigns are organized, how elections function, and how power is exerted (Benkler 2006). New political organizations have been birthed, political systems have changed, and politicians have risen and fallen. Some aspects of political communication, however, remain constant.

David Kaprf (2012) points out that "even the most radical changes to communications systems must be channeled through [pre-existing political] structural constraints in order to impact traditional political outcomes" (p. 10). He argues that political normalization has occurred alongside innovation and change. Newly powerful individuals and groups have ascended, but the global political elite remains the same—it has simply adjusted to, or co-opted, the altered state of political communication. Hindman (2008) takes this argument further, suggesting that democracy has not played out online as cyber-democracy optimists had hoped. Rather, he contends, the elite on the Internet are still elite and thus "online speech follows winner-take-all patterns" (p. 4).

In the face of large-scale changes to political communication, many actors therefore remain the same, albeit with alterations. Despite the rise of a multi-platform and multi-mediated experience, the traditional concerns of political communication (Chadwick 2013). Both older and newer media matter, but the distinctions between them, and between traditional and non-traditional political actors, are blurring (p. 4). However, the classical political communication study of the interactions between political actors, media, and public should provide a foundation for beginning to study both modified and new units of analysis such as political bots.

Units of Analysis: Old, New, and Aggregate

Contemporary political communication is currently in a transitional space, but perhaps this is simply a function of its being a field that focuses on the ever-evolving problem of interaction and power. The units of analysis for this subject are similarly in a state of revision and adaptation. Communication between political actors, media, and citizenry must still be interrogated, but it should be mapped onto new and emergent ways of interacting. Studying countries, individual political actors, and clusters of citizenry and politicians is still a worthwhile endeavor but such an exercise becomes more nuanced when considered alongside hybrid techno-social formations like bots. Political communication must therefore focus its attention on new units of analysis in order to better comprehend the current state of affairs. It is essential that the discipline also investigates the political role of networks, software, and algorithms.

To expand analyses in political communication as a field one must understand the interconnected nature of units of analysis—such as political actor groups, campaign managers, for instance—and move towards understandings of these groups as broader, and often digital, networks (Howard 2006). The study of computer code as politically and culturally imbued is crucial to building knowledge around both digital network functionality and digital network control (Gillespie 2011). Software and algorithms, as political units of analysis, offer new routes for understandings of power and control exerted at the level of code. Such research is crucial in developing understandings of political bots, and of the larger role of digital automation upon the digital political sphere can be analyzed as pure source code, as a formula for making decisions in a mediated environment, or as a code-driven product, like a computer bot. Geiger (2014) maintains, in fact, that the study of bots and "bespoke" code can aid researchers in understanding the material nature of software.

The battle for control of the online sphere, and the political power it affords, is taking place at all levels of the Internet (Benkler 2006). These levels include "the physical devices and network channels necessary to communicate; the existing information and cultural resources out of which new statements must be made; and the logical resources—the software and standards" (Benkler, p. 23). New units of political communication analysis, such as software and algorithms, are plotted along the lines of these physical, informational, and cultural layers. They are best interrogated alongside the traditional units of focus of political communication, such as countries, political parties, and political organizations.

Understanding Networks

Benkler (2006) argues that the rise of technical infrastructure and the networked society has the potential to revolutionize our economic, political, and cultural systems. Taylor Owen (2015) supports Benkler's assertions, arguing that the proliferation of networks and networked information has enabled a variety of "disruptive" power invested in nontraditional individual and loosely grouped political actors. He claims that networks have allowed individuals to "do things that replace and threaten existing institutions in all areas of international affairs, including: development, war, diplomacy, finance, international reporting, and activism" (p. 4). Ananny (2014), offering useful insight into the forth estate, analyzes how changes to networks have affected journalists, arguing that these alterations affect the way people in the news media interact with publics. Networks, then, complicate understandings of traditional units of political communication analysis, in this case, powerful institutions.

Networks can be examined in part or as sprawling units. The political communication researcher might, for instance, work to understand individual nodes (for example, politicians) and their relationships to one another. Castells (2007) argues that consistently changing networks of nodes have allowed for large-scale changes to communication and power, requiring scholars to dissect them in order to generate social understanding. Networks can also be understood as clusters (in various states of fragmentation and interconnection) that can be compared to one another. Himelboim (2008) conducts just such a comparative analysis and argues that the way networks are structured and related affects civic discussions and information sharing online (p. 156).

Scholars of political communication have made use of the wealth of research on social networks to begin to understand the Internet-mediated social sphere. Howard (2006) examines contemporary political campaigns by qualitatively studying the novel organizational formations functioning via and alongside new media. He calls this endeavor "networked ethnography" and argues that it is useful in developing understanding that is both rich in description and complex in scope. Kreiss et al. (2011) argue against the idea that "networked information exchange necessarily levels the social playing field, or that networked modes of social action are replacing their industrial antecedents." (244). Rather, they suggest that analyses of political networks reveal both normalization and the development of new power structures.

Bennett et al. (2011) assert that when a network of political agents takes on a particular argument or story "structural dynamics involving narrative choices, conflicts, and strategies can lead comparable networks to diverge even as they espouse the same cause" (p. 297). This means that online political groups, formal or informal, face particular challenges when attempting to make use of online means of organization. Bennett & Segerberg (2012) take this argument further, suggesting that new network forms make for a "connective" action that is more focused on the individual than the "collective" group. They contend that connective action is enabled by the rise of technological platforms like Twitter, Facebook, and YouTube, which allow for organization. This action prizes the rise of individual organization due to the de-anonymized and personally focused nature of such SMS.

Scholars of political communication can work to generate new knowledge on networked interaction by using established modes of social network analysis in tandem with qualitative means for interpreting the behavior of particular actors or groupings. Howard's (2006) hybrid method is useful to this end because it takes into account the dispersed and multifaceted nature of communication between and within campaigns. Owen (2015) develops an understanding of the networked power of "disruption" by situating new Internet-bred political groups like Anonymous—well-known users of political bots—in a broader history and context (pp. 18–20). Networks are, by nature, constantly changing formations, a fact that makes them difficult to analyze. But by making use of Chadwick's (2013) notion of media hybridity, alongside hybrid methods and contextual case study, political communication researchers can work to avoid the pitfalls of overgeneralization and determinism.

Understanding Software and Algorithms

Gillespie (2011) suggests that algorithms are not only inherently political but also that "we are coming to rely on these algorithms [and] we want them to be neutral." According to Gillespie, algorithms frame information in a particular way. The way this framing occurs is often complex and sometimes unknowable to even the coder who builds particular software. Despite this, algorithms and software code (generally) can work to enforce a particular politics. By analyzing source code, political communication scholars can understand aspects of the intent of the coder and possibilities for human–machine communication. More recently, Gillespie (2014) has suggested that algorithms are now a communications technology, like broadcast television or online newspapers, and are "caught up in the ways we ratify knowledge of civic life" (pp. 167–168). He makes a clear case for the role of algorithms in political communication.

The unit of analysis for computer code, software, or algorithms could be chunks of script, software units like social bots, or the front-end infrastructure of online social platforms, which are a manifestation of backend code. Woolley & Howard (forthcoming) are working to establish the social bot as a unit of analysis amongst scholars of communication. They argue that politicized social bots are amongst the most important developments in the technologically oriented study of political communication and suggest that scholars of all types should work to understand the ways these coded political actors are affecting public opinion worldwide. Gillespie's (2011) arguments on the political nature of algorithms suggest that scholars should study platforms' frontend interfaces, an arena often populated by bots, and backend code together in order to generate a more holistic understanding of the social or communicative role of code. According to Gillespie, algorithms control many facets of interaction on social media sites and therefore they can also affect public opinion and political discourse.

Literature from the computer and information sciences has developed practices for understanding the role of code in human–computer interaction. Preece et al. (1994) present an overarching explanation of human–computer interaction and offer a scholastic starting point for understanding the exchange between humans and code. Schneiderman and Plaisant (2009) offer a section on the potential social and individual implications of instituting particular software or design, acknowledging that outcomes are often hard to predict and can be reliant on the user (pp. 423–434). Suchman (2006) has built theory on understanding the social role of technology, explaining the inherent role designers and publics play in machine production. She also explains the ways in which machines can affect social formations. Nardi (1995) compares different theories of human–machine communication and suggests activity theory as a useful method for the naturalistic anthropological study of social machinery and people.

The work of Barocas et al. (2013) takes up the notion of algorithms as a unit of analysis and points to the algorithmic categories established by Gillespie (2014) as useful in differentiating between both type and potential effect. Gillespie works to map out a six-part overview of what he calls "public relevance algorithms," that is, those that affect socio-political processes via the particular framing of information. Geiger (2014) advocates for building understandings of online and offline coordination and socio-computational practices by analyzing the source code that runs alongside social platforms. He argues that this code, often in the form of automated bots, plays a functional and constructive role on websites like Wikipedia and should thus be analyzed in, at least partially, social, and communicative terms. Berry (2012) takes ideas of code-driven political even further. He suggests that, in order to appreciate contemporary international political economy, one must first comprehend computer code. For Berry, an understanding of code is integral to developing understandings of the financial systems, multinational companies, government organizations, and non-governmental organizations that now so thoroughly rely on computational infrastructure.

Communication or socially-based analysis of scripts or algorithms is methodologically tricky in that it requires (a) access to scripts that are often proprietary or hidden and (b) a background in computer coding. The first problem suggests that building an understanding of algorithms should be supplemented with or methods and units of analysis in order that piece together a secondary picture of code functionality in order to offset the problem of copyrighted code (Gillespie 2009). The need for an understanding of code can be tackled by building interdisciplinary teams of social scientists and computer programmers. The United States Nation Science Foundation has begun supporting such hybrid projects with the intention of developing knowledge of this nuanced phenomenon (US NSF—Transformative Research, n.d.). Barocas et al. (2013) call for a more critical interrogation of the way in which social scientists understand and attempt to study algorithms. They acknowledge the effect of algorithms on politics, culture, and society but argue that algorithms must be demystified as an analytical category.

Understanding Modifications to Classical Units of Analysis

Political communication research that analyzes interactions between and within traditional units such as countries/nation-states, non-governmental organizations, and political parties remains relevant but should work in tandem with analyses of new digital units in order to gain nuance. Chadwick (2013) argues, to this end, that "in our analyses we should try as far as possible to integrate the roles played by older and newer media systems" (p. 4). The ways people communicate politics between one another are changing but so are the ways political actors are making use of seemingly mundane technology. Mukerji (2003) contends that the ability to exert a Weberian notion of power has changed in the age of technology. Engineered locales are, according to Mekerji, "places where power and its legitimacy can be questioned and undermined" (p. 656). Therefore, in order to understand traditional state power, we must understand reconceptualizations of technology both new and old.

Turner (2006) notes the radical politics of early cyber-utopianists and developers and describes how these beliefs affected the subsequent development of the Internet. For Turner, there has always been a cyber-optimist undercurrent to social understandings of the online sphere. This research prizes the histories and perspectives of people, of individuals and social groups, in understanding the politics of technology. This hybrid approach to understanding political sociology analyzes the politics of groupings of people as well as the ways in which people might use the Internet towards political ends.

According to Karpf (2012), new organizations have risen to online to both supplement and substitute those formally in power. Karpf argues that the Internet has changed the middle organizational layer of politics, but that the political role of publics and other political actors largely remains the same in the Internet age. He contends that sustained political action still requires organization and also that Benkler's (2006) individualized notion of networked freedom doesn't note this (p. 8). Hindman (2008) takes the argument about contemporary online political normalization further. He contends that "direct political speech on the Internet—by which I mean the posting of political views online by citizens—does not follow egalitarian patterns" and goes as far as to deem digital democracy a "myth" (p. 17). Manovich (2001) provides a framework for thinking about new media, arguing that all media were once new. This perspective maintains a historically contingent understanding of all media and notes the interesting ways different media are used to further the social, cultural, and political goals of people.

Methodologically, scholars of political communication are equipped to study the classical aspects of interaction between political actors, media, and publics. There is methodological work to be done, however, when it comes to understanding communication driven by complex—digitally based and human created—actors like political bots. Chadwick's (2013) argument that all communicative power players be situated within a larger hybrid media environment is useful in its nuance. His argument is neither about how the Internet has provided "revolutionary chance" nor "politics as usual" but something in-between, something hybrid (p. 5). This project suits the analysis of communication amongst classical political actors and of those enabled and defined by new and emergent technology.

Automation, Qualitative Methods, and STS

An air of both "innovation and anxiety" has surrounded the theorization and formulation of qualitative methods for studying digital society (Hine 2005, 3). Scholars have been concerned with whether old methods of research can be re-envisioned or if they needed to create entirely new methods to capture online interaction. Because methods are largely based upon precedent, the idea of generating brand new ways to carry out research is a daunting endeavor. Markham & Baym (2008) suggest that qualitative methods can be "reconsidered without reinventing" (p. xv). They believe that the foundational aspects of qualitative methodology remain useful, particularly naturally based ethnography. After all, scholars in the past were still interested in studying human interaction, even if it occurred online. "Deep hanging out," it seems, is still possible on the Internet, with new and researchable social venues materializing with frantic regularity (Geertz 1973).

A similar feeling of anxiety and innovation surrounds the literature focused around the rise of what some scholars call "ethnography of information." This emerging branch of qualitative inquiry is focused on the sociality of technology. In other words, it asks the question: what happens when human agency is not necessarily the main force of social interaction and organization? This research closes the gap between "online" and "offline" worlds, arguing that much of the world is online is some way, shape or form. Hine's (2005) argument that "an anthropological perspective on mediation is largely concerned to understand why some media are perceived as mediating and others are not" has been extended to accept the interconnected nature of mediation and sociality via the uses of, and interaction with, wired technology (p. 14). Ethnographers of information find understanding today, more than ever, in Hine's notion that the "real" is a colloquial and not an epistemological term. For this variety of qualitative researcher, the complex, contextual, and multiple nature of reality is further affirmed by the increasingly social nature of "intelligent" technology.

Reconsidering Without Reinventing: Key Concepts in Ethnography of Information

Ethnography of information moves away from a methodological focus on the primacy of human interaction and towards a framework for qualitatively understanding the intersection of social technology and human communication. The rise of "intelligent" technology has transformed the way humans interact with one another but also, and more singularly, the way humans interact with technological actors. Social bots, learning algorithms, the Internet of things, and cloud-enabled robotics are all examples of technology that have ever increasing, and ever more sophisticated, interactions with people. Advances in parallel computation, big data and increasing algorithmic nuance have driven investment in artificial intelligence from the business side of technology, and brought about an urgent need for new methods and theories of understanding amongst scholars (Kelly 2014).

Emerging methodological concepts in ethnography of information, such as trace ethnography and networked fieldsites, are tied to the ever-evolving social scientific practices of mixed methodology and creative qualitative inquiry. Each of these research practices has its background in the digital research methods represented by digital anthropology, virtual data gathering and Internet research. These ideas about human social interactions in the digital sphere are, in turn, drawn from the even larger compendium of research on qualitative methodology from the social sciences and, most specifically when discussed ethnography, sociology and anthropology. This trajectory, from field methods to digital methods to ethnography of information, must be understood fully in order to capture the new concepts that are emerging. New works concerned with the sociality of technology and techno-human relationships place importance on the concepts of materiality, representation, and sense-making, but so do traditional texts on qualitative method.

As Markham and Baym (2008) note, the process of generating new methods should be a process of reconsidering without reinventing. For them, qualitative researchers should make sense of present techniques via those that came before. Scholars should still work to produce clear questions regarding the culture of the site, group, and individuals of study, but they must be adaptable in how they answer them.

The work of Lindlof & Taylor (2010) is foundational to qualitative communication researchers. They place primacy on the processes of sense-making and research production while paying heed to both historical methodology and new tools. Lindlof and Taylor are, notably, concerned with the communication between people, and their overview of communication methods doesn't include much space for research on the social interaction capabilities of technology. For Lindlof and Taylor, media and technology studies is an "informal" field of research in communication and Actor Network Theory is quite "controversial." Their brief overview of Actor Network Theory notes its role in analyzing technical materiality. They explain that ANT-oriented scholars hold "that in social interaction, nonhuman actors play as significant a role as do nominally human actors" (pp. 51–52). Beyond contributing a richly structured sourcebook for understanding qualitative communication methods, Lindlof and Taylor have aided the growing field of ethnography of information by promoting an approach to qualitative methodology that is theoretically flexible and empirically rigorous.

Miles et al. (2013) provide an approach to qualitative data analysis at large, with less of a focus on designing research and one that is more targeted towards understanding the product (data) of inquiry. They present a "sourcebook" targeted at any scholars of any field or discipline hoping to conduct qualitative analysis, though they present a highly formalized and orderly approach to understanding data. This treatment, focused on discovery and prediction, places high importance on display via network organization and matrices. This precise method for gathering and displaying data is useful for scholars studying the social intersections of technology and humans in that it provides a thorough presentation of the fundamentals of data-oriented methods, most of which are translatable to documenting the sociality of technology.

Though ethnographers of information work to understand an emerging phenomenon, the social role of intelligent technology, they must still work to capture interaction. Qualitative research has been built to understand human society and culture, and the communication therein, but smart tools are only understood as such because they have emerged via interrelations with humankind (Latour and Venn 2002). Van Maanen (2011) offers all ethnographers an introduction to building the socio-cultural narratives that are so central to the method. Van Maanen's survey of ethnographic voices reveals the open-ended nature of recording and presenting fieldwork, thus revealing its adaptability for reconceptualized (or mostly new) methods for studying novel social interaction. In fact, he advocates for experimenting with and reflecting on the ways social reality is presented—methods and theory should not be static (pp. 1–7). For van Maanen representation via ethnography has always been tricky because "no transparency theory can be confirmed" (p. 6). The primacy he places upon the reflexive nature of research gives power to the scholar observing sociality but also makes clear the subjective nature of presenting the observed. This point, regarding representation, is particularly crucial when discussing ethnography of social technology because the concept of artificial intelligence is still just that, a concept. Until machines can fully speak for themselves (and, if this ability is realized, than certainly after as well) ethnographers must take care in presentations of machine intent and causality.

Today's proliferation of digital qualitative methods has grown out of the work undertaken by the social scientists and humanistic scholars throughout academic history. The more recent collection of texts on "virtual" ethnography, however, have had an even more concerted effect on the budding literature on ethnography of information. Scholars have struggled with the proper terminology for studying online culture, vacillating between methods that are "digital," "virtual," "net-based," "netnographic," and "data-centric," but also with how to represent online culture (Bowler 2010). Those interested in Internet methods early on were particularly wary of the separation between the "real" world offline and the artificial one online. This separation has since been criticized itself as artificial, especially in light of the ever-increasing "wired" nature of society.

Hine (2005) presents a collection on virtual methods that asks, and works to answer, the pressing questions faced by scholars at the time. What does Internet mediation do to human interaction? Do old methods of qualitative inquiry have any use? If so, how can they be amended to better facilitate online research? Are new methods needed? Ultimately, Hine suggests that the energy and excitement surrounding the study of interaction via a new medium be grounded in the idea that all qualitative methods are contextual to the site of study and must be adapted in situ. She argues that methods, far from being apolitical, are encoded with meaning by both researchers and publics. This representational move makes important distinctions about the way qualitative scholars choose to speak about sociality, whether technical or human. Hine contends that qualitative social researchers should be open to the ways technology might alter the social sciences, both epistemologically and methodologically (p. 9). This move creates a path for reflexive theorizing about how humans affect technology but also about how technology effects humans.

Markham and Baym (2008) argue that "qualitative research requires a tolerance for chaos, ambiguity and inductive thinking" (p. ix). By being flexible qualitative researchers are able to work towards contextual understandings of new phenomena or new ways of thinking about existing phenomena, such as is the case with theorizing the social capabilities of technology. It is they who suggest that scholars should place faith in the historical and situated methodologies built by those who came before. While perhaps imperfect given certain circumstance, these methods were also often constructed with adaptability and subjectivity in mind. They argue that research requires the ongoing balance of different dialectical tensions and that researchers must work to articulate what is gained and lost by particular arguments (p. xix). With this in mind, the polarization of the technical and the social can be complicated by way of the process of reflexive identification and comparison.

The argument that "technology is not one more iota mediated by the digital" is central to the work of Horst & Miller (2013) and suggests that the notion of greater pre-digital authenticity is a fallacy (pp. 3–4). For Horst and Miller that interaction which occurs via technology is still fundamentally social, pointing towards the idea that the social permeates all facets of the world as humans see it, mediated or no. This ontological point is important in theorizing the role machines play in knowledge generation but also in directing qualitative methods for understanding technical society towards greater reflexivity. While Horst and Method advocate for a reconsidered multi-methods approach, Kozinets (2010) suggests that digital methods need to be standardized via the creation of a "new" methodology called netnography. This approach is limiting, however, in that it provides less interpretive flexibility for those seeking to study social technology.

The concept of trace ethnography, presented by Geiger & Ribes (2011), is based on the idea that ethnographic researchers ought make use of the multitude of data online. Geiger and Ribes argue that this tactic "combines the richness of participant observation with massive amount of data in logs in order to reconstruct patterns and practices of users in sociotechnical systems" (p. 1). Ethnographers of information must find ways to work with trace social data in tandem with traditional fieldwork practices in order to comprehend the heterogeneous nature of the networked environment (p.1). Howard (2002) proposes the idea that ethnography itself be networked in nature. Howard's study of multi-site political campaigns and constantly traveling campaign staff highlights the role of understanding social interaction via networked means. The work of Burrell (2009) takes this argument further to suggest that fieldsites can be realized as occupying virtual, physical, and imagined spaces.

Vertesi (2008) has done significant work around the socially imbued concept of techno-scientific representation. She argues that formal and practical representations of objects (for instance, the London underground map) often diverge from the colloquial and enculturated view of said objects. In other words, technological mediation can change the way people experience things in their everyday life but so too can people remix the "planned" meaning of producers.

Vertesi (2015) continues the work of speaking to critical issues of scientific representation in her ethnographic work with the Mars Rover team. Not only is this work a multi-site and multi-method endeavor to capture a complex social arrangement of professionals from different backgrounds but is also an interrogation of the interactive role of scientific visuals during the Rover mission. Vertesi places importance on the analytic framework of "drawing as," in which scientists use digital technologies to draw the red planet in ways that have implication for future research and viewings (22). Vertesi argues that "drawn" images of Mars must adhere to communal norms and "constraints" in ways that shape both the subjects and objects of analysis (pp. 191–193).

Coleman (2014) conducts multi-year participant observation amongst the activist/hacking group Anonymous in a way that underscores the relationship between the both the loose collective of individuals/groups that form the larger organization and the digital technology that facilitates their practices. For Coleman, Anonymous would not exist without technological platforms like 4chan and indeed the larger array of computational tools. She suggests, however, that technology, and especially the infrastructure of Internet security, has been significantly changed by Anonymous. This relationship is indicative of the larger goal of ethnography of information in that it reveals the interactive social relationship between both machines and humans.

The "Technically" Social: Challenges to Theory and Method

As Nagy & Neff (2015) point out, scholars of science and technology studies face theoretical challenges in developing an understanding of the relationship between society and technology, especially a contemporary space where the two are often so intrinsically interconnected. Neff et al. (2012) argue that agency can be viewed as "technical in two senses, as in the agency that is possible by systems of technology and as a limiting description of that agency or latent agency, a not quite, but "technically," agency" (p. 300). For them, the politics of technological systems should be reorganized to better capture the social role of technology.

The perspective of technological "affordance" has been used in attempts to begin a theorization of this relationship as existing between technological and social determinism. Wellman et al. (2003) suggest that the social affordance of wireless technology and "ubiquitous connectivity" allow people the possibility to link "at any place." However, Nagy and Neff argue that the term "affordances" has often been used inappropriately as a cursory catch all for a more human-centric notion of, what is actually, a relationship between both human and smart technology. They suggest that deeper and clearer theorizing is needed to generate knowledge on the "materiality of technology" (p. 2). They put forth the idea of "invisible affordances" in order to challenge academic consensus around the term, remind scholars about the user's role in imagination of use, and to trouble the "static," human dependent, notion of affordance (p. 1–2). Deeper inquiry into the relationships between society and technology, and an understanding of these relations as both complex and symbiotic, will result in better theory and methods.

Hogan (2015) identifies the ideology implicit in "personalized" algorithmically sorted lists, arguing for a rethinking of single ranking and towards a more user centric perspective. He suggests that moves like this create a better relationship between machine and human user and result in less ideological power input at the site of coding. The work of Star & Ruhleder (1994) supports the assertion that computational design has particular organizational and social effects. By accepting the cultural role that both coders and algorithms play, scholars will have more success in understanding techno-social relationships from a view beyond the human-centric and towards the human–machine.

Many academics argue that "simplistic technological determinism" is still a major theme of discussion amongst media practitioners, administrators and pundits (Lee and Streeter 2012, 95–96). In his interview with Lee, Streeter argues, in fact, that the Internet embodies a clear set of social choices. Streeter, as a "fully throated social constructionist," is way off the affordances perspective and more concerned with historically situating change than arguing for technological causality (p. 96). This view, not uncommon to scholars throughout the social sciences, must be avoided in order to understand the ways in which non-human actors act upon humans.

Latour & Venn (2002) provide a more nuanced way to think about the role of nonhuman agents (from tools to animals to architecture) in social interaction. They argue that using tools is what endows people with humanity; that tools and people have coevolved. This approach, neither socially constructive nor technologically deterministic, acknowledges that all types of agents generate difference. Latour and Venn argue that this decentralizing move is beneficial to the social sciences, in that in enriches understandings of networks by understanding the ways in which nonhuman actors are connected to humans.

The question of labor must also be considered by critical scholars hoping to understand the rise of "technically" social non-human agents. Neff (2012) builds the understandings of the way people working in the technology industry accept risk in order to seek the elusive success of start-ups. Like Vertesi (2008), Neff suggests that culture is co-created between both those who create narratives about particular social artifacts and constructions (be they images or jobs) and those who consume or perform them. For Neff risk is socially constitutive because concepts are built complex networks of actors. Such understandings of social representation are useful in generating theory about, and methods for understanding, technological actors and humans because they place an emphasis on the complexity of meaning-making.

In order to fully comprehend the way meaning is constructed, scholars must continue to theorize about the materiality of non-human actors. By accepting the notion that technology can alter humans as much as humans alter technology the social sciences will build better understandings of the socio-informatic landscape. Vertesi's (2015) analytic framework of drawing, the process of interrelations between images of Mars and scientists, is useful in this regard. As machines become more intelligent, as algorithms improve and data processing is streamlined, the complexity of techno-social interactions will only grow. It's the role of science, technology, and society scholars to continue the work of asking hard questions and accepting methodological flexibility in order to more fully comprehend these interactions.

When Substrate Becomes Substance: Implications for Communication

Communications scholars should heed Hine's (2005) reminder "to capitalize on the potential that new technologies provide for social science itself, in examining epistemological and methodological commitments afresh, opening up possibilities of new research designs and new approaches, and seizing on the opportunity for reflexive engagement." (9) The implications of studying bots, automation, and algorithm are inherent in the emergent nature of research in ethnography of information. That said the precedent set by traditional methods and the skills within each method are useful to the communicative study of techno-sociality and human technology relationships. Many of the methods used for studying problems of communication including, but not limited to, ethnography will also prove useful in studying social aspects of information society. After all, as Hine notes "communication is intrinsic in all social science" and the discipline of communication is itself adaptable to change.

The study of smart technology—of bots, automation, and algorithms—opens up the study of communication to various theoretical and methodological challenges, but it also presents opportunities. Communication researchers are no longer restricted to studying human interaction but are instead faced with the creative task of building understandings of nonhuman communication. While there has been much work on animal communication, few scholars have examined the communication habits of intelligent technology. Latour (2012) has generated theory that allows for the social abilities of nonhuman actors. It is time, however, for communication scholars to start studying these social abilities by examining the interaction between human and non-human actors, such as via empirical ethnographic and mixed qualitative research into the communications between humans, algorithms, bots, and other automated technology.

The emergent nature of hybrid methods like trace and network ethnography provide a further challenge to communication scholars hoping to advance their enquiries via the ethnography of information. Because methods are built upon precedent there is still much work to do in generating new techniques and incorporating tested ones. Jankowski & van Selm (2008) suggest that communication scholars should work to build understandings across multiple methods, and they encourage cross-national research because, they argue, such dialogue tends to result in innovation. These methodological endeavors are the future of research on socio-technical questions. The subject matter, like that of all humanistic research and social science, is concerned within interaction and thus lies firmly within the purview of communication studies.

The work of Geiger and Ribes (2011) presents a mixed methodological template for understanding the complex interactions between code, bots, and humans. By collecting trace data through net logs, they argue, scholars can piece together a rich interactive history of individuals, platforms and organizations. Beyer (2014) provides scholars of communication, sociology, and beyond with a means for generating comparative understandings of groups formed and acting via technical means. According to Beyer, larger, diffuse, and anonymous online organizations can facilitate civic engagement in ways that are often unreachable within net-based platforms that prioritize individual or small group interaction. This point suggests that scholars ought to focus attention on specific aspects of particular technologies and examine the dual nature of cause and effect between such tools and human Internet users.

Conclusion

Political communication scholars must focus their research on new units of analysis, but they should also hybridize this research by examining the continued role of traditional political actors, media, and citizenry. The converse is also true. The classical study of communication and politics must work to understand the ways new technologies, and aspects of these technologies, are used by emerging political actors. Chadwick (2013) argues that "hybrid thinking rejects simple dichotomies, nudging us away from 'either/or' patterns of thought and towards 'not only/but also' patterns of thought" (p. 4). This inclusive notion of political communication can only be useful to those attempting to understand today's complex communication landscape.

In order to understand the role of complex mediated networks, computer code, and elements of Internet infrastructure, political communication, researchers must pay attention to political normalization and the continued exertion of power. This phenomenon plays out via interconnected and "wired" spaces and will continue to do so. Interaction, mediated or otherwise, between the new and old guard of politics is a site for hybridized study and theorizing.

A variety of traditional methods and theories are available to scholars hoping to develop the study of new "technical" political actors. New methodological tactics, combined with the efficacy of mixed methods, present scholars with even more tools for generating data and knowledge. The challenge now is to move beyond the primacy of human interaction and towards an understanding of the symbiotic nature of relationships between humans and "technically" social actors like algorithms, bots, and those within the Internet of things.

Scholars will best be served by flexible approaches to building and understanding theory and methods. The interaction that occurs between technical actors, systems, and people is complex, situated, and requires reflexivity in those who study it. As technology becomes more intelligent, the social relationship between technological and human actors will become more complex. Ethnographers of information are well-poised to generate the understandings of this communicative phenomenon.

References

Alexander, L. (2015, April 2). Social Network Analysis Reveals Full Scale of Kremlin's Twitter Bot Campaign. *Global Voices*. Retrieved from https://globalvoices.org/2015/04/02/analyzing-kremlin-twitter-bots/

Alfonso, F. (2014, April 16). Reddit cofounder abandons r/technology following banned terms drama. Retrieved August 26, 2016, from http://www.dailydot.com/layer8/alexis-ohanian-reddit-technology-banned-words/

Ananny, M. (2014). Networked Press Freedom and Social Media: Tracing Historical and Contemporary Forces in Press–Public Relations. *Journal of Computer-Mediated Communication*, *19*(4), 938–956. https://doi.org/10.1111/jcc4.12076

Barocas, S., Hood, S., & Ziewitz, M. (2013, March 29). Provocation piece | Governing Algorithms. Retrieved from http://governingalgorithms.org/resources/provocation-piece/

Benkler, Y. (2006). *The Wealth of Networks: How Social Production Transforms Markets and Freedom*. New Haven, CT: Yale University Press.

Bennett, W. L., Foot, K., & Xenos, M. (2011). Narratives and Network Organization: A Comparison of Fair Trade Systems in Two Nations. *Journal of Communication*, *61*(2), 219–245. https://doi.org/10.1111/j.1460-2466.2011.01538.x

Bennett, W. L., & Segerberg, A. (2012). The Logic of Connective Action. *Information, Communication & Society*, *15*(5), 739–768. https://doi.org/10.1080/1369118X.2012.670661

Berry, D. M. (2012). The relevance of understanding code to international political economy. *International Politics*, *49*(2), 277–296. https://doi.org/10.1057/ip.2011.37

Beyer, J. L. (2014). *Expect Us: Online Communities and Political Mobilization*. New York, NY: Oxford University Press.

Boshmaf, Y., Muslukhov, I., Beznosov, K., & Ripeanu, M. (2011). The socialbot network: when bots socialize for fame and money. In *Proceedings of the 27th Annual Computer Security Applications Conference* (pp. 93–102). ACM. Retrieved from http://dl.acm.org/citation.cfm?id=2076746

Bowler, G. (2010). Netnography: A Method Specifically Designed to Study Cultures and Communities Online. *Qualitative Report*, *15*(5), 1270–1275.

Burrell, J. (2009). The Field Site as a Network: A Strategy for Locating Ethnographic Research. *Field Methods*, *21*(2), 181–199. https://doi.org/10.1177/1525822X08329699

Castells, M. (2007). Communication, Power and Counter-power in the Network Society. *International Journal of Communication*, *1*(1), 29.

Chadwick, A. (2013). *The Hybrid Media System: Politics and Power*. Oxford, UK: Oxford University Press.

Chu, Z., Gianvecchio, S., Wang, H., & Jajodia, S. (2010). Who is tweeting on Twitter: human, bot, or cyborg? In *Proceedings of the 26th annual computer security applications conference* (pp. 21–30). Austin, TX: ACM. Retrieved from http://dl.acm.org/citation.cfm?id=1920265

Chu, Z., Gianvecchio, S., Wang, H., & Jajodia, S. (2012). Detecting automation of Twitter accounts: Are you a human, bot, or cyborg? *IEEE Transactions on Dependable and Secure Computing*, *9*(6), 811–824. https://doi.org/10.1109/TDSC.2012.75

Coldewey, D. (2012). Romney twitter account gets upsurge in fake followers, but from where. *NBC News*.

Coleman, G. (2014). *Hacker, Hoaxer, Whistleblower, Spy: The Many Faces of Anonymous*. London ; New York: Verso.

Cook, D. M., Waugh, B., Abdipanah, M., Hashemi, O., & Rahman, S. A. (2014). Twitter Deception and Influence: Issues of Identity, Slacktivism, and Puppetry. *Journal of Information Warfare*. Retrieved from http://works.bepress.com/david_cook/15/

Dubbin, R. (2013, November 15). The Rise of Twitter Bots.

Forelle, M. C., Howard, P. N., Monroy-Hernandez, A., & Savage, S. (2015). *Political Bots and the Manipulation of Public Opinion in Venezuela* (SSRN Scholarly Paper No. ID 2635800). Rochester, NY: Social Science Research Network. Retrieved from http://papers.ssrn.com/abstract=2635800

Geertz, C. (1973). *The Interpretation of Cultures: Selected Essays*. Basic Books.

Geiger, R. S. (2014). Bots, Bespoke Code, and the Materiality of Software Platforms. *Information, Communication & Society*, *17*(3), 342–356.

Geiger, R. S., & Ribes, D. (2011). Trace Ethnography: Following Coordination through Documentary Practices (pp. 1–10). Presented at the 44th Hawaii International Conference on System Sciences (HICSS), IEEE.

Gillespie, T. (2009). Wired Shut, Copyright and the Shape of Digital Culture. *Journal of Information, Communication and Ethics in Society*, *7*(2/3), 213–218. https://doi.org/10.1108/14779960910955927

Gillespie, T. (2011, October 19). Can an algorithm be wrong? Twitter Trends, the specter of censorship, and our faith in the algorithms around us. Retrieved from http://culturedigitally.org/2011/10/can-an-algorithm-be-wrong/

Gillespie, T. (2014). The Relevance of Algorithms. In T. Gillespie, P. J. Boczkowski, & K. A. Foot (Eds.), *Media Technologies: Essays on Communication, Materiality, and Society*. Cambridge, MA: MIT Press.

Himelboim, I. (2008). Reply distribution in online discussions: A comparative network analysis of political and health newsgroups. *Journal of Computer-Mediated Communication*, *14*(1), 156–177. https://doi.org/10.1111/j.1083-6101.2008.01435.x

Hindman, M. (2008). *The Myth of Digital Democracy*. Princeton, NJ: Princeton University Press.

Hine, C. (2005). *Virtual Methods*. Oxford ; New York: Bloomsbury Academic.

Hogan, B. (2015). From Invisible Algorithms to Interactive Affordances: Data After the Ideology of Machine Learning. In E. Bertino & S. A. Matei (Eds.), *Roles, Trust, and Reputation in Social Media Knowledge Markets* (pp. 103–117). Springer International Publishing. Retrieved from http://link.springer.com/chapter/https://doi.org/10.1007/978-3-319-05467-4_7

Horst, H. A., & Miller, D. (2013). *Digital anthropology*. A&C Black. Retrieved from http://books.google.com/books?hl=en&lr=&id=9KuPzBgus7oC&oi=fnd&pg=PR5&dq=digital+anthropology&ots=uzgYGqVIHd&sig=OHQMy5JPTRN_zeAov_Jj0mi_3Gk

Howard, P. (2002). Network ethnography and the hypermedia organization: new media, new organizations, new methods. *New Media & Society*, *4*(4), 550–574.

Howard, P. N. (2006). *New Media Campaigns and the Managed Citizen*. Cambridge ; New York: Cambridge University Press.

Jankowski, N. W., & Selm, M. V. (2008). Internet-Based Political Communication Research. *Javnost - The Public*, *15*(2), 5–16. https://doi.org/10.1080/13183222.2008.11008967

Karpf, D. (2012). *The MoveOn Effect: The Unexpected Transformation of American Political Advocacy*. Oxford, UK: Oxford University Press.

Kelly, K. (2014, October 27). The Three Breakthroughs That Have Finally Unleashed AI on the World. Retrieved October 14, 2015, from http://www.wired.com/2014/10/future-of-artificial-intelligence/

Kozinets, R. V. (2010). *Netnography: Doing Ethnographic Research Online*. Thousand Oaks, CA: SAGE Publications Ltd.

Kreiss, D., Finn, M., & Turner, F. (2011). The limits of peer production: Some reminders from Max Interneter for the network society. *New Media & Society*, *13*(2), 243–259. https://doi.org/10.1177/1461444810370951

Latour, B. (2012). *We Have Never Been Modern*. Cambridge ; New York: Harvard University Press.

Latour, B., & Venn, C. (2002). Morality and Technology The End of the Means. *Theory, Culture & Society*, *19*(5–6), 247–260. https://doi.org/10.1177/026327602761899246

Lee, H. J., & Streeter, T. (2012). Romancing the Internet An Interview With Thomas Streeter. *Journal of Communication Inquiry*, *36*(2), 95–110. https://doi.org/10.1177/0196859912445067

Lindlof, T., & Taylor, B. (2010). *Qualitative Communication Research Methods*. Thousand Oaks, Calif: SAGE Publications, Inc.

Manovich, L. (2001). *The Language of New Media*. Boston, MA: MIT Press.

Markham, A. N., & Baym, N. K. (2008). *Internet Inquiry: Conversations About Method*. Los Angeles: SAGE Publications, Inc.

Medeiros, J. de. (n.d.). Turkey's Twitter-Bot army and the politics of social media. Retrieved July 24, 2014, from http://Internetmagazin.de/en/Turkey-Twitterbot-army-and-social-media-politics-174524

Metaxas, P. T., & Mustafaraj, E. (2012a). Social media and the elections. *Science*, *338*(6106), 472–473.

Miles, M., Huberman, A. M., & Saldaña, J. (2013). *Qualitative Data Analysis: A Methods Sourcebook* (Third Edition). Thousand Oaks, California: SAGE Publications, Inc.

Motti, J. (2014, August 12). Twitter acknowledges 23 million active users are actually bots. Retrieved August 26, 2016, from http://www.techtimes.com/articles/12840/20140812/twitter-acknowledges-14-percent-users-bots-5-percent-spam-bots.htm

Mukerji, C. (2003). Intelligent Uses of Engineering and the Legitimacy of State Power. *Technology and Culture*, *44*(4), 655–676.
Nagy, P., & Neff, G. (2015). Imagined Affordance: Reconstructing a Keyword for Communication Theory. *Social Media + Society*, *1*(2), 2056305115603385. https://doi.org/10.1177/2056305115603385
Nardi, B. (1995). Studying Context: A Comparison of Activity Theory, Situated action Models, and Distributed Cognition. In *Context and Consciousness: Activity Theory and Human-computer Interaction* (pp. 69–96). Boston, MA: MIT Press.
Neff, G. (2012). *Venture labor: work and the burden of risk in innovative industries*. Cambridge, Mass.: MIT Press.
Neff, G., Jordan, T., McVeigh-Schultz, J., & Gillespie, T. (2012). Affordances, Technical Agency, and the Politics of Technologies of Cultural Production. *Journal of Broadcasting & Electronic Media*, *56*(2), 299–313. https://doi.org/10.1080/08838151.2012.678520
Orcutt, M. (2012, June 21). Twitter Mischief Plagues Mexico's Election. Retrieved June 20, 2014, from http://www.technologyreview.com/news/428286/twitter-mischief-plagues-mexicos-election/
Owen, T. (2015). *Disruptive Power: The Crisis of the State in the Digital Age*. Oxford University Press.
Preece, J., Rogers, Y., Sharp, H., Benyon, D., Holland, S., & Carey, T. (1994). *Human-Computer Interaction*. Essex, UK, UK: Addison-Wesley Longman Ltd.
Qtiesh, A. (2011, April 18). Spam Bots Flooding Twitter to Drown Info About #Syria Protests [Updated]. Retrieved from http://advocacy.globalvoicesonline.org/2011/04/18/spam-bots-flooding-twitter-to-drown-info-about-syria-protests/
Schneiderman, B., & Plaisant, C. (2009). *Designing the User Interface: Strategies for Effective Human–Computer Interaction* (5th edition). Boston: Pearson.
Star, S. L., & Ruhleder, K. (1994). Steps Towards an Ecology of Infrastructure: Complex Problems in Design and Access for Large-scale Collaborative Systems. In *Proceedings of the 1994 ACM Conference on Computer Supported Cooperative Work* (pp. 253–264). New York, NY, USA: ACM. https://doi.org/10.1145/192844.193021
Stone-Gross, B., Cova, M., Cavallaro, L., Gilbert, B., Szydlowski, M., Kemmerer, R., … Vigna, G. (2009). Your Botnet is My Botnet: Analysis of a Botnet Takeover. In *Proceedings of the 16th ACM Conference on Computer and Communications Security* (pp. 635–647). New York, NY, USA: ACM. https://doi.org/10.1145/1653662.1653738
Suchman, L. (2006). *Human–Machine Reconfigurations: Plans and Situated Actions* (2nd edition). Cambridge, UK: Cambridge University Press.

Turner, F. (2006). *From Counterculture to Cyberculture*. Chicago, IL: University of Chicago Press.

Urbina, I. (2013, August 10). I Flirt and Tweet. Follow Me at #Socialbot. *The New York Times*. Retrieved from http://www.nytimes.com/2013/08/11/sunday-review/i-flirt-and-tweet-follow-me-at-socialbot.html

US NSF - Transformative Research. (n.d.). Retrieved October 13, 2015, from http://www.nsf.gov/about/transformative_research/submit.jsp

Van Maanen, J. (2011). *Tales of the Field: On Writing Ethnography, Second Edition* (Second Edition). Chicago: University of Chicago Press.

Vertesi, J. (2008). Mind the Gap: The London Underground Map and Users' Representations of Urban Space. *Social Studies of Science*, *38*(1), 7–33. https://doi.org/10.1177/0306312707084153

Vertesi, J. (2015). *Seeing Like a Rover: How Robots, Teams, and Images Craft Knowledge of Mars*. Chicago ; London: University of Chicago Press.

Wellman, B., Quan-Haase, A., Boase, J., Chen, W., Hampton, K., Díaz, I., & Miyata, K. (2003). The Social Affordances of the Internet for Networked Individualism. *Journal of Computer-Mediated Communication*, *8*(3). https://doi.org/10.1111/j.1083-6101.2003.tb00216.x

Woolley, S., & Howard, P. (2016, May 15). Bots unite to automate the presidential election. Retrieved from http://www.wired.com/2016/05/twitterbots-2/

Woolley, S., & Howard, P. N. (forthcoming). Social Media, Revolution, and the Rise of the Political Bot. In *Routledge Handbook of Media, Conflict, and Security*. New York, NY: Routledge.

CHAPTER 19

Outlook for Prosperity and Peace in the Emergent Global Political Economy of Robots

Ryan David Kiggins

On October 7, 2016, the US government announced that Russia was utilizing information technologies to directly interfere with the US 2016 presidential election.[1] The purpose, from the view of US officials, was to directly affect US voter participation and voter choice in a cynical ploy to ensure that the democratic candidate for president lost the 2016 election (Mission accomplished). According to Wood (2017), the Russian operation was comprised of three interrelated and mutually supporting operations. The first operation illicitly gained access to information about

[1] *Joint Statement from the Department Of Homeland Security and Office of the Director of National Intelligence on Election Security*, 7 October 2016. Available at: https://www.dhs.gov/news/2016/10/07/joint-statement-department-homeland-security-and-office-director-national. Last accessed 31 March 2017.

R. D. Kiggins (✉)
Department of Political Science, University of Central Oklahoma, Edmond, Oklahoma, USA
e-mail: rkiggins@uco.edu

R. Kiggins (ed.), *The Political Economy of Robots*, International Political Economy Series, https://doi.org/10.1007/978-3-031-95171-8_19

the 2016 democratic candidate for president and her senior campaign staff from their personal email accounts in order to glean any potentially damaging information that could be incorporated into negative news stories and advertisements about the democratic candidate for president, her campaign, and senior campaign staff.

The second operation distributed the negative news stories and advertisements utilizing automated bots, what Woolley's contribution to this volume characterizes as political bots, to make a large number of posts on various social media outlets for each negative news story and advertisement. The use of automated bots is a timeworn practice in the political economy of information used by email spammers, financial and betting market arbitragers, and hackers conducting cyberattacks. In this application, allegedly by Russia, automated political bots were employed to create an online viral effect for the negative news story, supported by Russian government-controlled news organizations, by repeatedly reporting the negative news story. A viral effect occurs as the number of clicks and reposts of a news story or advertisement exponentially increases. Automated bots are designed to produce such quantifiable clicks and reposts that are relied upon as empirical proof that online views of a news story or advertisement are rapidly increasing. As the viral effect took hold, negative news stories were re-reported by rightwing news sites in the US, then Fox News, followed by the rest of mainstream media outlets in the US. Negative advertisements largely remained confined to social media posts and re-posts, though, Russian government-controlled media organizations broadcast negative advertisements within the US that were later picked up by rightwing websites, such as infowars.com.

The second operation was to sway US voter opinion regarding the democratic candidate for president, through constant negative news and advertising throughout the 2016 campaign. The efficacy of negative news and advertising on US voter choice is a much-researched topic (Mattes and Redlawsk 2015; Soroka and McAdams 2015; Soroka 2014; Johnson-Cartee and Copeland 2013; West 2013; Kahn and Kenny 1999; Lau et al. 1999). An interesting recent finding within this research is that humans tend to listen to and remember information framed in negative terms over information framed in positive terms. Whether or not voter mobilization or demobilization follows remains an open empirical question, with evidence supporting both propositions. However, conventional wisdom among political consultants—those who work for and plan political campaigns—is that negative advertising and negative news stories do have effects on voter choice. With these views of negative news and political advertising in mind, if the objective is to influence voter choice in an election, developing, distributing, and repeating negative news stories and advertisements during a political campaign, could achieve that objective. The advent of social media, big data analytics, automated bots, the internet, digitized and stored voter lists sharply lowers the costs, complexity, and time–space barriers to surreptitiously influence an election outcome by any actor, let alone a state actor such as Russia.

The third objective of the allegedly Russian operation was to illicitly acquire access to US voter information databases which, Director of the Federal Bureau of Investigation, James Comey, confirmed occurred during testimony, given on 10 January 2017, before the Intelligence Committee of the US House of Representatives. These databases of US voters are maintained on poorly secured county and state, and, apparently, the democratic national party computer networks. Making for easy hacks by (allegedly) Russian government agents or affiliates. The purpose of acquiring this information of US voters was to support micro-targeted negative advertising about the 2016 democratic presidential candidate through social media and web advertisements. Microtargeting is a political campaign advertising technique that leverages big data analytics to develop voter profiles that segment the voter population, enabling more directed advertising to voters (see West 2013; Nielsen 2012; Murry and Scime 2010). Big data-supported microtargeting means that campaigns may know voters better than voters know themselves. Relying on microtargeting allows a political campaign to identify the crucial campaign issue for segments of the voter population, design specific ads tailored to each voter segment and profile, and then distribute those ads to voter segments and individual voters using internet browser ad technologies, social media ads, likes, posts, and reposts, mailers, and campaign canvassers. The purpose of microtargeting, as with any political communication, is to influence voter opinion, participation, and, ultimately, choice.

This recent example of alleged Russian shenanigans in the 2016 US presidential elections underscores the purpose for this book: to broaden the conversation about digital automated information technologies beyond killer robots to other areas of human endeavor affected by digital automation. So, focused on killer robots and the implications to ethics and war, scholars may have elided how digital automation alters the prospects for peace and prosperity in the global political economy. In terms of peace, current and former US elected and public officials, including Senator John McCain[2] and former US Vice President, Dick Cheney,[3] have each characterized Russian use of automated technologies to effect a desired US presidential outcome, as an act of war. The Obama Administration, prior to leaving office, imposed economic sanctions on Russia as a result of its belief that Russia did in fact interfere with the 2016 US presidential election. Tensions between Russia and the United States were recently characterized by President Vladimir Putin's spokesperson as worse than the Cold War.[4] Tensions between Russia and the US are no longer confined to balance of power, nuclear deterrence, or great power rivalry. Significantly, the use of political bots rather than killer bots contributed to furthering the tension between Russia and the US.

Russian political bots targeted US political institutions for the purpose of sowing mistrust in information received by US voters through US news sources, undercut US voter perceptions of US government efficacy to secure against foreign intervention, and provoke uncertainty as to whether or not future US elections will be free and fair, that future elected officials were not indirectly (directly?) assisted in their campaigns by foreign powers attempting to influence the election of a preferred candidate who will implement policies in office less detrimental to the interests of that foreign power. The US, is the latest democracy to experience Russian electoral process meddling, democracies in Europe have also experienced similar Russian electoral interference using information and automated technologies as far back as 2004.[5] The upshot is that information and automated technologies do not have to be militarized to be used in a manner that threatens the prospects for domestic tranquility and global peace. Political bots are as potentially lethal and destabilizing as killer bots. Prospects for global peace are not solely based on a future of militarized bots, rather the prospects for global peace may also rest on

[2] CNN news report available at http://www.cnn.com/2016/12/30/politics/mccain-cyber-hearing/. Last accessed 31 March 2017.

[3] NBC News report available at http://www.nbcnews.com/politics/white-house/dick-cheney-russian-election-interference-could-be-seen-act-war-n739391. Last accessed 31 March 2017.

[4] ABC News report available at http://abcnews.go.com/Politics/kremlin-spokesman-russia-us-relations-worse-cold-war/story?id=46486662. Last accessed 31 March 2017.

[5] LA Times news report. http://www.latimes.com/world/europe/la-fg-russia-election-meddling-20170330-story.html. Last accessed 31 March 2017.

whether or not the global community creates global agreements, regimes, and institutions to regulate how automated technology is utilized in pursuit of political ends.

Regarding the prospects for prosperity in the emergent global political economy of robots, this book highlights automation of global trade, labor, health care, and finance. The automation of global value chains contributes to the divorce of humans from their labor, from their capacity to earn wages for sustaining human life. In addition, automated hiring practices raise concerns about individual privacy and the possibility that intangible talent and skill, unrecognizable by current hiring data analytics technologies, may contribute to chronic human unemployment. Combined, automated technologies may negatively affect workers through replacement and hiring discrimination. Another threat to workers arises from the automation of financial innovation which increases the likelihood of excessive risk-taking giving rise to speculative bubbles which, threaten economic stability and employment. Automated global trade combined with automated financial innovation undermine the prospects for sustaining the embedded liberalism compromise,[6] by which governments implemented welfare policies to protect workers from economic dislocations attendant from a global open economy. Moreover, the precarity of human prosperity in the political economy of robots may be amplified by automated healthcare that fails to individualize diagnosis and treatment. However, all may not be lost. The ultimate outcome humans may experience, from the emergent global political economy of robots, assuming the advent of singularity, as Walsh and Sculos note in this volume, may be a labor-less future in which humans are free from the oppression of the wage-labor contract, free to pursue individual progress. Failing that, while recognizing that information and automated technologies present unique challenges to democratic governance, humans must cultivate practical wisdom in order to preserve human cognitive faculties. Part of this effort to retain and cultivate our cognitive capacity is to develop policy-relevant frameworks that enable human political, economic, and social institutions and practices to incorporate and leverage information and automated technologies for the betterment of global society. For, in the end, digital automated information technologies were conceived, designed, and built by humans to better society, to

[6] Ruggie, J. G. (1982). International regimes, transactions, and change: embedded liberalism in the postwar economic order. *International organization*, *36*(02), 379–415.

enhance human productivity, learning, and communication. Our choice is whether or not to institutionally constrain the political economy of robots to support global prosperity and peace. Our hope is that this collection provokes reflection, conversation, and action to do so, in order to ensure that any favorable promise attendant the political economy of robots delivers global prosperity and peace in the twenty-first century and beyond.

Works Cited

Johnson-Cartee, K. S., & Copeland, G. (2013). *Negative political advertising: Coming of age*. Routledge.

Kahn, K. F., & Kenney, P. J. (1999). Do negative campaigns mobilize or suppress turnout? Clarifying the relationship between negativity and participation. *American political science review*, *93*(04), 877-889.

Lau, R. R., Sigelman, L., Heldman, C., & Babbitt, P. (1999). The effects of negative political advertisements: A meta-analytic assessment. *American Political Science Review*, *93*(04), 851-875.

Mattes, K., & Redlawsk, D. P. (2015). *The positive case for negative campaigning*. University of Chicago Press.

Murray, G. R., & Scime, A. (2010). Microtargeting and electorate segmentation: data mining the American National Election Studies. *Journal of Political Marketing*, *9*(3), 143-166.

Nielsen, R. K. (2012). *Ground wars: Personalized communication in political campaigns*. Princeton University Press.

Soroka, S., & McAdams, S. (2015). News, politics, and negativity. *Political Communication*, *32*(1), 1-22.

Soroka, S. N. (2014). *Negativity in democratic politics: Causes and consequences*. Cambridge University Press.

West, D. M. (2013). *Air wars: Television advertising and social media in election campaigns, 1952-2012*. Sage.

Wood, P. (2017). "Trump Russia dossier key claim 'verified'." *BBC News*, published 30 March 2017. Available at: http://www.bbc.com/news/world-us-canada-39435786. Last accessed 3 April 2017.

Index

R. Kiggins (ed.), *The Political Economy of Robots*, International Political Economy Series, https://doi.org/10.1007/978-3-031-95171-8

The manufacturer's authorised representative in the EU is Springer Nature Customer Service Centre GmbH, Europaplatz 3, 69115 Heidelberg, Germany. If you have any concerns regarding our products, please contact ProductSafety@springernature.com

Printed and bound by CPI Group (UK) Ltd, Croydon, CR0 4YY
01/10/2026
02225954-0009